PROPAGATION OF HORTICULTURAL CROPS
(PRINCIPLES AND PRACTICES)

PROPAGATION OF HORTICULTURAL CROPS

(PRINCIPLES AND PRACTICES)

DR. R.R. SHARMA, FHSI, FBRS

M.Sc., Ph.D. (Hort.)

Principal Scientist (Fruit Science)

Division of Food Science and Postharvest Technology

Indian Agricultural Research Institute

New Delhi-110 012

KALYANI PUBLISHERS

LUDHIANA • NEW DELHI • NOIDA (U.P.) • HYDERABAD • CHENNAI
KOLKATA • CUTTACK • GUWAHATI • KOCHI • BENGALURU

KALYANI PUBLISHERS

Head Office

B-1/1292, Rajinder Nagar, **Ludhiana**-141 008 • Ph : 0161-2760031, 2745756
E-mail : kalyanibooks@yahoo.co.in **Website :** www.kalyanipublishers.co.in

Administration Office

4779/23, Ansari Road, Daryaganj, New Delhi-110 002
Ph : 011-23271469, 23274393 **E-mail :** kalyani_delhi@yahoo.co.in

Works

B-16, Sector-8, **NOIDA** (U.P.)

Branches

1, Mahalakshmi Street, T. Nagar, **Chennai**-600 017 • Ph : 044-24344684
Gopabandhu Lane, Behind Govt. Bus Stand, Badambadi,
Cuttack-753 012 (Odisha) • Ph : 0671-2311391
3-5-1108, Narayanaguda, **Hyderabad**-500 029 • Ph : 040-24750368
10/2B, Ramanath Mazumdar Street, **Kolkata**-700 009 • Ph : 033-22416024
Arunalaya, 1st Floor, Saraswati Road, Pan Bazar, **Guwahati**-781 001 • Ph : 0361-2731274
Koratti Parambil House, Convent Road, **Kochi**-682 035 • Ph : 0484-2367189
No. 24 & 25, 1st Floor, Hameed Shah Complex, Cubbonpet Main Road, **Bengaluru**-560 002

R-35334 05

First Edition, 2002
Second Revised Edition, 2013
Third Revised Edition, 2016

Reprinted, 2022

Typesetting at
Times Printographic

ISBN 978-93-272-6880-5

PRINTED IN INDIA
Kalyani Printings, B-15, Sector 8, NOIDA
and published by Mrs. Usha Raj Kumar for
Kalyani Publishers, New Delhi-110 002

Dedicated to my reverend mother

(SMT. VIDYA DEVI SHARMA)

and affectionate father

(LATE SH. GANDHI RAM SHARMA)

FOREWORD

The field of horticulture is like an ocean, which includes fruits, vegetables, ornamentals, plantation crops, condiment and spices, mushrooms, root and tuber corps. It has emerged as one of the most important sectors for the diversification of agriculture. Today, India is the 2nd largest producer of fruits and vegetables in the world, but yet, we are targeted to achieve the rainbow revolution, which would only be possible after achieving the golden revolution. Golden revolution can only be achieved if the production and productivity of horticultural crops is increased to its potential. There are many constraints, which limit the productivity of horticultural crops but the major is the non-availability of genuine planting material. Thus, plant propagation is the backbone of horticulture industry.

A course on plant propagation and nursery management is taught to UG and PG students in almost all the agricultural universities. However, the books covering the whole syllabus are limited, and if available, these don't contain latest information in a simple and understandable language. Moreover, there are about 50,000 nurseries in India and the nurserymen do require some readymade material for the production of quality material. I am happy that Dr. R.R. Sharma has come up with the idea of writing a book as per the needs of the students, teachers, nurserymen and plant propagators.

The latest available information has been put systematically in 36 chapters, covering all aspects of sexual and asexual methods of propagation of horticultural crops. It is interesting that special emphasis has been given on seed propagation and nursery management. Similarly, a due attention has been paid to include some important chapters such as use of plastics in plant propagation, rejuvenation of old orchards, practical use of plant bioregulators with some chapters exclusively on commercial methods of propagation of all the horticultural crops. In my opinion, it is a complete treatise on the theory, principles and practices of plant propagation.

I am sure that the book will attain great popularity among students, teachers, researchers, extension workers, nurserymen and alike in the field of horticulture.

Place : I.A.R.I., New Delhi-12
Date : May 3, 2002

(Dr. S. NAGRAJAN)
Director
Indian Agricultural Research Institute
New Delhi-110 012

PREFACE

THIRD EDITION

The second edition of the book entitled ***'Propagation of Horticultural Crops'*** received appreciation from several readers and I am hopeful that it might have proved very useful to the students, teachers, researchers, extension personnel in India and abroad. I received several queries and suggestions for its improvement since its 2nd edition was published. Considering these requests, I have revised and enlarged this version and two more chapters have been included in the revised version. In addition, several new information have been incorporated in the revised edition. Besides, other information on different aspects of plant propagation of horticultural crops of the second edition have been revised and updated.

I am not only hopeful but fully confident that the revised version of the book will be much more useful and beneficial to the readers. However, still there may be some omissions, errors and mistakes in this revised version. Hence, I request all the readers to be critical and invite them to suggest modifications for the next edition.

2015 **R.R. SHARMA**

PREFACE

FIRST EDITION

The field of horticulture is divided into three main branches *viz.* pomology, olericulture and floriculture. Pomology is the study of different aspects of fruit growing; olericulture deals with vegetable growing and floriculture with all aspects related to flowering and ornamental plants. In the production of these horticultural crops, one of the fundamental problems is that of propagation, which is the reproduction of plants. Among different horticultural crops, raising of new plants is not difficult in certain vegetables and flowering annuals because they are self-pollinated and do not show much heterozygosity. However, in most of fruit plants and some ornamental and plantation crops, many different propagation methods have been developed, which create addition complications because one method standardised for a particular crop may not be ideally suited for the other crop in the same or different environment.

A course on plant propagation and nursery management is taught to the U.G. and P.G. students of horticultural colleges in all the agricultural universities. Some books are available on plant propagation, which covers the syllabus of this course but these are either very costly or their language is difficult enough to make the students understand the exact principles and procedures involved in raising the horticultural crops. Further, there are about 50,000 nurseries in India and about two lakhs of people are dependent on them directly or indirectly. People engaged in this profession propagate plants with traditional methods, because they don't have access to the easily available literature on plant propagation. Production of quality material should be the first and foremost aim of a nurseryman but due to lack of sufficient knowledge and literature, most nurserymen in our country are not in a position to prepare and supply the quality planting material to the indentors. Keeping these facts in mind, the idea of writing a small but useful book on propagation of horticultural crops was conceived.

I have consulted many books, research papers and review articles for incorporating the available information in this book. Though, I have tried to incorporate the maximum possible information briefly but still I feel that there may be many omissions and errors. I request all the readers to please pin point the omissions and errors and write to me without any fear, which would act as a guiding factor in organizing the 2nd edition.

This book has been written in a very simple and understandable language. It is hoped that it will be of great help to the researchers, teachers, extension workers and students related to the discipline of horticulture. It covers almost all aspects *i.e.,* theory, principles and practices of sexual and asexual propagation of different horticultural crops, which are required for their multiplication. Further, special emphasis has been given on nursery management. Similarly, the commercial methods of propagation of different horticultural crops have been dealt in separate chapters.

My dream of writing a book could not have true without the help of many professionals, teachers, students and alike in the field of horticulture. For which, I acknowledge with thanks the help and inspiration rendered by Dr. Room Singh, Head and Dr. A.M. Goswami (Ex-Head), Division of Fruits and Horticultural Technology, IARI, New Delhi-110 012. Dr. S.K. Saxena, Dr. P.C. Jindal, Dr. Sanjay Kumar Singh, Dr. S.R. Sharma, Dr. Netra Pal, Dr. K.P. Singh, Dr. V.B. Patel, Dr. Manish Srivastav, Scientists and many students of Discipline of Horticulture, IARI, New Delhi have helped me from time to time during the preparing of this manuscript by way of giving useful suggestions, for which I will remain beholden to all them throughout my life. I am also thankful to Mr. Surinder Kumar for taking beautiful photographs. Mrs. Karuna Dixit and Mr. K.S. Negi for drawing beautiful sketches.

Each and every member of my family including *Gainthu* (Abbu), my little cousin, have helped me in every possible manner during the preparation of this book and I thank them all from the core of my heart.

In the last but the least, I am especially thankful to my wife, whose filial attachment did not come in my way during the preparation of this book.

2002 **R.R. SHARMA**

CONTENTS

Pages

1. INTRODUCTION 1—5

Major methods of propagation, genetical concepts concerning plant propagation.

2. PLANT STRUCTURE 6—20

Different parts of a plant, inflorescence and its types, pollination, kinds of fruits, fertilization, life cycle of horticultural plants.

3. SEXUAL PROPAGATION 21—25

Advantages and disadvantages of sexual propagation, development of seed, different kinds of seeds.

4. SEED DORMANCY 26—36

Types and phases of dormancy, hormonal control of dormancy, measures to control seed dormancy.

5. SEED GERMINATION 37—45

Stages and types of dormancy, factors and treatment to facilitate seed germination.

6. SEED VIABILITY 46—52

Seed types, factors affecting seed viability, methods of testing seed viability.

7. SEED PRODUCTION 53—56

Attributes of a good quality seed and hybrid seed production.

8. SEED TESTING AND CERTIFICATION 57—62

Seed testing methods, phases and stages of seed certification.

9. SEED VIGOUR 63—65

Vigour testings, constraints in vigour testing, tests for measurement of seed vigour.

10. SEED STORAGE 66—68

Factors affecting seed viability and points for safe storage of seed.

11. NURSERY ESTABLISHMENT 69—75

Pre-requisites for nursery establishment, importance of nursery, factors affecting nursery establishment, parts of a nursery.

12. MEDIA FOR PROPAGATION 76—81

Characteristics of an ideal medium, criteria for selection of a medium, characteristics of commonly used media.

13. PROPAGATION STRUCTURES 82—90

Greenhouse, hotbeds, cold frames, mist propagation units, net house, lath house etc.

14. RAISING PLANTS IN THE NURSERY 91—94

Seed sowing, hardening of seedlings and transplanting

15. NURSERY MANAGEMENT 95—101

Nutrient, insect-pest and disease management in nursery.

16. WATER MANAGEMENT IN NURSERY 102—103

Quality of Irrigation water, methods of irrigation in nursery.

17. NURSERY REGISTRATION AND CERTIFICATION 104—106

Licensing requirements for the nursery responsibilities of applicant and civil penalties.

18. MARKETING OF NURSERY PLANTS 107—110

Lifting, packing, storage and sale of nursery plants.

19. TOOLS AND ACCESSORIES REQUIRED IN NURSERY 111—115

Important tools (secateurs, forks, pruning shears, knives, saws) and accessories (tying and wrapping material grafting wax, polythene tapes, supports, baskets, pots, pesticides etc.)

20. HORMONAL POWDERS, SOLUTIONS AND PASTES 116—119

Types, method of preparation and application of hormonal powders, solutions and pastes.

21. ROUTINE NURSERY PRACTICES 120—124

Clipping, defoliation, pinching transplanting, pricking, disbudding, deshooting, staking, soil sterilization etc., nursery practices.

22. CHEMICALS AND PLANT GROWTH REGULATORS IN PLANT PROPAGATION 125—127

Use of different chemicals and plant growth regulators in sexual and asexual propagation of plants.

23. ROLE OF PLASTICS IN PLANT PROPAGATION 128—131

Uses in greenhouse, polytunnels, polyhouses, typing and wrapping material etc.

24. VEGETATIVE PROPAGATION 132—134

Advantages and disadvantages, general methods of vegetative propagation.

25. CLONE : ORIGIN, MANAGEMENT AND VARIABILITY 135—144

The clone, production and maintenance of a clone, variability in a clone, topophysis, chimera etc.

26. PROPAGATION BY CUTTINGS : PRINCIPLES 145—156

Advantages and disadvantages, anatomical and physiological basis of rooting in cuttings, factors affecting rooting in cuttings.

27. TYPES OF CUTTINGS 157—163

Stem, root, leaf, bud and herbaceous cuttings etc.

28. PROPAGATION BY GRAFTING : PRINCIPLES 164—170

Formation of graft union, factors affecting the healing of graft union, limitations of grafting etc.

29. GRAFT INCOMPATIBILITY 171—175

Symptoms, types and causes of graft incompatibility.

30. METHODS OF GRAFTING 176—186

Advantages and disadvantages of grafting, attached methods (inarching, tongue, cleft, bridge) and detached methods (splice, whip, side) of grafting etc.

31. SPECIAL GRAFTING TECHNIQUES 187—194

Root grafting, nurseroot grafting, green wood grafting, cutting grafts, softwood grafting, epicotyl grafting, micrografting etc.

32. STOCK-SCION RELATIONSHIPS 195—198

Effects of rootstock on scion, scion on stock and interstocks etc.

33. BUDDING : THEORY AND PRINCIPLES 199—202

Advantages of budding, times of budding, bud union formation, and factors affecting budding etc.

34. METHODS OF BUDDING 203—208

T-budding, inverted 'T' budding, chip budding, patch budding, flute budding, microbudding etc.

35. BUD-WOOD CERTIFICATION IN CITRUS 209—213

Biological indexing and sesological methods of virus detection.

36. SHOOT-TIP GRAFTING AND MICRO-BUDDING IN CITRUS 214—217

Procedure of STG, laboratory and serological tests, procedure for micro-budding.

37. REJUVENATION OF OLD ORCHARDS 218—221

Selection of plants, heading back, grafting/budding techniques, after care of grafted/budded plants.

38. APOMIXIS AND POLYEMBRYONY 222—224

Types of apomixis, polyembrony and its significance.

39. PROPAGATION BY LAYERING 225—232

Advantages and disadvantages, methods of layering like tip, simple serpentine, trench, air, mound etc.

40. PROPAGATION BY SPECIALIZED VEGETATIVE STRUCTURES 233—241

Tuber, bulb, rhizome, suckers, slips, pseudobulbs, runners, stolon etc.

41. MICROPROPAGATION : PRINCIPLES AND PRACTICES 242—254

Basic requirements, advantages and limitations and methods of *in vitro* propagation etc.

42. MAJOR PROPAGATION METHODS OF FRUIT PLANTS 255—264

Different methods, ease and success, commercial feasibility etc. of different fruit plants.

43. PROPAGATION METHODS OF VEGETABLE CROPS 265—271

Seed propagation and vegetative propagation of major vegetables

44. PROPAGATION OF PLANTATION CROPS 272—275

Brief procedures for commercial propagation methods of important plantation crops.

45. PROPAGATION METHODS OF CONDIMENTS, SPICES, MEDICINAL AND AROMATIC PLANTS 276—278

Brief introductory information on propagation methods of spices, medicinal and aromatic plants.

46. PROPAGATION METHODS OF FLORICULTURAL PLANTS 279-289

Brief procedures of commercial propagation methods of different ornamental plants.

47. MODERN TECHNIQUES OF RAISING ANNUALS 290—292

Plug tray nursery raising : Advantages and requirements.

GLOSSARY 293—300

SELECTED REFERENCES 301—304

LIST OF TABLES

Table No.	Title	Page No.
2.1.	Botanical classification of fruits	16
4.1.	Kinds of seed dormancy	30
5.1.	Stratification requirement of seeds of some temperate fruit crops	44
6.1.	The longevity and storage conditions for seeds of different fruit and plantation crops	47
6.2.	The longevity and storage conditions for seeds of some vegetables and annual flowers	48
7.1.	A list of some F_1 hybrids of vegetable crops released in India	55
8.1.	The size of sample to be submitted for testing to seed testing laboratory	58
8.2.	Minimum seed certification standards for some vegetable crops	61
10.1.	Maximum safe seed moisture contents of some vegetable during storage at different temperatures	67
15.1.	Deficiency symptoms of some important nutrients in the nursery plants	96
15.2.	Common insect-pests and diseases of nursery and their management	98
35.1.	Indicator plants for virus and virus-like diseases in citrus	211
35.2.	Serological and molecular techniques for detecting viruses and virus-like diseases in citrus	213
41.1.	Chemicals used for sterilizing explants in micropropagation	245
41.2.	Composition and preparation of the stock solutions for MS basal medium	246
41.3.	Composition and preparation of the stock solutions for B_5 basal medium	247
41.4.	Micropropagation technology developed for some horticultural crops in India	253
43.1.	A list of some important vegetables propagated by seeds	266
43.2.	A list of vegetable crops raised by vegetative methods of propagation	270
45.1.	Common propagation methods of major condiments and spices	276

45.2. Common propagation methods of seed spices 277

45.3. Important herbal spices and their propagation methods 277

45.4. Methods of propagation of medicinal and aromatic plants 278

46.1. Major bulbous plants and their commercial propagation methods 283

46.2. Common indoor plants and their commercial propagation methods 285

46.3. Important shrubs and trees and their commercial propagation methods 286

LIST OF FIGURES

Fig. No.	Title	Page No.
2.1.	Common root systems	7
2.2.	Different types of leaves found in horticultural plants	9
2.3.	Phase changes in the life cycle of a seedling plant	10
2.4.	Parts of a flower	10
2.5.	Whorl of sepals and petals of a flower	11
2.6.	Internal structure of some perfect flowers	12
2.7.	Different types of inflorescences found in horticultural plants	14
2.8.	Structure of a flower pistil showing the process of fertilization	15
2.9*a*.	Fleshy and dry fruits	18
2.9*b*.	Female cone of *Pinus roxburghii*	18
2.10.	A coniferous plant	18
2.11.	Angiospermic plants	19
3.1.	Different kinds of seeds produced by horticultural plants	23
5.1.	Seed germination in dicots	39
8.1	Different types of tiers used for seed sampling	58
11.1.	Lay-out plan of an ideal nursery	75
13.1.	A cold frame widely used propagating structure in temperate climate	84
13.2.	Plants growing in a cold frame	85
13.3.	A small mist propagation unit	86
13.4.	Components of an intermittent mist propagation unit	87
13.5.	A small poly tunnel	88
13.6.	A commonly used electric propagator	89
19.1.	Different tools and instruments used in the nursery	111
23.1.	A polytunnel widely used for raising flower and vegetable seedings.	129
25.1.	Phase variation in a mature seedling plant	141
25.2.	Different types of chimeras observed in horticultural plants	142
25.3.	Different stages of development of graft chimera in nightshade and tomato	143
26.1.	Rooting co-factors in rooting process of cuttings	150
27.1.	Preparation of softwood cutting from the plants	159
27.2.	Propagation by root cuttings	159

27.3. Root cuttings (a) thinner roots, (b) Thicker roots 160
27.4. Different types of cuttings and their relative position on the plant 163
28.1. Components of a grafted plant 164
28.2. Polarity in grafting 168
29.1. Development of an outgrowth at graft union, a symptom of graft incompatibility 172
30.1 Different methods of approach grafting 178
30.2. Different steps involved in bridge grafting 180
30.3. Different steps involved in whip grafting 181
30.4. Different steps involved in tongue grafting 182
30.5. Different steps involved in stub grafting 183
30.6. Methodology of cleft grafting 185
30.7. Bark grafting 186
31.1. Components of a double worked plant 189
31.2. Different steps involved in nurse-seed grafting 190
31.3. Softwood grafting 191
31.4 Epicotyl or stone grafting 192
31.5. Renewal grafting 194
34.1. Position of "T" cut on the stock 204
34.2. Method of insertion of shield on the stock in "T" budding 204
34.3. Different types of budding 205
34.4. Prepared bud-stick 208
35.1 Citrus budwood certification programme 210
38.1. Development of nucellar embryos in citrus 223
39.1. Tip layering 227
39.2. Methodology involved in simple layering 228
39.3. Serpentine layering 229
39.4. Different steps of trench layering 229
39.5. Methodology of stooling 231
40.1. Tubers of dahlia 233
40.2. Formation of offsets in (*a*) Chives, (*b*) Lily 234
40.3. Different parts of a gladiolus corm 236
40.4. Lily plant showing bulbil production 238
40.5. (*a*) Tuberous roots of dahlia (*b*) Tuberous begonia plant 239
40.6. Runner production in strawberry plant 240
41.1. Different types of micropropagation systems 250
41.2. Different steps involved in meristem tip culture 251
41.3. Flow diagram for *en masse* production of grape under *in vitro* conditions 254

LIST OF PLATES

Plate No.	Title
1.	A view of series of glass houses
2.	A side view of a poly house
3.	National Phytotron Facility built up at IARI, New Delhi
4.	Tomato seedlings in the nursery beds ready for transplanting
5.	Poly bags filled with soil mixes for seed sowing
6.	Mango rootstocks seedling in poly bags ready for grafting
7.	Earthen pots widely used for container grown plants in India
8.	Mango shoot etiolated with black polyethylene sheet
9.	Unrooted and rooted grape cuttings
10.	Cured scion sticks of mango for side veneer grafting
11.	A veneer grafted mango plant
12.	Methodology of stone grafting in mango commercially employed in *Konkan* region of India
13.	Air layered shoot of lemon
14.	Profuse rooting in stooled plant of apple
15.	A view of rooting in guava stools
16.	Runner formation in strawberry
17.	A view of flowering annuals raised in a plastic house
18.	Different steps involved in raising grape through micropropagation
19.	A mango explant growing in MS medium supplemented with activated charcoal
20.	Raising citrus plants through micropropagation
21.	A view of Shimla mirch in a green house
22.	Papaya seedlings in plug trays
23.	Seed sowing in plug trays
24.	A view of cucurbits seedlings raised in plug trays
25.	Plug tray raised seedlings ready for transplanting

Introduction

The existence of plants on the earth is highly essential for the survival of mankind because plants not only provide the basic needs of animals but also release oxygen, which is much more essential to the mankind. Man is continuously disturbing the nature since ancient times for his gains. Among different agricultural crops, horticultural plants are known to provide almost all the basic needs of man. For his survival, much progress has been on the domestication of wild types and improvement in the existing types. The whole system of domestication and improvement could have collapsed if there would have not been multiplication of the plants. Basically, all plants in this universe multiply themselves both by sexual (seed) or asexual (vegetative) means. However, scientists have now developed several new techniques for the speedy and better multiplication of different plants.

The propagation of plants has been one of the fundamental occupations of mankind since civilization begins. Ancient man learnt about growing of some cereal crops for his survival and with advancement, many more cereal crops, medicinal and horticultural plants were also added. Later, progress was made on the selection of the desirable types from the wild population (tomato, barley, rice etc.), natural hybrids (wheat, strawberry) and natural population (Cole vegetable crops). But this progress would have of no significance if man could not have any method to maintain them. Most of the selected plants would have lost or could have attained undesirable forms if they were not propagated under strict environmental control. This basically forced man to learn about new techniques or methods of propagation for perpetuation of desirable types. Initially, ancient man learnt to propagate potato and sweet potato by division of tubers; banana and ginger by division of rhizomes and fruit crops like grape, fig, pomegranate, olive etc., by hardwood cuttings and pineapple and date palm by suckers. Interestingly, these crops are still propagated commercially by these methods initiated by man at that time, although a lot of progress has been made by the scientists in the field of propagation of fruit crops. After the above mentioned techniques, many new methods of budding and grafting have been standardized with the progress in time. Now, every day, new dimensions are made in the field of plant propagation in general and horticultural plants in particular. For example, invention of glass house; mist propagation; hybrid seed production technology; micro-budding; micro-grafting; rooting co-factors in the rooting of cuttings; micropropagation; epicotyl grafting in mango; stooling in guava, mango and litchi; side-veneer grafting and softwood grafting in mango; bottom heat technique for rooting of cuttings of some difficult-to-root species are few new entries in the field of propagation of horticultural crops in the last century.

CHOICE OF THE PROPAGATION METHOD

The first and foremost objective of a propagation method should be to produce individuals that are identical of mother or original plant. Thus, a successful propagation method is the one, which transmits all the desirable characters of a mother plant to the offspring. If a method is unable to do it, then it is considered as an undesirable method. Hence, only desirable or recommended method of propagation should be tried. Most of the present day horticultural plants have been selected from the wild species, which are the resultant of natural hybridization and mutation, and if these are not multiplied by the suitable methods, these may lose their identity and all efforts made by the breeder will go futile. In general, the self-pollinated plants, which are considered as homozygous, are propagated by seeds. Most vegetables and flowering annuals fall in this group, which are propagated by seeds and they maintain all the desirable characters of the mother plants in their offspring by self-pollination. However, the cross-pollinated plants (mostly fruit and nut crops) unlike self-pollinated plants, show heterozygosity, if propagated by seeds. As a result, seeds do not carry the true characteristics of the mother plant and hence seed propagation of these plants is not desirable. Therefore, such plants should be propagated by vegetative means because the vegetatively propagated plants are phenotypically and genotypically similar to the mother plant. Further, such plants don't show any variation until and unless some mutations take place in them. In general, the choice of propagation method for a specific crop is different owing to genetic, climatic and other variations. Listed below are some methods of propagation of major horticultural plants :

I. Sexual Propagation

Seeds: Most annuals, biennials and some perennial fruit plants; vegetables crops; plantation, aromatic and medicinal plants; ornamental, flowering and shade providing shrubs and trees.

II. Asexual Propagation

(i) **Propagation by apomicitic seedlings :** Citrus, mango, apple etc.

(ii) **Propagation by vegetative structures**

(*a*) ***Propagation of division***

Division of stem tubers : Potato, Jerusalem artichoke, caladium etc.

Division of tuberous roots : Sweet potato, canna, turmeric, dahlia, bigonia etc.

Division of rhizomes : Ginger, banana, blueberry, bamboo, iris, oxalis etc.

Division of stock : Gooseberry, current, strawberry, etc.

(b) Propagation by separation

Bulbs: Onion, daffodil, Bird of Paradise, tulip, tuberose, lily, narcissus, hyacinth etc.

Corms: Glodiolus, freesia, crocus etc.

(c) Propagation by suckers

Pineapple, banana, chrysanthemum, date palm, blackberry, raspberry, *kair, pilu,* etc.

(d) Propagation by runners

Strawberry

(e) Propagation by stolon

Doob grass (*Cynodon dactylon*), *Oxalis* etc.

(f) Other vegetative structures

Bulbils : Lilies

Offsets : Pineapple, banana, cacti and succulents, tulip, iris, lily etc.

(*iii*) Propagation by cuttings

(a) Stem cuttings

Hardwood cuttings : Grape, fig, pomegranate, mulberry, kiwifruit, olive, quince, rose, privet, honeysuckle, wisteria, weeping willow, poplar etc.

Semi-hardwood cuttings : Mango, guava, jackfruit, camcllia, coleus, geranium, lemon, sweet potato, rhododendron etc.

Softwood cutting : Juniper, hollyhock, azalea, lilac etc.

Herbaceous cuttings : Chrysanthemum, dahlia, defenbacia, begonia, coleus, carnation, geranium etc.

(b) Leaf cuttings :African violet, sansevieria, bigonia, bryophyllum etc.

(c) Leaf bud cuttings : Blackberry, lemon, cammelia, hydrangea, kalanchoe etc.

(d) Root cuttings *:* Blackberry, fig, cherry, apple, pear etc.

(*iv*) Propagation by layering

(*a*) *Tip layering* : Trailing blackberry, raspberry etc.

(*b*) *Simple layering* : Ornamental climbers, bougainvillea, grape, guava, pomegranate, lemon etc.

(*c*) *Stooling/mound layering :* Clonal apple rootstocks, guava, mango, litchi, gooseberry etc.

(*d*) *Serpentine layering :* Clematis, American grape etc.

(*e*) *Trench layering* : Clonal apple rootstocks, cherry, plum etc.

(*f*) *Air layering (Gootee or Marcotage)* : Litchi, guava, lemon, pomegranate, Macadamia nut, rubber plant, croton, rose etc.

(v) Propagation by grafting

(*a*) *Whip grafting* : Several fruit plants

(*b*) *Cleft grafting* : Avocado, apple, pear, plum, camelia etc.

(*c*) *Bark grafting* : Several fruit plants

(*d*) *Side grafting* : Mango, avocado, junipers etc.

(*e*) *Inarching* : Mango, sapota, guava, litchi etc.

(*f*) *Veneer grafting* : Mango, cashewnut etc.

(*g*) *Stone/epicotyl grafting* : Mango, cashewnut *etc*.

(*h*) *Top working :* Mango, *ber*, cashewnut, mulbery etc.

(*i*) *Softwood grafting* : Cashewnut, mango etc.

(*j*) *Bridge grafting* : Used for repairing of the damaged fruit plants.

***(vi)* Propagation by budding**

(*a*) *T-budding* : Citrus, plum, peach, cherry, *ber*, *aonla*, rose etc.

(*b*) *Patch budding* : Walnut, pecannut, *bael*, Indian rubber tree etc.

(*c*) *Chip budding* : Mango, grape etc.

(*d*) *Ring budding* : Peach, plum, *bael*, *ber*, mulberry etc.

***(vii)* Micropropagation**

(*a*) *Shoot tip culture* : Citrus, papaya, grape, orchids, asparagus, lily, gladiolus, strawberry, carnation, chrysanthemum etc.

(*b*) *Embryo culture* : Citrus, rhododendron etc.

(*c*) *Single cell culture* : Carrot, sunflower etc.

(*d*) *Embryo rescue* : Grape, peach, citrus etc.

(*e*) *Meristem culture* : Banana, strawberry, guava, rose etc.

(*f*) *Anther culture* : Datura

(*g*) *Spore culture* : Ferns

(*h*) *Seed culture* : Orchids

GENETICAL CONCEPTS CONCERNING PLANT PROPAGATION

For proper understanding of the science and technique of plant propagation, one should have basic knowledge about the plant growth, structure of plant (root, stem, leaves, flower, seed etc.), different kinds of plants and various possible methods by which different plants can be propagated. For basic knowledge on genetic processes and structure, one should have idea about the inheritance pattern and genetic manipulations that maintains the characters of a plant from generation to generation.

In fact, cell is the basic unit of life and propagation also has cellular bases. As stated earlier, plant propagation involves the control of two different types of life cycle, the sexual (seed) or asexual (vegetative) but both should have the basic capacity to reproduce offspring similar to the mother plant.

MEIOSIS AND SEXUAL REPRODUCTION

Sexual reproduction involves the union of male and female sex cells, the formation of seeds and creation of a population of seedling individuals with new and different genotypes. Meiosis cell division, which is also known as reduction division, takes place, which basically involves sex cells and reduction in number of chromosomes to half each in female and male gamate. They unite together, giving rise to new offspring cells. The offspring so produced may resemble either or none of the parents. The inheritance of characters from the parents takes place from generation to generation through the action of gene present on the chromosomes of the sex cells. Some traits are controlled by single and some by multiple genes. In such cases, two terms *viz.* homozygous and heterozygous are used for describing the genotype of a seed or plant produced by it. As stated earlier, a population is usually homozygous if produced by self-pollinated plant and heterozygous, if produced by cross pollinated plant.

MITOSIS AND ASEXUAL PROPAGATION

Asexual propagation is possible because each cell of a plant contains all the genes necessary for the growth and development (called as totipotency of a cell) and during cell division (mitosis), these are replicated in daughter cells. In mitosis cell division, the chromosomes divide into two daughter cells. As the chromosomes produced are the same as in mother cells, the daughter cells also have all the characteristics similar to the mother cells and ultimately the new plants also have genotype and phenotype similar to mother plant. Mitosis occurs basically in specific growing plant parts like shoot apex, root apex, cambium, intercalary zone and callus mass, which results in the growth and development of vegetative parts of a plant. Basically, mitosis cell division is responsible for wound healing and normal vegetative growth, and regeneration, which make the possibility of raising many horticultural plants by cuttings, layering, separation or division. These methods of vegetative propagation are of commercial importance because they permit large-scale multiplication of desirable plants. However, the most important aspect is that the plants produced by these methods are identical to the mother plants except in some cases, where sudden heritable changes (mutations) or chimeras may alter the genotype and phenotype of new individuals as well.

• • •

CHAPTER 2

Plant Structure

A person who is engaged in plant propagation work must have fair idea about the structure of a plant, because, in principle, plant propagation methods are based to a considerable extent on the structure of a plant and the functions of plant parts. The major parts of a plant are the roots, stem, buds, leaves, flowers, fruits and seeds. These parts differ widely in respect to shape, size and colour among different horticultural plants.

Root

Root is an essential underground part of a plant, which in addition to anchorage, absorbs water and nutrients for the soil. In some plant species, it acts as an important organ for food storage also (*e.g.*, potato, sweat potato). Mostly, two root systems exist in horticultural plants *i.e.* tap root system and fibrous root system (Figure 2.1). In general, a root has four parts *viz.* region of conduction, region of absorption, region of cell elongation and root cap. All parts of a root, except root cap, make secondary growth. The shape and size of these root systems is different in different plant species. For example, the taproots of young pecan plant develop more prominently than the lateral roots. In peach and apple, there is very limited development of the taproot and it gives rise to laterals, which constitutes the main root system. Similarly, the root system of the monocot and dicots is different and accordingly their external as well as internal structure is also entirely different. Thus, it is quite possible that the method of propagation is considerably influenced by the root system of a plant.

Stem

The stem is an aerial part of a plant, which provides the framework to plant. It supports the branches, leaves, flowers and the fruits. It also serves as a conducting system between the roots and the leaves. Like roots, the stem of dicots and monocots has different structure. Stem of dicots has xylem, cambium, phloem, cortex and epidermis in successive order outwards from the pith in the centre. Meristematic cells located between the xylem and phloem are called as cambium, cambium layer and the cambial region. Cambium region is vital in plant propagation as it plays three important roles *i.e.*, (*i*) it is responsible for the development of new cells, thereby growth, healing of wounds, (*ii*) development of callus, and (*iii*) rooting primordia. All these are essential for the success of a propagation method. In monocots, the cambial strands and vascular bundles are isolated in masses of other tissues. Thus, the presence of cambial layer in dicots make the rooting process easy in such plants and absence of this layer is primarily responsible for failure of vegetative methods in monocots.

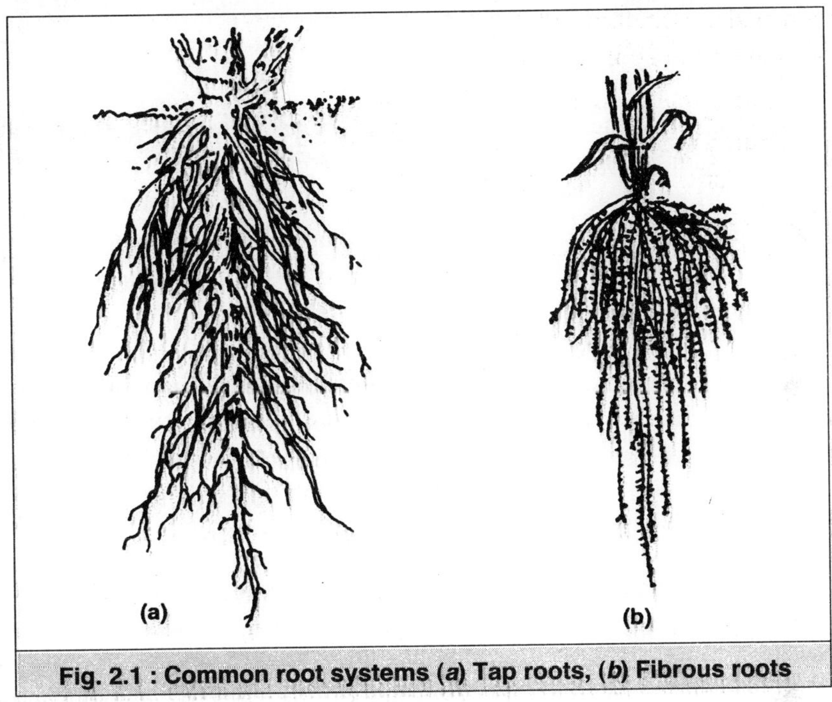

Fig. 2.1 : Common root systems (*a*) Tap roots, (*b*) Fibrous roots

Buds

A bud is a condensed, immature, young shoot, having undeveloped internodes because of which, the leaves become overlapped and the bud acquire a compact structure. Usually, young stem develops some scaly structures (buds) at points called as *nodes*. The area between nodes is called as *internode*. A bud is a growing point surrounded by embryonic leaves or floral parts. In fact, it is a rudimentary stem, protected by an envelope of scales. Buds have been classified on the basic of the mode or time of origin; position on the stem, node, and time at which they begin to grow and function.

On stem, these may be *lateral* or *terminal*. Most lateral buds are borne in the axils of leaves, called as *axillary buds*. However, sometimes no bud occurs (as in peach, tung etc.) in leaf axils, which are called as *blind buds*. In some species, only one bud develops, and in others, even 2, 3 or 5 buds may develop. For example, in pecannut and walnut, 2-4 buds develop at one node, and one nearest to the terminal of the shoot, is the largest, called as *primary bud*. Others buds are called as *secondary* or *reserve buds*.

On the basis of function, the buds are classified as *leaf buds* and *flower buds*. Leaf buds contain a growing point of undifferentiated tissue and they give rise only to new vegetative shoots and thus plant height and spread are mostly due to these buds. On the other hand, fruit buds contain various rudimentary parts of a flower. These buds give rise to flower and thereby to fruit. The vegetative buds are pointed, while the floral buds are plumpy. In some fruit plants, the leaf buds may change to flower bud (as in peach). Similarly, some fruit plant produce mixed buds, containing both flowers and vegetative buds in axis within the same cluster of bud scale, as in apple.

On the basis of growth, the buds may be *dormant* and *latent* as in temperate fruit crops, wherein, most fruit plants develop and mature in a given season and remain dormant in winter. Such buds start growing in the following spring, whereas, some remains inactive for a long time and become *latent*. Sometimes, the buds may arise from the young shoots or twigs, which emerge from a point other than branches. These buds are known as *adventitious buds*.

Water Sprouts and Suckers

In some fruit and ornamental plants, some shoots have very rapid and luxuriant growth as compared to the other shoots on the same plant. These shoots may arise from the base or from side branches on the stem. Such shoots, which grow from adventitious buds on the roots, are called as *suckers*. Formation of suckers is quite common it some ornamental plants, blackberries, dwarfing apple rootstocks, pineapple and date palm.

Vigorous growing shoots that arise from the trunk or large limbs of the plant are called as *water sprouts*. Water sprouts may arise from adventitious buds but usually come from latent buds that remain dormant for quite some time and start growing. Such sprouts are usually unproductive and should not be used for propagation. Occurrence of water sprouts is quite common in apple, citrus, bougainvillea, rose, grape and chrysanthemum.

Leaves

Leaf is the main photosynthetic part /organ of a plant. Leaves are the lateral appendages formed by the stem in elongation. These are required by the plants for the manufacture of carbohydrates – the food material, which is highly essential for growth. Different plants have leaves of different shapes and sizes (Figure 2.2) but their functions are the same. Some plants retain their leaves season after season and remain green throughout, called as *evergreen plants*. In evergreen plants, some are *broad-leaved evergreen* (citrus, mango, litchi, avocado etc.) and some are *narrow-leaved evergreen* (leaves commonly called as needles) as in coniferous trees like *Pinus roxburghii, Cedrus deodara, Pinus wallichiana, Pinus gerardiana* etc. Some plants shed their leaves at certain seasons, called as *deciduous plants*. For example apple, pears, *phalsa*, grape, apricots, pecans, walnut etc., shed their leaves during winter while *ber* sheds its leaves during summer.

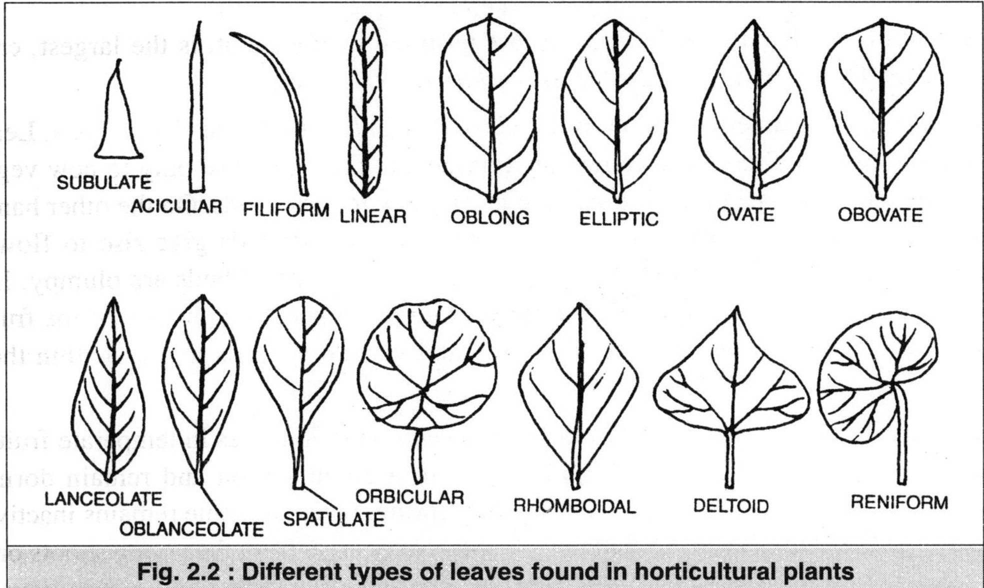

Fig. 2.2 : Different types of leaves found in horticultural plants

Flower

It is the reproductive part of a plant and is a fore-runner of the seed. It is very essential to study the structure of a flower, which basically plays vital role in the formation of fruits and seed.

Most horticultural plants except some ornamental foliage plants, produce flowers. For flower induction, two distinct morphological stages *i.e.*,vegetative and reproductive take place in the life cycle of a plant (Figure 2.3). For reproductive stage to occur, a plant has to complete certain period of juvenile phase. Vegetative phase continues for a certain time during which plants do not respond to flowering inducing stimulus. In the reproductive stage, the vegetative growing points differentiate to become reproductive and thus develop into flowers. The initiation of flowers makes the end of the vegetative stage of that particular plant part. In a particular plant, only some parts develop flowers and others may not. Thus, some parts remain vegetative for some time but in the years to come, they may also become reproductive.

The vegetative and reproductive stages of plants are different and vary widely among different horticultural plants. For example, annual plants (mostly flowering annuals and vegetable crops) have very short vegetative growth (3-4 months,) whereas, woody perennials have usually very long (6-8 years) pre-bearing period (juvenile phase) or even more. Similarly, the bud differentiation also varies among different horticultural plants. In some plants, flowering takes place in summer, in others, it takes place during spring and still in others during winter, depending on species and environment in which they are growing. The conversion of vegetative phase of a plant into reproductive phase, is genetically controlled. However, it can be mediated by environmental conditions of a particular locality, cultivar, exogenous application of hormones or other biological manipulations. But, production of a flower is most important for perpetuation of that plant in nature as the flower after fertilization leads to the production of seeds.

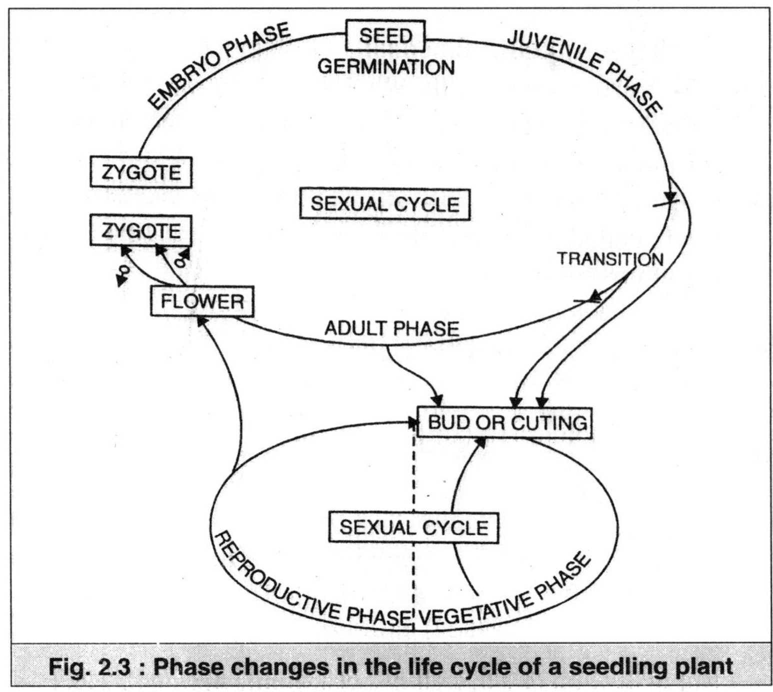

Fig. 2.3 : Phase changes in the life cycle of a seedling plant

The flower may be borne singly or clusters in the axils of leaves or at the apex of a stem. Their shape, size and colour vary widely amongst different horticultural plans. Usually, a flower has four parts, the sepals, the petals, the stamens and the pistils, which are borne on a receptacle (Figure 2.4). The sepals are usually green leaf like structures, collectively called as

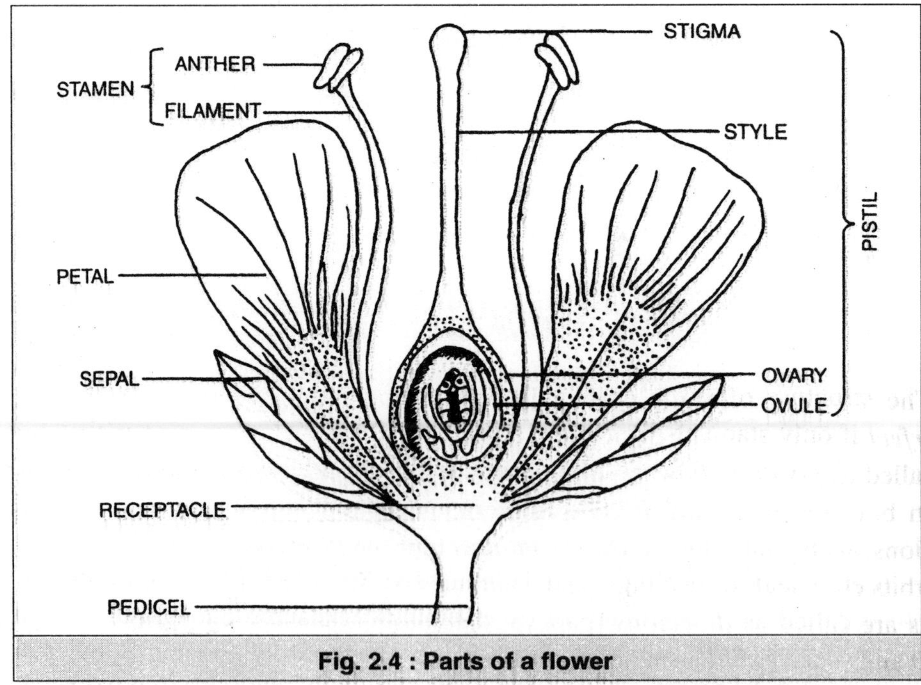

Fig. 2.4 : Parts of a flower

calyx (Figure 2.5). They enclose the other parts of a flower and protect it from waggaries of climate or otherwise. Next to sepals, is the whorl of coloured floral parts (petals), collectively called as *corolla* (Figure 2.5). The colour, shape, size and arrangement of corolla is different in various kinds of plants. It is the most beautiful part of a plant and it chiefly attracts the insects towards it, which help in pollination. The male part of a flower is called as stamen, which are collectively called as *androecium*. Each stamen has a filament and anther. Anther produces the pollen grains for pollination. The pistil is the female part of a flower, which has three distinct parts *viz.* the stigma, the style and the ovary. The pistils are collectively called as *gynoecium*. Two classes of flower are generally recognized. The flowers are called as *perfect* or *hermaphrodite* if both stamens and pistils are present in it. Such flowers are capable of self-pollination. Perfect flowers are found in some fruit plants (apple, pear, peach, orange), most vegetables (tomato, peas, lettuce, onion) and flowering annuals.

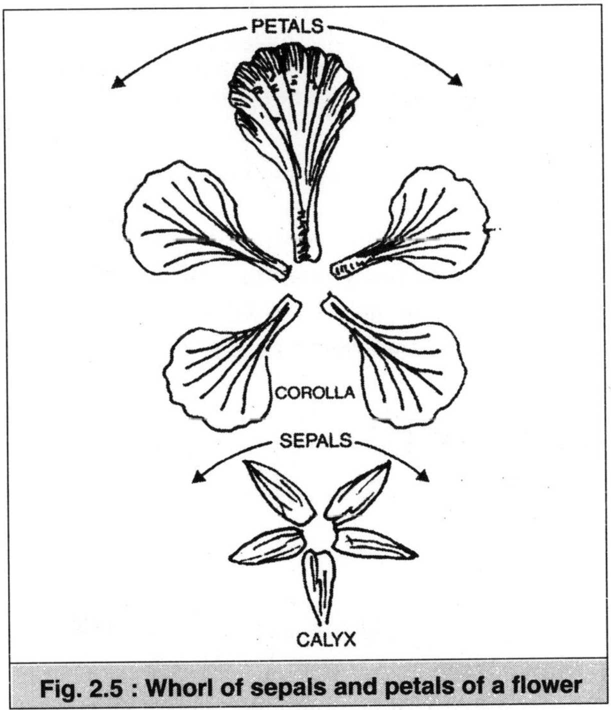

Fig. 2.5 : Whorl of sepals and petals of a flower

The structure of some perfect flowers is given in figure 2.6. A flower is called as *imperfect* if only stamens or pistils are present in it. Flowers having pistils but not stamens, are called as *pistillate flowers,* and *staminate,* if only stamens are present but not the pistils. When both the male and female flowers are present on the same plant but at different locations, such plants are called as *monoecious* (walnut, pacannut, hazelnut, tung, coconut, cucurbits etc.) and if pistillate and staminate flowers are present on different plants, such plants are called as *dioecious* (papaya, date palm, pistachio nut, Mascadine grape, pointed gourd and snake gourd, asparagus etc.)

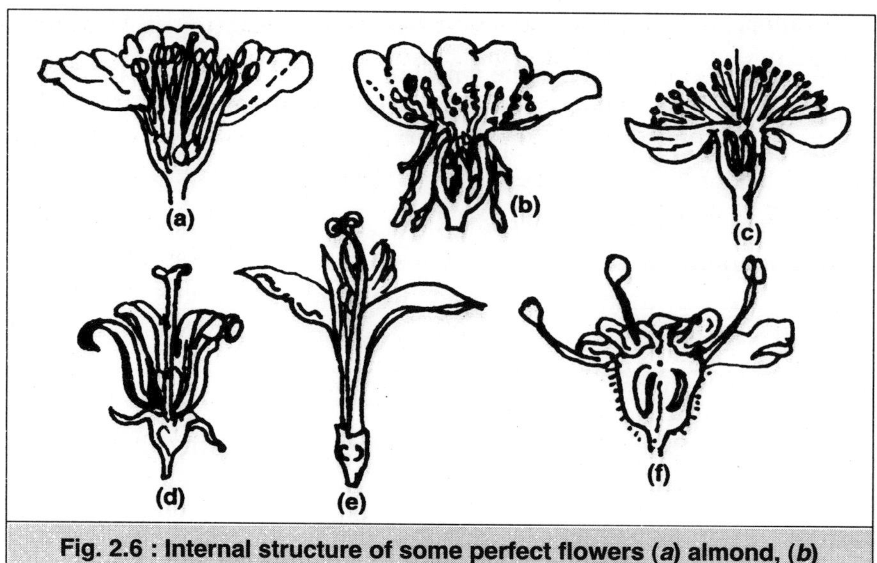

Fig. 2.6 : Internal structure of some perfect flowers (*a*) almond, (*b*) bluebell, (*c*) rose, (*d*) coffee, (*e*) myrtle, (*f*) carrot

Inflorescence

The arrangement of flowers on the stem axis is called as *inflorescence*. The stem of plant that bears the individual flower of the inflorescence is called as *pedicle* and the main stem of the inflorescence from which the pedicels arise is called as the *peduncle*. The stem of the flower, borne singly is also known as the *peduncle*. A flower attached to peduncle without stalk is called as *sessile flower*. The flattened structure to which flower parts are attached, is called as *receptacle*.

TYPES OF INFLORESCENCE

The different types of inflorescences found in horticultural plants have been described briefly hereunder :

A. Racemose Inflorescence

In this type of inflorescence, older flowers are at the base and younger one at top or near to growing points. It is of five types :

Spike : In this type, the main axis is elongated and flowers are sessile. It is commonly found in banana, gladiolus and snapdragon. *Catkin* is a spike, which bears either pistillate or staminate flowers and usually falls of the plant. It is found in pecan nut, oak, mulberry, dates and walnut.

Raceme : In this type, the main axis of the inflorescence is elongated and flowers are borne on pedicles of about equal length at maturity. It is common in cabbage, cauliflower and water hyacinth.

Corymb : In corymb, main axis is elongated but the pedicels of the older flowers are longer than the younger ones and thus, it has a flat top. Cherry, pear and candytuft have this type of inflorescence.

Umbel : In umbel, the main axis is short and knob like. The flowers are borne on pedicels, which are of equal length. Onion and carrot are important examples of umbel.

Head : In head, the main axis is very short, which bears sessile flowers. Sometimes, whole head is treated as one flower but in reality, it is an inflorescence, which contains numerous separate flowers. Sunflower, lettuce, marigold, and chrysanthemum are some examples of this type of inflorescence.

B. Cymose Inflorescence

In cymose inflorescence, the main axis ceases to grow due to formation of flowers at the top and other flowers are borne on lateral branches. It is very interesting to note that in this type of inflorescence, the youngest, and last, flower to bloom is the farthest from the tip. It is of the following three types :

Solitary : Some plants produce only one terminal flower on the peduncle. For example, rose, tulip and poppy.

Fascicle : In this type, flowers are closely crowded. Phlox and apple have fascicle type of inflorescence.

Glomerecule : In this type, the flowers are very closely crowded, more than fascicle. However, flowers of centre first open, progressively outward. Dogwood plant produces this type of inflorescence.

C. Compound Inflorescence

Some horticultural plants produce both determinate and indeterminate type of inflorescence which is called as compound inflorescence. There are different types of compound inflorescence but the simplest type is *panicle* in which, a group of flowers like an ordinary raceme is borne as in mango, grape and litchi. Compound umble inflorescence is found in *parsnip* in which many umbels are produced on the central axis. Some common types of inflorescence found in horticultural crops are given in Figure 2.7.

Pollination

Pollination may be defined as the process of transfer of pollen grains from anthers to the stigma of a flower. It is the first and foremost step in the process of fertilization and for seed formation. For proper pollination to occur, the pollen should be mature and stigma should be receptive. The process by which pollen is shed is known as *anthesis*. On the basis of pollination requirements, the horticultural plants may be self-pollinated or cross-pollianted. In self-pollination, which is also called as *autogamy,* the pollen comes from the same flower to pollinate the stigma of the same flower. Self-polliantion, in general, is a rule in plants having perfect flowers, although chances of cross-pollination are never overruled due to one or the other peculiarity present in the flower. Cross-pollination, which is also known as *allogamy,* is the pollination of a flower with pollen from any other flower. Cross-pollination between two flowers of the same plant is called as *geitogamy*. Most of the fruit plants and their cultivars are naturally cross-pollinated but several vegetables and flowering annuals are self-pollinated.

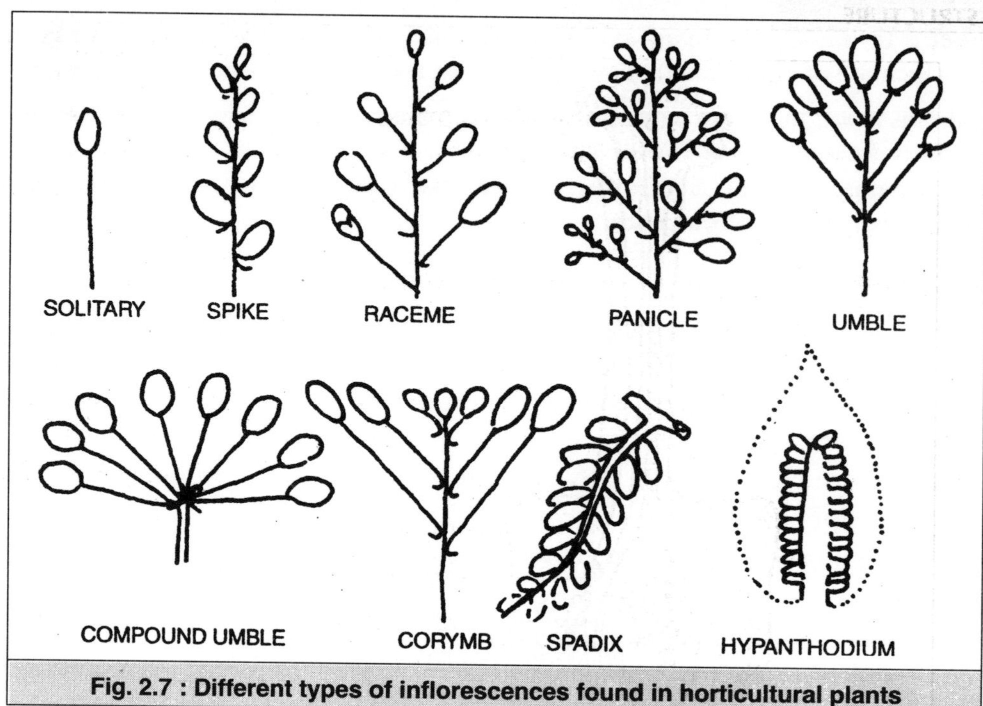

Fig. 2.7 : Different types of inflorescences found in horticultural plants

Some mechanisms inherited in the plants like dioecy (as in date palm, pistachio nut and asparagus), monoecy (as in pecans, walnut, filberts, cucurbits etc.), dichogamy (as in walnut, pacannut) macadamia nut etc.), self sterility (as in some cultivars of almond and other stone fruits, onion, marigold etc.) and incompatibility (as in mango, *ber*, loquat, apple cole group of vegetables, petunia, lily, etc.) favour cross pollination in them. Cross-pollination is accompanied in nature by various agencies like wind (as in walnut, pecannut and other temperate zone nuts, coconut, date palm), insects like honeybees, moths, wasps and syrphid flies (as in apple, mango, peach, fig, coconut, cole crops, onion, cucurbits etc.), birds (as in pineapple and banana), and water (as in water lily, water chestnut etc.). Hand pollination is generally done in hybrid seed production of many vegetables and flowering annuals and sometimes in date palm. For producing hybrids in horticultural plants, hand pollination is usually done.

The Process of Fertilization

After successful pollination, the pollen grains germinate on the stigma of the flower and reach the ovule. The two nuclei of the pollen grains are discharged into the embryo sac. In the mean time, there is development of eight nuclei in the ovule, consisting of an egg cell, two synergists at one end, three antipodal nuclei at the other end and two polar nuclei near the centre. One of the nuclei of the pollen tube, fuses with the egg cells, producing the embryo of the seed and, the other with two polar nuclei to produce endosperm. This complete process is called as *double fertilization* (Figure 2.8). After proper fertilization, the fruit grows. Many factors affect fruit set in different horticultural crops but defective floral

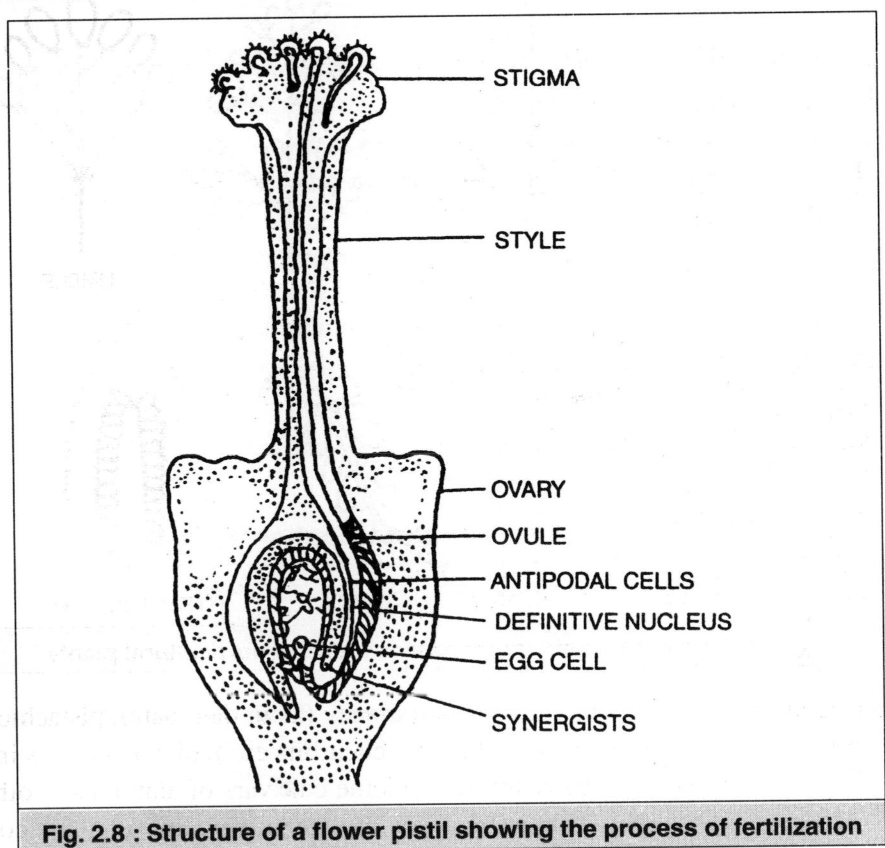

Fig. 2.8 : Structure of a flower pistil showing the process of fertilization

parts, inadequate pollination, incompatibility, dichogamy, dioecy, monoecy, and several physiological, environmental and nutritional factors are known to affect it invariably. After successful fertilization, fruit setting takes place.

FRUIT

Botanically, fruit is a mature ovary. However, in some fruits crops, other floral parts are also involved in the fruit development process. A fruit is called as a '*true fruit*', if it is developed fully from cells or tissue of one or more ovaries but as *false fruit,* if it is composed of tissues other than that of an ovary as in apple, pear, and pineapple. In general, fruit has one or more seeds. However, seedless fruits are also produced by some plants (as in banana pineapple etc.). For the development of a fruit, proper pollination and fertilization is required but parthenocarpic fruits (seedless) don't require even pollination or fertilization.

Kinds of Fruits

Although, most fruits consist of only a single ovary, there are some, including a number of the most important cultivated ones, which consist of or include several ovaries. In some of these cases, the several ovaries are from a single flower, and in others, they are from several very closely clustered flowers. Hence, the fruits have been classified into three major groups (Table 2.1) on the basis of the number of ovaries involved in their formation, as follows :

Simple fruits : Consisting of a single enlarged ovary, with which some other flower parts may be incorporated. Most common fruits are simple fruits except those listed below as examples of aggregate and multiple fruits.

Aggregate fruits : Consisting of a number of enlarged ovaries belonging to a single flower and massed on or scattered over the surface of a single receptacle. The separate ovaries are spoken of as fruitlets (as in raspberry, blackberry and strawberry).

Multiple fruits : Consisting of the enlarged ovaries of several flowers, more or less coalesced into one mass (as is mulberry, fig and pineapple).

On the basis of their consistence, structure and dehiscence, simple fruits have been divided into the following different kinds (Table 2.1).

TABLE 2.1 : Botanical classification of fruits

Simple fruits			*Aggregate fruits*	*Multiple fruits*	*Accessory fruits*
1. Fleshy fruits	2. Dry fruits				
	Dehiscent fruits	Indehiscent fruits			
(*a*) *Berry* (banana, grape, papaya, tomato, pepper, brinjal etc.)	(*a*) *Capsule* (iris, lily, gladiolus, azalea, rhododendron, portulaca, poppy etc.)	(*a*) *Achene* (butter cup, sun flower etc.)	Strawberry, custard apple, raspberry	Pineapple, mulberry, fig etc.	Dewberries, black berries etc.
(*b*) *Hesperidium* (orange, mandarin, lime, lemon, grape-fruit, pummelo, etc.)	(*b*) *Follicle* (delphinium, marigold, larkspur, peony etc.)	(*b*) *Caryopsis* wheat, rice, maize, barley, oats, most grasses etc.			
(*c*) *Pepo* (watermelon, muskmelon, cucumber, squashes, pumpkins, gourds, melons etc.)	(*c*) *Siliqua* (mustard, stocks, cauliflower, cabbage, radish turnip, rutabaga rapeseed, knol khol etc.)	(*c*) *Nut* (acorn, pecannut, walnut, hazelnut, oak, cashewnut etc.)			
(*d*) *Pome* (apple, pear, quince, loquat etc.)	(*d*) *Pod or legume* (peas and beans etc.)	(*d*) *Samara* (maple, ash elm, dioscorea etc.)			
(*e*) *Drupe* (*stone*) (mango, peach plum, cherries, apricot, almond, olive etc.)		(*e*) *Schizocarp* (parsnip, celery, mallow, carrot, parsely etc.)			

1. Fleshy fruits

Fleshy fruits have been divided into following categories :

(A) Berry: In berry, the ovary wall is fleshy and enclosing one or more carpels and seeds as in grape, banana, papaya, brinjal, pepper, tomato, and date. It is of two types :

(*a*) *Papo:* A type of berry with hard rind. For example, squash, watermelon, muskmelon, pumpkin, bottle gourd, snake gourd, bitter gourd, and cucumber.

(*b*) *Hesperidium* : A type of berry with a leathery rind. It is common in mandarin, orange, lemon and limes.

(B) Drupe or stone fruits : Such fruits are derived from a single carpel and are one-seeded; exocarp thin, forming the peel; mesocarp fleshy; endocarp stony. Examples of this group are cherry, peach, mango almond, plum and olive.

(C) Pome fruits : Such fruits are derived from several carpels; receptacle fleshy; outer portion of pericarp fleshy; inner portion papery, forming 'core'. Important examples of this type are pear, apple, loquat, and quince.

2. Dry fruits

A. Dehiscent fruits

(*a*) **Legume or true pod :** In this type, carpel is one; splitting along two sutures. For example, pea, beans, vetch etc.

(*b*) **Follicle :** In this type, carpel is one; splitting along one suture. For example, delphinium, marigold, larkspur, columbine, peony etc.

(*c*) **Capsule :** In this type, carpels are two or more; dehiscing in one of four different ways:

1. Along the line of union of carpels (septicidal dehiscence) : *e.g.* azalea.
2. Along the middle of each carpel (loculicidal dehiscence): *e.g.* iris, lily.
3. By pores at the apices of carpels (poricidal dehiscence): *e.g.* poppy.
4. Along a circular horizontal line (circumscissile dehiscence): *e.g.* purslane and plantain.

(*d*) **Siliqua :** In this type, carpels are two; separating at maturity, leaving a partition wall persistent as in mustard, cabbage, cauliflower, turnip, radish, knol khol etc.

B. Indehiscent fruits

(*a*) **Achene:** Such fruits are one-seeded; seed attached to ovary wall at one point only. Its examples are buckwheat, sunflower, buttercup etc.

(*b*) **Caryopsis (grain) :** Such fruits are one-seeded; pericarp firmly united all around to testa or seed coat. Its examples are wheat, corn, rice, barley, broom corn, oats, and all other grasses.

(*c*) **Samara or 'key' fruit :** Such fruits are one or two-seeded; pericarp, bearing a wing-like outgrowth. Its examples are ash, elm, maple, and dioscorea etc.

(*d*) **Schizocarp :** In such fruits, carpels are two or more; united splitting apart at maturity. Its examples are carrot, parsnip, parsley, celery, mallow etc.

(*e*) **Nut :** Nut is a hard, one-seeded fruit, generally produced from a compound ovary. Its examples are acorn, peanut, walnut, hazelnut, cashewnut etc.

Some common types of fruits are shown in figure 2.9*a* and 2.9*b*.

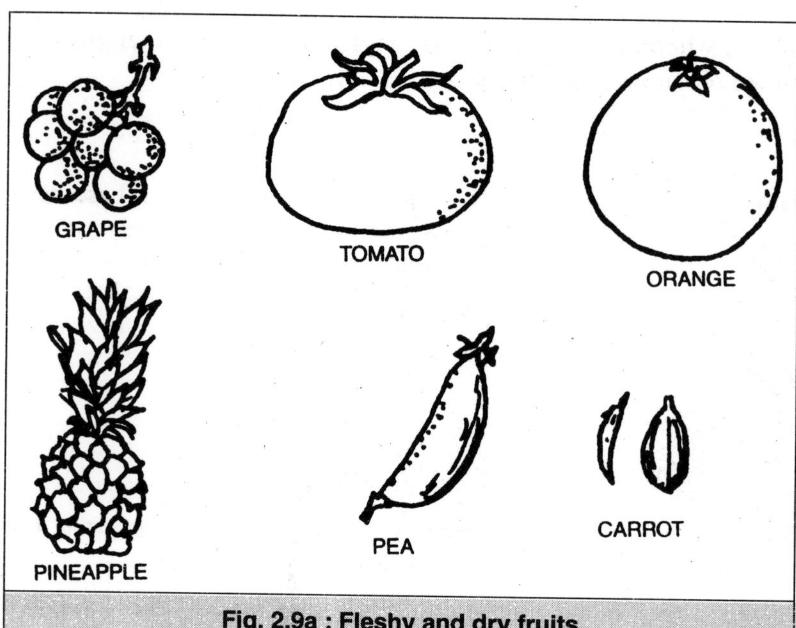

Fig. 2.9a : Fleshy and dry fruits

Fig. 2.9b : Female cone of *Pinus roxburghii*

Seedless Fruits

Some horticultural plants produce seedless fruits. The seedlessness may be due to parthenocarpy and embryo abortion. In parthenocarpy, fruit setting takes place without fertilization. Parthenocarpy is of two types. (*i*) When plants set and mature fruits without pollination, it is called as *vegetative parthenocarpy* as in banana, Japanese persimmon, fig, orange, grapefruit and some cucumber cultivars (*ii*) In some instances, the stimulus of pollen is required, though, the ovule develops without fertilization, it is called as *stimulative parthenocarpy*. This type is common in Black Corinth grape and Jerusalem cherries. Interestingly, in some fruits, there is development of seedless fruits even if successful pollination and fertilization has taken place. It is common in grapes and is known as *stenospermocarpy*. In such fruits, there is development of embryo up to certain stages but it aborts at later stages. This phenomenon has been observed in plum and cherries, although it is of no special significance in these fruit crops because of presence of bony endocarp.

Fig. 2.10 : A coniferous plant

Seed

A seed may be defined as the mature ovule, which contains a miniature plant (embryo), storage tissue (cotyledon or endosperm) and seed coat. In some plants, the seeds are enclosed and are not visible (angiosperms) and in some, these are naked (gymnosperms) (Figure 2.10). The seeds of some plants have only one cotyledon, called as *monocots*, whereas, some has two cotyledons, called as *dicots* (Figure 2.11). Seeds are the major source of propagation of several horticultural plants.

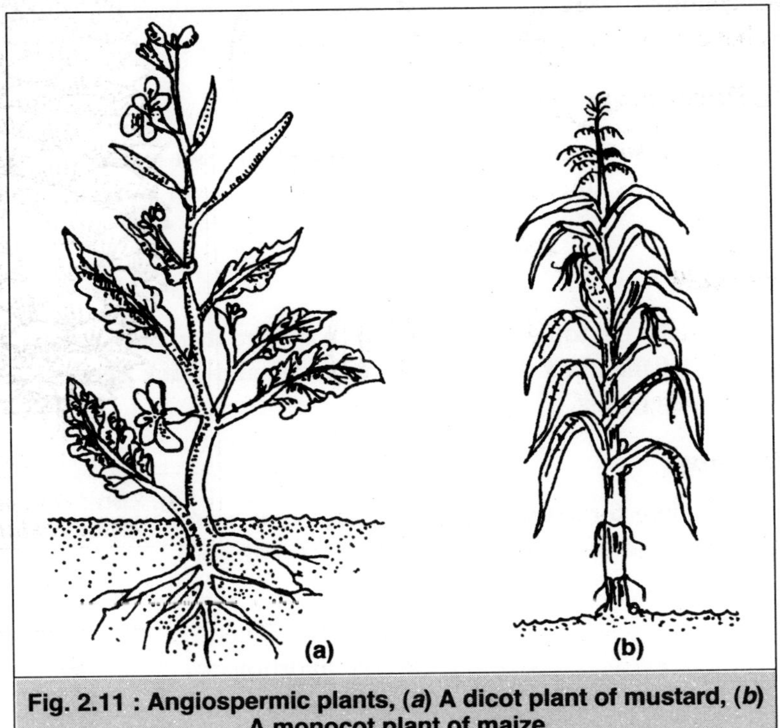

Fig. 2.11 : Angiospermic plants, (*a*) A dicot plant of mustard, (*b*) A monocot plant of maize

LIFE CYCLE OF HORTICULTURAL PLANTS

The life cycle of a plant is the entire process of growth that is involved from germination of a seed to the production of a crop of seed by the new plant. This duration is variable among horticultural plants as some plants complete this process in very short period and others take years together. For example, some flowering annuals complete their life cycle within a few weeks, whereas, others may take even 8 to 10 years to form seed. One group of plants produces one crop of seed and then dies. These are called as *monocarpic* plants. Other plants although complete their life cycle by producing seeds but don't die and continue to live and produce seeds for many years. These are called as *polycarpic* plants. However, on the basis of life cycle, the horticultural plants have been classified as annuals, biennials and perennials as under.

Annuals

Such plants complete their life cycle in one season and die afterwards. Most of the vegetables and flowering annuals fall in this group.

Biennials

Such plants make vegetative growth in one season and produce flowers and seeds in the next season. Cabbage, cauliflower and hollyhock are some examples of biennials.

Perennials

Such plants continue to grow for years together. They produce fruits and seeds but don't die. Perennials have been classified into following three types :

Herbaceous Perennials

In such type of plants, the top dies in winter but root persists. Under favourable environmental conditions, these again putforth new growth next season. For example mallow, asparagus etc.

Woody Perennials

In woody perennials, the top of the plant lasts for more than one season. Like herbaceous perennials, the roots of the plant don't die. Dewberries and black berries are two important examples of woody perennials.

Trees and Shrubs

Trees and shrubs live for several years and add some wood every year. Most fruit crops and ornamental shrubs and ornamental trees fall in this category.

• • •

CHAPTER 3

Sexual Propagation

Raising plants by means of seeds is called as 'sexual propagation'. Without doubt, there are numerous plants raised for our gardens from seeds than by any other method of propagation. Flowering annuals and vegetables are almost exclusively raised through seeds. The plants produced through seeds are called as *seedlings*. In sexual method of propagation, the sex organs of flower (androecium and gynoecium) are involved in processes like pollination and fertilization, resulting in the formation of seeds. The seeds on germination give rise to individuals with a new genotype.

The plants are propagated by seeds for three basic reasons, (*i*) for producing commercial crops, (*ii*) to develop new varieties, and (*iii*) for raising rootstocks for budding and grafting purposes etc. Raising of plants through seeds has certain advantages over vegetative propagation as follows :

ADVANTAGES OF SEXUAL PROPAGATION

The following are some advantages of sexual propagation :

1. Seed is relatively a cheaper and simple way of obtaining large number of plants as compared to vegetative propagation.
2. Seedlings have usually long life as compared to vegetatively propagated plants.
3. Seedling plants have better root system and therefore provide better anchorage than the vegetatively propagated plants.
4. The hybrids are first raised through seeds.
5. It is the only practical method of propagation for most of vegetables, annual flowers and fruit plants like papaya, *phalsa*, mangosteen, which can't be easily propagated successfully through vegetative methods.

6. Most of the rootstocks (except clonal rootstocks) for budding and grafting purposes in fruit plants are raised through seeds only.
7. Sexual propagation may sometimes leads to the production of chance seedlings, which are sometimes superior to the parent plant.
8. Seedling plants are comparatively resistant to insect-pests and diseases than the vegetatively propagated plants.
9. Seedlings are usually hardy and can tolerate waggaries of adverse climatic conditions in a better way than the vegetatively propagated plants.
10. Seeds can be stored for longer time and can be transported easily to distant places than vegetative propagation.
11. No special technical skill is required for raising plants through seeds.
12. Most viruses are not transmitted through seeds. Thus, seed propagation is useful in producing virus-free plants.
13. The mangosteen (*Garcinia mangostana*) fruit is developed through parthenogenesis. Seedlings obtained from the seeds of such fruits are similar to the mother plant and thus, propagated commercially through seeds.
14. Polyembryony is a rule in some varieties of citrus and mango. Their seeds on sowing give rise to more than one seedling. Thus, seed is the most favourite propagation method in polyembryonic varieties.

DISADVANTAGES OF SEXUAL PROPAGATION

Propagation through seeds has certain disadvantages also. Some of the disadvantages are enumerated below :

1. Seedling plants are highly heterozygous in nature. Owing to segregation, the seedling trees are not uniform in growth, yield and fruit quality parameters than vegetatively propagated plants.
2. Seedling plants are usually tall and spreading types and thus the management cost involved in operations like harvesting, pruning and plant protection etc., is comparatively higher.
3. Seedling plants have usually long juvenile phase and take more time to come into bearing.
4. Usually, seeds in most of fruit plants loose viability very soon after extraction and thus, have very low germination rate.
5. The beneficial effects of rootstock on scion variety can not be exploited in sexual propagation.
6. Some fruit plants like pineapple, banana etc., do not produce viable seeds and propagation through seeds is not possible in them.
7. Seed-borne viruses exists in a number of fruit plants like psorosis in citrus and mosaic in peach, cherry and almond. Thus, multiplication of these plants by seed is not feasible.
8. It is not possible to perpetuate the exact characters of any superior selection through seeds.

DEVELOPMENT OF SEED

Successful pollination and subsequent fertilization are the two most important events for formation of viable seeds. During the process of pollination, pollen grains are transferred from anthers to the stigma. The pollen grain germinates and grows downward through the style and reaches embryo sac. Out of the two male gametes released by pollen tube, one fuses with female gamete to form the zygote (2N) and other with two polar nuclei to form the endosperm. The zygote develops and gives rise to embryo. The embryo undergoes many morphological and physiological changes during seed development. Thus, a seed basically consists of three parts namely embryo, endosperm/cotyledon and seed coat (testa). Out of these, embryo is the basic part, which carries miniature plant and gives rise to new individual under favourable conditions. It is formed within the seed from a male and female gamete during fertilization. The size of embryo varies from species-to-species. In some species, it is rudimentary, and in others, it is differentiated into distinct growing points, the *radicle,* giving rise to root system and the *plumule,* giving rise to the shoot system. On the root-shoot axis, one or two cotyledons are also present between the growing points. The transition zone between plumule and radicle is called as *hypocotyl*. The endosperm of the seed act an as food reserve. The seed coverings are usually tough, which protect the embryo from injury. The seed coat usually develops from the outer layers of the ovule, called *integument*. The scar that is visible after breaking the seed from the stalk is called as *hilum* and the small point near the hilum is called as *misropyle*. The seeds vary in shape, size, form, colour and composition (Figure 3.1).

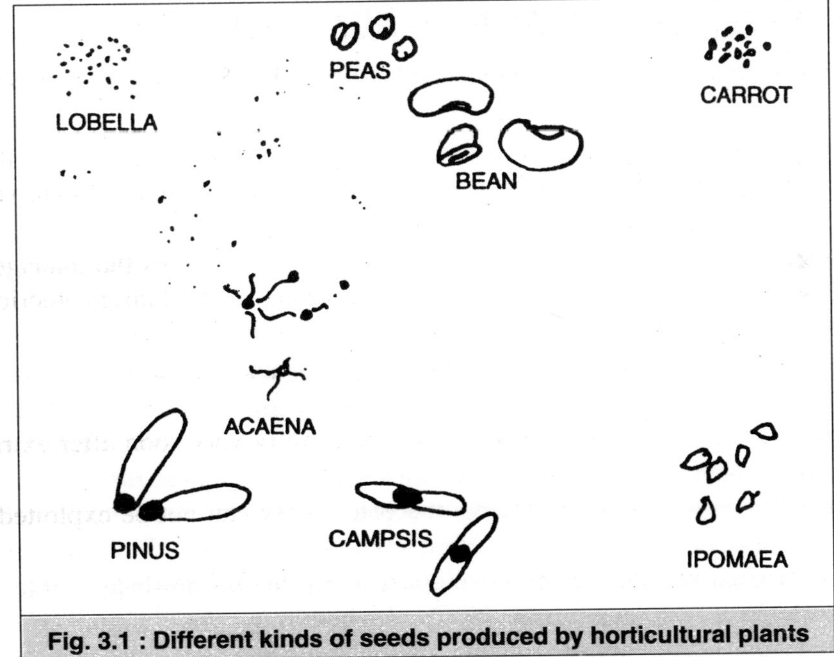

Fig. 3.1 : Different kinds of seeds produced by horticultural plants

Developmental Stages of Seed

In general, three physiological stages of seed development *viz*. embryo differentiation, cell expansion and maturation of seed have been recognized. These stages have been discussed briefly hereunder.

Stage I : Embryo differentiation

This stage is characterized by the differentiation of the embryo and endosperm mostly due to cell division. In it, the embryo reaches the beginning of the cotyledon stage of development. During this stage, there is a rapid increase in both fresh and dry weight. Many changes occur during this stage, which are different in dicots, monocots and gymnosperms.

Stage II: Cell expansion

It is a period of rapid cell enlargement due to the accumulation of food reserves. There is accumulation of carbohydrates, fats, oils, proteins in the storage organs of seeds such as cotyledons, endosperm, nucellus etc. The food reserve manufactured in the seeds from photosynthate is loaded into seed from the mother plant. In seeds of dicots, there is direct vascular connection between mother plant and the seed through *funiculus*. However, in monocots, no such connection exists but special cells, usually called as *'transfer cells'* facilitate the passage of photosynthates from mother plant to the endosperm of seed.

Stage III : Maturation drying

In this stage, seeds reach physiological maturity. However, before this stage, embryo must have reached maximum dry weight. Thus, there is greater water loss in this stage and there is no longer vascular connection between mother plant and funiculus. This stage may be considered as an end to the seed development and there is synthesis of 'Lea proteins' in response to water loss. Following maturation, seed acquires the ability to germinate. However, sometimes seeds may attain dormancy or may become *quiescent*. Such seeds fail to germinate under unfavourable conditions.

STRUCTURE OF SEED

Seeds of different plants vary greatly with respect to size, form, external colour and markings, internal structure, and amount and nature of stored food, but most of the seeds consist of seed coat or coats, remains of nucellus, endosperm (sometimes lacking), plumule, embryo, cotyledons (sometimes only one), hypocotyl and radicle.

Seed coat : The seed coat is developed from the integument or integuments of the ovule. When two seed coats are present, the inner one is usually thin and delicate, and the other one is thick and tough. The outer seed coat is a very effective protective structure. The principal external features, which may be found in mature seeds are the *hilum, raphae, micropyle,* and *chalaza*. The hilum is a scar, which marks the place where the seed breaks from the stalk (funiculus). It is always present, but it is often small and indistinct. In those plants in which the ovules become curved during the development, so that the micropyle lies close to the hilum, there may be a ridge composed chiefly of a bundle of vascular elements. This ridge, formed by the stalk of the ovule, and present only in ovules in which the funiculus is sharply bent at the base of the ovule, is called the *raphae*. In such ovules, the term *chalaza* is applied to that region at the upper end of the raphae where the funiculus spreads out and unites with the base of the ovule.

Endosperm : In some seeds, which when mature, are without endosperm, because the endosperm has been used up by the developing embryo. In seeds of this type, such as, the common bean, pea, alfalfa, clover etc., the stored food is mostly in the cotyledons. Conspicuous examples of seeds with endosperm are castor oil seed, corn, wheat and seeds of some other cereals.

Embryo : The rudimentary plant or embryo varies greatly in appearance in different seeds, owing generally to differences in the form and relative development of the parts of the embryo. However, with few exceptions, all embryos are made up of the same organs. The organs of the embryo in most seeds are:

The plumule—a rudimentary shoot.

The cotyledons—seed leaves (sometimes only one).

The hypocotyl—part of the embryo between the attachment of the cotyledons and the upper end of the rudimentary root, usually very short in some seeds.

The radicle—rudimentary root.

In few plants, such as most orchids, the embryo is very rudimentary and consists merely of a small mass of undifferentiated cells. In such cases, the ripening of the seed interrupts the development of the embryo much earlier than in most other plants. In such rudimentary embryos, when development is resumed at the time of germination, the embryo passes through developmental stages corresponding to those through which the embryos of most seeds pass before the seeds mature.

TYPES OF SEEDS

Seeds have different shapes and sizes, ranging from the minute, almost microscopic seeds of orchids to the massive 18 kg seeds of double coconut. There are five natural groups of seeds : dust-like seeds, hard-coated seeds, fleshy seeds, oily seeds and winged or plumed seeds. Some consider another category of seeds as 'pelleted' seeds. Pelleted seeds are usually very small seeds, which are coated in an inert substance to make them larger and/or a more even shape. This enables them to sow easily by machine or hand. As a general rule, the pelleted seeds should be sown less deep than to be normal seed. Similarly, the soil or other growing medium used should be kept in an evenly moist condition until germination takes place.

The various types of seeds have different properties and requirements. Dust-like seeds, such as those of begonias, don't have long storage life. They are sown just at the top layers of soil. Some seeds have hard seed coat and often don't germinate until protective coverings are not broken down. Some seeds are prone to being chipping. Some seeds as those of temperate fruits require stratification to overcome dormancy and to facilitate proper germination. In contrast, seeds of some horticultural plants (eucalyptus, brooms etc.) should be subjected to the fire and heat in some degree to facilitate seed germination. Fleshy seeds, as in peas and beans, require soaking in water for higher germination. There is always difficulty in germination of oily seeds as in carrot and magnolia. Such seeds are usually shriveled and should be sown immediately after the ripening.

• • •

Seed Dormancy

Seeds of some fruit plants (*e.g.*, mango, citrus) germinate immediately after extraction from the fruits under favourable conditions of moisture, temperature and aeration. However, in others (*e.g.*, apple, pear, cherry), germination of seeds does not take place even under favourable conditions. This phenomenon of seed is called as *dormancy*. Nature has gifted these species by way of preventing germination prior to adverse weather conditions that would otherwise kill young and tender seedlings after germination. However, it is really a problem for the plant propagators. Thus, the plant propagators and nurserymen should have fair knowledge of dormancy, seed germination and measures to overcome seed dormancy.

TYPES OF DORMANCY

Various types of dormancy is encountered in seeds primarily because of different germination controlling mechanisms present within the seed and outside growing environment. Depending upon different species, the dormancy may be of different types. Different investigators have studied the phenomenon of dormancy in different species and the following different categories of dormancy are found in different species :

1. Seed Coat Dormancy

Many times, the dormancy is due to seed coat because of physical, mechanical or chemical barriers, which are described briefly hereunder :

(*a*) Physical dormancy : In this type of dormancy, the seed coat and sometimes the hardened section of the seed covering becomes impermeable to water and gases, which prevents the physiological processes leading to the germination process. The seeds having seed coat dormancy do not germinate until impermeable barrier is weakened or replaced. Although, the embryo is quiescent (non-dormant) but covered by a water impermeable covering. This type of dormancy is common in members of Leguminaceae, Convolvulaceae,

Malvaceae, Chenopodiaceae and Solanaceae families. Among cultivated crops, hard seediness is found chiefly in the herbaceous (*e.g.* pulses) as well as woody legumes like robinia and acacia, which is considered as a genetic character of a species, variety or a result of environmental conditions. It is considered that impermeability of a seed is due to the layer of palisade like macrosclereid cells, coated with a thin layer of wax during seed maturation and storage.

In nature, the hard seed coat or seed coverings are softened by various agents of environment like continous weathering, mechanical abrasion, alternate freezing and thawing, attack by soil microorganisms (*e.g.*, fungi and bacteria) or through passage of the seeds through the digestive tracts of the animals. Softening of seed coverings by fungi and bacteria is more effective above 50°F. As a result of these actions, the seed coat softens and becomes permeable to water and gases, which facilitates easy germinations of the seeds.

(*b*) Mechanical dormancy : In this type of dormancy, the seed coverings become so hard that they do not allow the embryos to expand during germination. Hardening of seed coverings is never a sole cause of dormancy in any species but the hardening in combination with other factors, may delay seed germination. For example, the seed coat (as pits of stone fruits and shells of nuts) are apparently permeable to water and gases, but have very high mechanical resistance to allow expansion of embryo, as a result, the germination does not occur unless the seed coats are softened. Such mechanical resistance of seed coverings is decreased by creating moist and warm conditions during storage by microbial activity.

(*c*) Chemical dormancy : The presence of certain chemicals in the seed may also inhibit germination. Many chemical inhibitors have been extracted from various plant parts, which are responsible for inhibiting seed germination. The inhibitory chemicals are produced and accumulated in the fruit as well as in the seed coverings. Some of the substances associated with inhibition of germination are various phenols, coumarin and abscsic acid etc. The fleshy fruits in juices can inhibit seed germination. It is common in citrus, stone fruits, apple, pears, grapes and tomatoes. Similarly, the fruit coverings like capsule of brassicas can inhibit seed germination. Such inhibitors have also been extracted from the seeds of Chenopodiaceae (*e.g.*, atripex), Polygonaceae, Portulaceae (*e.g.*, portulaca) families and in some other species in which embryo is located on the periphery (*e.g.*, beet seed). Similarly, the seeds in some species of Cruciferae (*e.g.*, mustard), Linaceae (*e.g.*, flax) Violaceae (*e.g.*, voilet), Labiateae (*e.g.*, levendra) families have thin seed coat with a mucilaginous inner layers that contain inhibitors. Though, it is very difficult to establish a direct relationship between inhibitors and seed germination but there are always instances (*e.g.*, iris), where the germination of seeds can be improved or stimulated by leaching of inhibitors with water, or removing seed coverings or both.

2. Morphological Dormancy

Morphological dormancy is common in plant families and species in which the embryo within the seed is not fully developed but is redimentary or undeveloped at the time of ripening or maturity. Such seeds do not germinate, if planted immediately after extraction/harvesting. Plant with rudimentary embryos produce seeds with little more than a pro-embryo embedded in a massive endosperm (as in magnolia) at the time of fruit maturation. Formation of rudimentary embryo is common in herbaceous flowers crops such as anemone, ranunculus, poppy, fastisa etc. Some temperate zone plants like holly and snowberry have also rudimentary embryos. Sometimes inhibitors also become active at high temperature, making the problem more complex. The germination in such seeds can be improved by (*i*) exposing the seeds to 15°C or below, (*ii*) exposing seeds to alternate temperatures, and (*iii*). treating seeds with KNO_3 or GA_3.

In some plants, the morphological dormancy is due to the partially developed or torpedo shaped embryos also. It is common in carrot, rhododendron and cyclamen. Such seeds should be subjected to a temperature of 20°C or treated with GA_3 to facilitate germination in them. In many tropical species, the seeds with rudimentary and undeveloped embryo require longer period at a high temperature for germination to take place. For example, the seeds of palm generally require storage for several years to germinate but this period can be reduced to 3 to 4 months by keeping them at a temperature of 38 to 40°C. Similarly, GA_3 (1,000 ppm) also accelerates seed germination in palms.

3. Internal Dormancy

In several seeds, the dormancy is controlled internally within the living tissues of the seed. It may be exerted by semi-permeability of the seed covering and secondly due to dormancy present within the embryo. Freshly harvested seeds of some cereals, grasses, vegetables and flower crops have internal dormancy. Internal dormancy is of the following two types:

(*i*) Physiological dormancy : Most temperate fruits such as apple, pear, cherry, peach and apricot, vegetables and flowering annuals have physiological seed dormancy, lasting for one-to-six months. Such type of dormancy usually disappears during seed storage. The seeds of these species have dormant embryos and they do not germinate even under fovourable conditions. This type of dormancy may be due to heat sensitivity (thermo-dormancy) in which temperature has direct influence on seed germination or due to light sensitivity in which seeds of some species require light or darkness to germinate. The control of physiological dormancy lies in physiological activities and living seed coverings. Removal of seed coat and its abrasion by some means, may bring about germination in many species. This mechanism is purely due to semi-permeability of seed coverings, which allows water to enter on removal of coverings but not the gases. The world known horticultural practice of stratification has been developed to overcome physiological dormancy of embryo in temperate fruits. In stratification, the seeds are placed between layers of moist sand or soil in boxes or in ground and are exposed to chilling temperature either out doors in the field or in the refrigerators.

(*ii*) Embryo dormancy : Embryo dormancy is commonly found in seeds of trees, shrubs and some herbaceous plants of temperate region, where seeds overwinters in soil and germinate in spring under natural conditions. To overcome embryo dormancy, the following pre-requisite conditions are necessary :

1. Imbibition of moisture by seed;
2. Exposure of seeds to chilling temperature;

3. Proper aeration, and
4. Certain time period.

Moisture : For proper after-ripening of dormant embryos, it is important that the seeds have imbibed sufficient moisture. Usually, the process of imbibition is fast if there is no barrier (like hard endocarp) for it. On imbibition, the seeds swell and coverings split open. If there is no proper imbibition, it may cause an injury to the embryo and after-ripening process is affected adversely.

Low Temperature : It is the most important factor affecting the process of after-ripening of seeds. Temperature above freezing *i.e.* 2-7°C is the most effective for after-ripening process. Temperature below or above the critical limit may adversely affect the after-ripening rate. At a higher or lower temperature, seeds usually fail to germinate and temperature may induce secondary dormancy in the seeds.

Proper Aeration : Proper supply of O_2 during seed stratification is absolutely necessary for proper after-ripening process of the seeds. The amount of O_2 required for after-ripening process depends on temperature, moisture content of imbibed seeds and type of the seed. In general, the O_2 requirement at chilling temperature is low but if the temperature is more, the O_2 demand also increases.

Time : The time required for the after-ripening process of seed is dependent on species and chilling requirement of the seeds. The seeds of most woody plants require 1 to 3 months time for after-ripening, although seeds of certain species require even 5 to 6 months also.

4. Epicotyl Dormancy

Interestingly, the seeds of some species have different stratification requirement for their radicles, hypocotyl and epicotyl and this type of dormancy is called as *epicotyl dormancy*. On the basis of chilling requirement for epicotyl, plant species have been divided into two sub-groups. In one group, seeds germinate initially during warm period of 1-3 months and produce roots and hypocotyl but require 1-3 months chilling temperature for epicotyl to germinate as in lily and peony etc. In second group, seeds require a certain amount of chilling period to after-ripen the embryo, followed by a warm period for the roots to grow and again a chilling period to stimulate shoot growth as in trillium and certain perennials of temperate climate.

5. Double Dormancy

In certain plant species, seeds have dormancy due to hard seed coats and dormant embryos. This is called as *double dormancy*. For proper germination of such seeds, both the blocking conditions must be modified to allow water to penetrate the embryo and after-ripening of the embryos. This type of dormancy is found in trees and shrubs producing seeds with hard seed coats, grown primarily in cold winter area. To overcome the problem of double dormancy, the seeds are first treated to soften the seed coat and then subjected to cold stratification treatment.

6. Secondary Dormancy

Seeds of some plant species acquire dormancy after the seed has been separated from the plant, this type of dormancy is called as '*secondary dormancy*'. This is basically a term applied to a type of embryo dormancy, which develops gradually if the intact seeds are exposed to some conditions that allow imbibition by water but prevent germination. This

TABLE 4.1. Kinds of seed dormancy

Kind of dormancy	*Indicators of dormancy*	*Conditions necessary to break dormancy*
1. Exogenous dormancy	Imposed by factors outside the embryo	
Physical	Impermeable seed coat	Scarification
Mechanical	Seed coverings restrict radicle growth	Removal of seed covering; warm or cold stratification
Chemical	Inhibitors in seed coverings	Removal of seed coverings (fruits) or leaching of the seeds with water
2. Endogenous dormancy	Imposed by embryo factors	
Morphological	Underdeveloped embryo	Warm stratification
Physiological	Factors within embryo inhibits germination	A period of dry storage
Non-deep	After ripening Photodormant	Exposure to red light
Intermediate	Embryo will germinate if separated from the seed coat	Moderate periods (upto 8 weeks) of cold stratification
Deep	Embryo will not germinate when removed from seed coat or will form a physiological dwarf	Long periods (>8 weeks) of cold stratification
Epicotyl	Radicle is non dormant and growth begins when temperature and water permit, but epicotyl is dormant	Warm followed by cold stratification
3. Double dormancy	Combinations of dormancy conditions that must be satisfied sequentially	
Morphophysiological	Combination of underdeveloped embryo and physiological dormancy	Cycles of warm and cold stratification
Exo-endodormancy	Combinations of exogenous and endogenous dormancy conditions	Sequential combinations of dormancy releasing treatments.
4. Secondary dormancy	Germination conditions induce dormancy in seeds	
Thermodormancy	High temperature induces dormancy	Growth regulators or cold stratification
Conditional dormancy	Change in the ability to germinate in relation to time of the year	Chilling or warm stratification

Source : Hartmann *et al.* (1997).

type of the dormancy develops after the imbibition of water by the seed but before the emergence of the radicle. The conditions like high temperature, low O_2 supply and lack of light etc., favour the development of secondary dormancy in seeds. However, it is also considered as an ecological adaptation for plants in nature. Seeds separated from plants remain viable for long periods if dried and loose their primary dormancy in storage but when imbibed with water, they germinate promptly. Thus, the induction of secondary dormancy allows the imbibed seeds to remain viable for longer periods. It is particularly significant for seeds buried in soil, which do not germinate until disturbed or exposed to light. Seeds can be released from secondary dormancy by chilling, light and in some cases by treating with gibberellic acid. Similarly, dehydration and dry storage also favour germination in seeds having secondary dormancy. Different types of dormancy have been depicted in table 4.1.

ADVANTAGES OF DORMANCY

Seed dormancy is an evolutionary adaptation of some plant species in which germination is delayed even under favourable conditions of aeration, moisture and temperature. Thus, nature has gifted these plants because it is advantageous to them in the following ways :

1. Due to dormancy, seeds germinate only when environmental conditions become favourable for seedling survival. For example, seeds or most of the temperate region plants are available as and when winter sets in. If they are sown during this time, they may germinate if not dormant but seedlings would die soon due to severe cold (snow). These species in dormancy (mostly in winters) require special treatments like moist chilling or stratification for germination. By the time, winter is over, their requirements are also naturally met and then they germinate freely as the spring sets in (February-March) and then there is no danger to the seedlings from vagaries of climate.

2. It helps in the creation of a natural gene bank. Some seeds remain dormant for years in seed bank, which is wonderful ecological adaptation, though it is a problem if seeds of the weeds have such a long dormancy. Further, some plants produce *polymorphic* seeds, which have different degree of dormancy. For example, cocklebur fruit produces two seeds, of which, one is dormant and the other is non-dormant. One seed is for immediate germination and the other for seed bank for future germination.

3. Seed dormancy facilitates dispersal of seeds effectively. It is particularly helpful in seeds having wings or special structural adaptations, which are only modified if they pass through the digestive tract of animals or birds.

4. Seed dormancy can also synchronize germination at a particular time of the year. For example, spring germinated seedlings of plants of temperate zone plants have sufficient time to grow and develop and that of summer germinated seedlings to acquire sufficient growth to face coming winter.

HORMONAL CONTROL OF DORMANCY

Endogenous hormones that may be growth promoting or inhibiting, mediate the seed dormancy and germination processes. The commonly associated growth promoting hormones are gibberellins, cytokinins and ethylene, whereas, abscisic acid acts as inhibitor. Hormonal application has been proven to be most successful aid in germination process of many seeds. The success, however, depends upon species, variety and the stage of dormancy or stage of germination. The role of different hormones in seed dormancy has been discussed here under.

***(a)* Gibberellins :** Among different hormones, gibberellins are directly related with dormancy or germination of seeds. Gibberellins are found in high proportion in developing seeds and usually drop to the lower level in mature dormant seeds. Among various GA's, GA_3 is the most widely used gibberellin for getting rid of dormancy in seeds of many plant species. Gibberellins play vital role in initiation of enzyme activity, when the quiescent embryo imbibes water and later it helps in the induction of α–amylase enzyme activity, which converts the starch reserves of the embryo into sugars. These sugars are then translocated to the growing points of embryo to provide energy for further growth of the seedling. In this way, GA application enhances the seed germination process.

***(b)* Abscisic acid (ABA) :** Abscisic acid is the most common inhibitor found in dormant seeds. Its level usually increases with the maturation of fruit and is involved in the prevention of vivipary and induction of dormancy in many plant species. Its level is quite high in seed coats of dormant peach, walnut, apple and plum, which usually disappears as and when the seeds are stratified. In general, application of ABA can inhibit germination in non-dormant seeds. It is also known for counteracting the effects of applied GA_3. However, these affects are temporary and disappear when seeds are placed in ABA free media.

***(c)* Cytokinins :** The cytokinin activity is quite high in growing fruits and seeds, but it disappears as the seed matures. It is believed that cytokinins counteract the effects of abscisic acid (ABA) during the seed germination process. Thus, cytokinins play a permissive role in germination process by allowing GA to function appropriately. Application of synthetic cytokinins like benzyladenine (BA) or kinetin (KIN) has been found useful in breaking seed dormancy of many plant species. However, the reversal of seed dormancy by cytokinin has also been reported. This reversal in dormancy is mainly in seeds in which inhibition occurs due to natural occurring inhibitors. In such cases, cytokinin application improves germination but GA does not. Hence, in such cases, application of both GA and cytokinin is required to overcome seed dormancy.

***(d)* Ethylene :** Ethylene gas is a naturally occurring hormone, which is involved in many aspects of plant growth. It has been demonstrated that ethylene acts as germination promoter hormone for some kind of seeds. Artificial application of ethylene has been proved to be effective in breaking certain kind of dormancy in seeds. However, ethylene is particularly effective to overcome high temperature dormancy in seeds of certain species. It is very effective if used in combination with kinetin, CO_2 or light treatments.

***(e)* Other compounds :** There are certain other compounds, which may stimulate the germination process. For example, thiourea and potassium nitrate are two commonly used compounds for breaking seed dormancy. The role of these compounds is not clear but it is assumed that the effects of thiourea may be due to cytokinin activity. Similarly, *Fusicoccin* and *cotylenin* are natural occurring compounds, which overcome seed dormancy, as does the combination of gibberellins and cytokinins.

PHASES OF DORMANCY

The period of dormancy in seeds follows four different phases *viz.,* induction phase, maintenance phase, trigger phase and germination phase, which are briefly described hereunder :

1. Induction phase : The dormancy in seeds is mainly induced by various environmental factors. Among different climatic factors, the temperature and light at the time of seed development and maturation are the first to induce the dormancy in many seeds. For example, hot and dry weather during seed maturation helps in the development of hard seed coat in clover seeds. However, the seeds have soft coats if they mature and ripen during rainy season.

The balance of promoter and inhibitors has also some control on induction of seed dormancy. It has been observed that seed dormancy in many species is induced if the level of inhibitors (ABA) is more than the level of promoters. As discussed earlier, auxins, gibberellins and cytokinins act as promoters and abscisic acid (ABA) as an inhibitor for seed dormancy.

2. Maintenance phase : In this phase, the promoter-inhibitors balance has shift towards inhibitors, which causes certain metabolic blocks. This phase can be shortened or lengthened artificially through exogenous application of growth promoters or inhibitors. Usually, pre-treatment of seeds either with a promoter or inhibitor at 40-50°C for 2 to 3 hours may shorten the rest period in some seeds to some extent.

3. Trigger phase : In trigger phase, the promoter-inhibitor balance shift towards promoters by the environmental triggers. Stratification, scarification, exposure of seeds to light and leaching of inhibitors are some treatments, which may trigger the dormancy in many seeds.

4. Germination phase : During this phase, the seeds germinate under favourable environmental conditions. Water is absorbed by seeds, they get imbibed, as result of which, the protoplasm becomes active. The metabolic block created by one or other factors is removed and the metabolic activity of the seeds increases due to the increased enzymatic and hormonal activities, which initiate the seed germination process.

MEANS TO OVERCOME DORMANCY

Various methods can be employed to break dormancy of seeds with variable success. A brief account of such methods is given below :

1. Softening Seed Coat

Softening of seed coat and other coverings of seed helps in better absorption of water and gases, which results in better germination of the seeds. It can be done by a process called a *scarification*.

(a) Scarification

It is the process of scratching, breaking, chipping, nicking or softening the hard seed coat and other coverings of the seeds to make them permeable to water and gases. Scarification can be done by various methods as follows :

(*i*) **Mechanical scarification :** This is a very simple and effective mean of scarification. In mechanical scarification, the seeds are scratched or injured by rubbing against a rough surface or sand-paper. However, for large-scale scarification, mechanical scarifires are used. The seeds should be dried after mechanical scarification and may be stored or planted immediately, as these are liable to be attacked by pathogens. Mechanical scarification is usually followed in breaking seed dormancy of legumes like alfalfa and clover.

(*ii*) **Acid scarification :** Treatment of dry seeds with concentrated sulphuric acid is an effective aid for softening seed coat and other coverings. In this practice, the dry seeds are placed in concentrated sulphuric acid in the ratio of 1:2 in a glass, plastic, non-metal or wooden containers. The seeds along with acid should be stirred continuously to avoid acid injury to the seeds. The treatment time depends upon the kind and amount of seeds, varying from 15 minutes in some species to 6 hours or even more in others. At the end of treatment, the acid should be poured off and seeds are washed thoroughly under running tap water for 10-15 minutes to remove the acid from the seeds. The treated seeds may be placed in a large amount of water with small amount of baking soda to neutralize the adhering acid. Afterwards, the treated seeds may be sown immediately when wet or dried and stored for later sowing. Seeds of most legumes have tough seed coats and they respond better to sulphuric acid treatment.

(*iii*) **Hot water scarification :** In hot water scarification, the dry seeds are placed in hot water, maintained at 77 to 100°C for 12 to 24 hours. The ratio of water to seed should be 1:5. After imbibition, the seeds would swell and the swollen seeds can be used immediately for sowing, whereas the unswollen seeds should be discarded.

(*iv*) **Warm moist scarification :** By placing the seeds in moist-warm medium for several months can also soften the seed coats and other coverings through microbial activity. This treatment is highly beneficial in seeds having double dormancy. In this treatment, the hard seeds are planted in summer or early fall when the soil temperature is still higher, which usually facilitates germination.

(*v*) **High temperature scarification :** Seed coats of certain species are modified only if subjected to high temperature. For example, the seeds of some gymnosperms usually respond to high temperature in fires. It is believed that the high temperature of fire is responsible for the melting of resin, which otherwise blocks the cone and hence breaks the dormancy.

(*vi*) **Harvesting immature fruits :** In certain plant species, seeds are extracted prematurely to avoid the development of hard seed coats. Such seeds should be sown immediately after extraction.

(b) Stratification

It is a world known horticultural practice in which imbibed seeds are subjected to a certain amount of chilling temperature (5-10°C) to after-ripen the embryo. Usually, the term *moist chilling* is used as a synonym to stratification. It can be achieved by different means as follows :

(*i*) **Refrigerated stratification :** In this practice, the seeds are first soaked in warm water for 12-24 hours and when seeds are fully imbibed, these are mixed in the ratio of 1:3 with some medium, which retains moisture, provides sufficient aeration and contains no harmful substances. The suitable media for stratification are peat, vermiculite, sand or sphagnum moss etc. The imbibed seeds, mixed with medium are kept either in boxes, cans or jars with perforated lids, and then subjected to a temperature of 0-10°C. The stratification times depends upon the kind of seed, yet for most of seeds, a time of 1 - 4 months is sufficient under low temperature stratification. During the process of stratification, one must examine the boxes periodically, and if the seeds are dry, the medium should be re-moistened immediately. Further, if the sprouting begins, the seeds should be shifted to lower storage temperature. This treatment is very helpful in overcoming seed dormancy in most of the temperate fruits like apple, pear, cherry, plum etc.

***(ii)* Out-door stratification :** If refrigerated storage facilities are not available, out-door stratification may be employed either by storing the seeds outside in deep pits or in raised beds, enclosed in wooden frames. Basically, the natural rainfall and winter chilling of the outside environment are used in a way to meet the temperature and moisture requirements of the seeds. In out-door stratification, it is likely that the seeds are destroyed by excessive rains, freezing, drying or by rodents and therefore these should be protected from these circumstances.

***(iii)* Out-door planting :** Seeds of different species require cold treatment for after-ripening of the embryos. Such seeds may be planted directly in the seedbeds, when the natural environmental conditions are favourable for them. The seeds should be sown in such a way that they get proper moisture and make use of all other natural conditions, particularly the winter chilling period. The seeds, in general germinate quickly during spring, when the soil temperature begins to rise. The seeds of stone fruits (*e.g.*, plum, peach and cherries) germinate promptly, if sown in early summer, mainly because, the seeds get enough period of warmth before the onset of chilling. Therefore, the seeds that require high temperature followed by chilling should be planted in late summer to fulfil their demand of warm temperature followed by subsequent winter period that satisfies their chilling requirement. In out-door plantings, if the seeds are to be kept there for longer period prior to germination, they should be protected from drying, adverse climatic conditions, rodents, birds, diseases and weeds.

2. Leaching of Inhibitors

Certain inhibitors and phenolic compounds are present in seed coverings of many species, which inhibit germination. Thus, soaking of seeds in the running water for 12-24 hours or placing them in water for few hours before sowing helps in leaching off the inhibitors and phenolic compounds, which facilitates easy germination of seeds.

3. Pre-chilling

In seeds of certain plant species, dormancy can be overcome by pre-chilling treatment. In this treatment, the soaked or imbibed seeds are kept at a temperature of 5-10°C for 5-7 days before sowing. After this, the seeds can be sown in the field immediately.

4. Pre-drying

Pre-drying is also a useful practice in some seeds to overcome seed dormancy. In this treatment, the dry seeds are subjected to a temperature of 37-40°C for 5-7 days before sowing. After the treatment, the seeds can be sown in the field.

5. Hormonal Treatments

Several hormones and plant growth regulators (PGR) are known to play vital role in dormancy release of the seeds. However, GA_3 is commercially used for breaking seed dormancy in various kinds of seeds. The concentration of GA_3 depends upon kind of seed but usually a concentration of 200-500 ppm is most widely used. Cytokinin is another group of hormones used for breaking physiological seed dormancy and for stimulating germination in seeds of many species. Kinetin and BA (6-benzy1 aminopurine) are commercial preparations of cytokinins used for overcoming seed dormancy. Soaking seeds in 100 ppm solution of kinetin for 3-5 minutes in highly effective concentration for overcoming dormancy in many seeds. The treatment time and concentration, however, vary from seed-to-seed. Ethylene also stimulates germination in seeds of some species.

6. Chemical Treatments

Some compounds other than hormones or PGRs are also known to break dormancy but their role is not clear. Thiourea is one example which known to stimulate germination in some kinds of dormant seeds. For this treatment, soaked seeds are put in 0.5 to 3.0 per cent solution of thiourea for 3-5 minutes. Afterwards, the seeds are ringed with water and then sown in the field. Similarly, potassium nitrate and sodium hypochlorite also stimulate seed germination in many plant species.

7. Seed Priming

Seed priming refers to the procedures followed to overcome dormancy in freshly harvested seeds. Most widely used seed priming procedures are osmo-conditioning, infusion and fluid drilling.

Osmo-conditioning : In this process, the seeds are placed in a shallow layer in a container having 20-30 per cent solution of polyethylglycol (PEG). The seeds are then incubated at 15-20°C for 7-21 days. Different hormones and fungicides can also added in PEG solution to protect the seeds from pathogens. After treatment, the seeds are washed and dried at 25°C and then stored until use.

Infusion : In this process, the hormones, fungicides or insecticides and antidotes are infused into dormant seeds through organic solutions. The seeds are placed in acetone or dichloromethane solution containing chemical to be infused for one-to-four hours. Afterwards, the solvent is allowed to evaporate and seeds are dried slowly in a vacuum desiccator for 1-2 hours. These seeds absorb the infused chemical directly in to embryo when soaked in water. The treated seeds may be directly sown in their permanent locations.

Fluid drilling : In this procedure, the seeds are suspended in a special gel before sowing. Nowadays different types of gels are available in the market but sodium alginate, guar gum, and synthetic clay are most widely used for this procedure.

• • •

Seed Germination

Seed germination is the process of reactivation of the metabolic activities of the seed embryo, resulting in the emergence of radicle (root) and plumule (shoot), thus leading to the production of a seedling or a young plant. The process of seed germination is very complex as it involves several biochemical, physiological and morphological changes within a seed. The initiation of germination basically requires three important conditions, (*i*). The seed must be viable *i.e.*, the embryo should be alive and capable of germination, (*ii*). The seed should be non-dormant and the embryo should be quiescent *i.e.*, there should not be any dormancy or any chemical barrier for germinations, and (*iii*). All appropriate environmental conditions like proper moisture, temperature, air (O_2) and light must be available. If all these conditions are fulfilled, the quiescent embryo in the seed will resume growth, thus initiating the process of germination. In early stages of germination, the embryo draws nutrients from stored food in cotyledons or endosperm of the seed. Later, new shoots and leaves are developed, which produce their own photosynthetic system.

STAGES OF GERMINATION

The process of seed germination involves several consecutive and overlapping events, like : (*i*) absorption of water, (*ii*) initiation of cell enlargement and division, (*iii*) increased enzymatic activity, (*iv*) food translocation to growing embryo, (*v*) increase in respiration and assimilation, (*vi*) increase in cell division and enlargement, and (*vii*) differentiation of cells into tissues and organs of a seedling. The sequence of these events is not specific and one event may overlap the other. However, the entire process of germination can be divided into following different stages and the sequence of these stages is discussed thereunder :

1. Activation Stage

The following series of events takes place in this stage for initiation of germination of a seed :

(*a*) Water absorption : It is also called as 'hydration phase' in which water is absorbed by the process of imbibition and osmosis by the dry seeds, which softens the seed coat and other coverings and causes hydration of the protoplasm. After imbibition of water, the seed swells and seed coverings rupture, which helps the protoplasm in resuming metabolic activity with the activation of enzymes. During hydration phase, the seed coat acts as a limiting factor and its rupture increase water uptake. Water enters the seed through micropylar pore and hilum.

(*b*) Synthesis and activation of enzymes : After hydration, enzyme activity begins very quickly. Activation of enzymes is partly from the reactivation of stored enzymes and partly by synthesis of enzymes during germination initiation process. The hydrolytic enzymes convert complex food material into simple forms, which can be readily translocated and absorbed by the embryo. The oxidative enzymes are involved in respiration and releasing energy for cell division and growth.

(*c*) Cell elongation : Hydration, synthesis and activation of enzymes help in the elongation of cells, which results in the emergence of radicle. Emergence of radicle is the first visible symptom of germination, which results from the elongation of cells rather than from the cell division. It is observed that under favourable conditions, the emergence of radicle may take place within a few hours as in non-dormant seeds or a few days after seed sowing. The emergence of radicle is considered as the end of 1st *i.e.,* activation stage.

2. Translocation Stage

Food material *i.e.,* fats, carbohydrates or proteins are stored in the endosperm or in cotyledons. These compounds are converted in to simple forms and are translocated to the growing points of the embryo. The process of conversion differs with the type of food material reserve in the seed. For example, fat and oils are converted enzymatically to fatty acid first and then to sugars. Proteins are converted to amino acids and nitrogen, and starch to simple sugars. All these conversions are regulated by metabolic activity of specific enzymes in a proper sequence.

3. Seedling Growth Stage

In this stage, the development of the seedling plant takes place by continued cell division in different growing points of the embryo, subsequently followed by the expansion of the seedling structures. However, the cell division in the growing points and subsequent cell elongations are two independent processes taking place in a seedling. As the germination proceeds, the structure of seedling soon becomes evident.

TYPES OF SEED GERMINATION

The radicle, *i.e.,* the growing point of root emerges from the base of the embryo axis and the plumule *i.e.,* the growing point of shoot is at the upper end of embryo axis, above the cotyledons. The section of seedling stem above the cotyledons is called as *epicotyl* and below the cotyledons, as *hypocotyl.* Two types of germination are commonly found in cultivated plants (Figure 5.1). In one type, the hypocotyl elongates and raises the cotyledons above the ground surface; it is called as *epigeous* or *epigeal germination*. This type of germination is very common in the seeds of beans, gourds, castor, tamarind and onion etc. In the other type, the epicotyl elongates and the hypocotyl does not raise the cotyledons above ground. It is called as *hypogeous* or *hypogeal germination*. This is common in the seeds of pea, gram, mango, custard apple, lotus and maize etc.

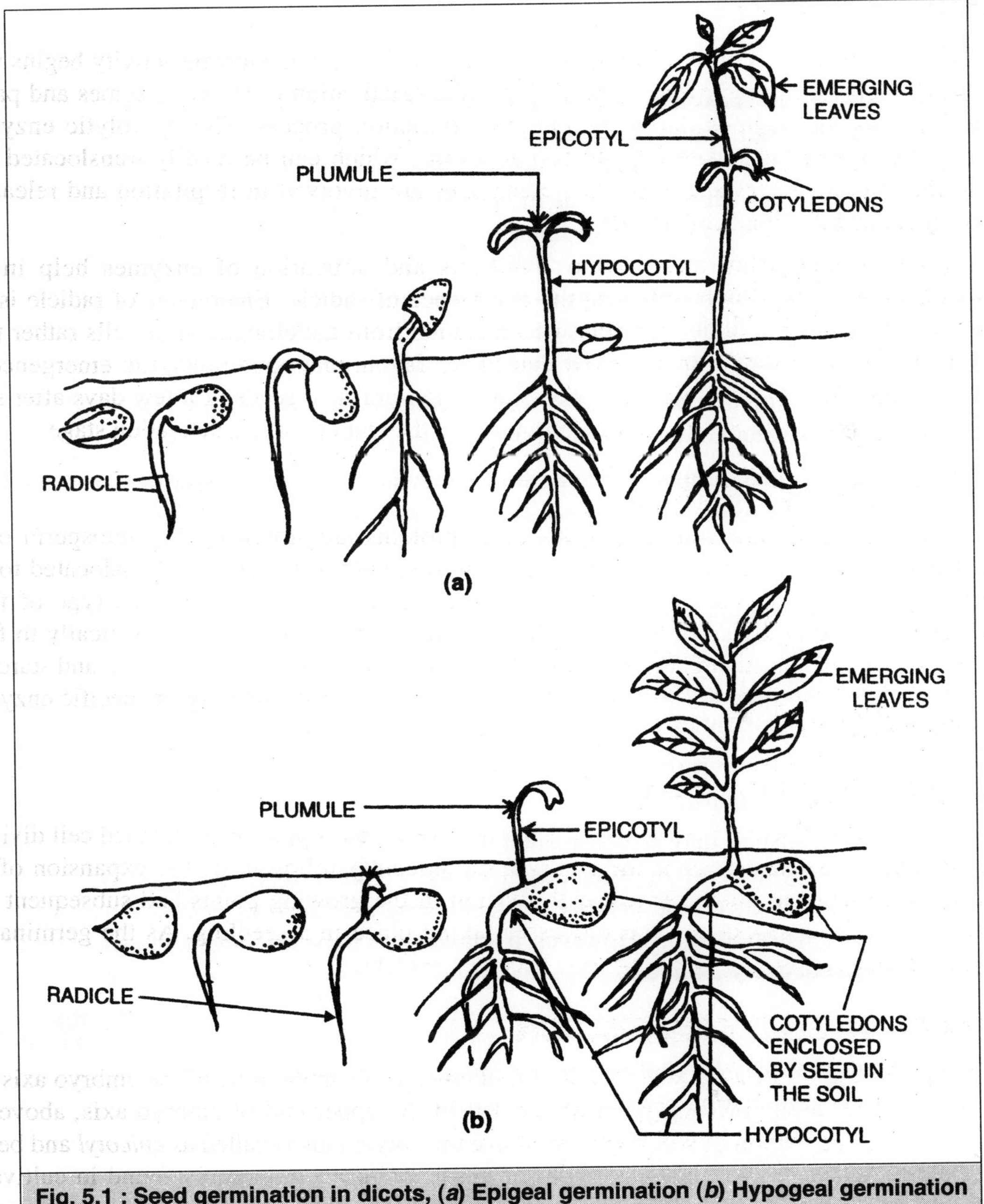

Fig. 5.1 : Seed germination in dicots, (*a*) Epigeal germination (*b*) Hypogeal germination

FACTORS AFFECTING SEED GERMINATION

Many factors affect seed germination, which can be broadly grouped as (*a*) external factors and (*b*) internal factors. A brief description of these factors is given in the following pages :

A. External Factors

Water

Water is one of the most important factors affecting seed germination. In general, seeds become active only after water absorption and the germination percentage or rate depends upon the water absorbed by the seed. Generally, seed germination does occur at moisture below 40 to 60 per cent. However, seeds of certain plants can germinate at a low moisture level and *vice versa*. The rate of water uptake is influenced by many factors like nature of seed coat, prevailing temperature, and available moisture contents etc. It is a known fact that impermeable seed coat more often acts as a barrier for water absorption. Similarly, water absorption is higher at a higher temperature but up to a certain limit. However, it is essential that the optimum moisture should be available to initiate the process of germination.

In general, the moisture requirement varies from species-to-species. On the basis of water requirement, vegetables have been classified into the following four different groups :

Group 1 **:** Seeds that germinate freely in soils having moisture content from permanent wilting percentage to above field capacity. Examples of this group include cabbage, cauliflower, carrot, turnip, radish, squash, onion, watermelon, muskmelon, cucumber, and tomato etc.

Group 2 **:** Seeds that germinate freely in soils having moisture content from intermediate to above field capacity. This group includes vegetables like lettuce, beet, bean, peas etc.

Group 3 **:** Seeds that germinate freely in soils having moisture content near to field capacity. The most important example being celery, a temperate vegetable.

Group 4 **:** Seeds that germinate at a low moisture level but have poor seed germination near field capacity as in spinach.

The excess of moisture may be detrimental too, as it may reduce seed germination due to poor aeration and by way of inviting the diseases. Soaking seeds prior to sowing is a universally followed horticultural practice for facilitating germination process in seeds. But, prolonged soaking is also harmful. However, if soaking has to be prolonged, the water should be changed atleast in every 24 hours.

Temperature

Temperature also plays a vital role in the initiation and regulation of seed germination process. It affects germination percentage and germination rate of the seeds. Usually, every plant species has certain optimum temperature requirement for germination, below and above of which, the germination process is inhibited. *Thus, the optimum temperature is the one at which the largest number of seeds germinate at the highest rate*. In fact, the temperature requirement of seeds varies from species-to-species. For example, the seeds of cool season vegetables (*e.g.*, cabbage, broccoli, celery etc.) may germinate at or below 4°C, whereas, the seeds of warm season vegetables (*e.g.*, cucumber, pumpkin, brinjal, tomato etc.) germinate freely above 10-15°C. However, the optimum temperature range for the seeds of most of the fruit plants is between 25 to 30°C. Freshly harvested seeds of temperate fruits like apple, pear, peach and vegetables like lettuce, celery do not germinate freely because of dormancy. However, if such seeds are stratified for some time, they also germinate freely. Temperature not only affects the germination process but also regulates the subsequent growth of the seedlings.

Aeration

Germinating seeds respire at a very higher rate. Thus, adequate supply of O_2 and exchange of gases between germinating medium and the embryo is necessary for rapid and uniform germination and subsequent seedling growth. However, the O_2 requirement of different seeds varies widely but some seeds germinate freely at O_2 tension below atmospheric concentration. The supply of oxygen to the germinating seed can be limited either by seed coat restrictions or excessive water in the growing medium. Similarly, sometimes hard soil crust, which is formed by excessive irrigation or heavy rains also limits O_2 supply and retards the seed germination process. Germination rate is always poor in over-watered or poorly drained seedbeds. However, seeds of celery, lettuce, beet, portulaca and petunia can germinate in water logged soils also because they have low O_2 requirements.

Light

Seeds of certain plants have specific requirement for light and will not germinate in absence of light. Light intensity have relatively less effects on seed germination but photoperiod and wavelength have more pronounced effects. In general, seed germination in blueberry, strawberry, raspberry, carrot, celery is influenced by light. In contrast, seed germination in nigella, allium, phlox and pansy is retarded by light. The seeds in which germination is promoted by light are called as *positively photoblastic* and the one in which germination is inhibited by light, are called as *negatively photoblastic*. Therefore, the seeds are planted in the top layer of soil in the seedbeds to meet their light requirements. In general, the light requirement of seeds is a very complex mechanism and it depends upon the age of seed, degree of imbibition, temperature, daylength, level of hormones etc. However, the red light is considered as stimulative and far-red as an inhibitory agent for seed germination.

The light requirement of seeds also depends on the amount of water imbibed by the seeds, type of the soil and treatment with the chemicals. For example, the light requirement is partially reduced if seeds are fully imbibed. Further, the light requirement for seed germination is less in sandy soils as compared to heavy soils. Similarly, light requirement can be partially reduced by treating seeds with GA_3, thiourea or KNO_3.

Temperature and Light Interaction

Temperature and light combination may have great influence on seed germination. It is particularly true with flowering annuals, which have been classified into nine different groups on the basis of temperature and light requirement as follows :

Group I **:** Seeds in which germination takes place over a wide range of temperature with no light requirement as in ageratum, alyssum, aster, candytuft and marigold etc.

Group II **:** Seeds in which germination takes place at cool temperature (5 to 7°C) with no light requirement. This is particularly true in cosmos and dimorphotheca.

Group III **:** Seeds in which germination takes place at warm temperature with no light requirement. For example, amaranthus, balsam, gaillardia and celosia etc.

Group IV **:** Seeds in which germination takes place in restricted temperature range without light requirement. For example, carnation, portulaca and sweet pea etc.

Group V **:** Seeds in which germination takes place over a wide range of temperature and if seeds are exposed to light, germination is enhanced as in begonia (fibrous roots) and saintpaulia.

Group VI **:** Seeds in which germination is enhanced over a wide range of temperature if seeds are exposed to light as in begonia (tuberous roots), salvia, primula and petunia, etc.

Group VII **:** Seeds in which germination takes place over a wide range of temperature and it is enhanced at warm temperature if exposed to light. Important examples in this group are cineraria and snapdragon etc.

Group VIII **:** Seeds in which germination takes place over a wide range of temperature and if held in dark as in calendula, phlox and vinca etc.

Group IX **:** Seeds in which germination takes place over a wide range of temperature, enhanced at warm temperature and if held in dark. For example, cyclamen, larkspur, pansy, viola etc.

Salinity

Salts present in the soil may also prevent or retard seed germination of some salinity-sensitive species. Salinity becomes more problematic if the growing medium is light, and irrigated with poor quality water. Frequent light irrigation results in salt deposits on the upper soil layers in which seeds are usually sown, thus inhibits seed germination process. Salinity may even damage the growing seed. The problem of salinity is acute in areas having high salts in water. This problem can be avoided by using soils and water with low salt content, withholding or using lesser amount of fertilizers and by irrigating less frequently, if water is salty.

Pathogens

Pathogenic organisms on the seed surface, in the growing medium or in the irrigation water may also inhibit the germination process and subsequent seedling growth. 'Damping off' of the seedlings is the most peculiar example in which pathogens completely destroy the young seedlings. Proper care must be taken to avoid the pathogenic organism in the growing medium by treating seeds or growing medium with suitable chemicals before use.

Tool injury

Sometimes, germinating seed or seedlings are injured by tools during intercultural operations, which may also result in poor germination of the seeds.

Seed coat barriers

Seed coat of some species is impermeable to water due to presence of palisade like macrosclereid cells. It inhibits the imbibition of water, as a result of which, non-dormant embryo is unable to develop and emerge due to moisture stress. Sometimes, seed coats are hard (*e.g*., leguminous vegetables, walnut and some stone fruits) and the tender embryo can not rupture the seed coat, which ultimately affects germination. Similarly, hard seed coat prevents the proper exchange of gases, which is a pre-requisite for seed germination. All these conditions of seed coat would affect seed germination.

Rest Period

In some plants, post extraction rest period is responsible for poor germination of their seeds as in fig.

TREATMENTS TO FACILITATE GERMINATION

There are several physical and chemical means to stimulate the germination process in seeds. However, the efficiency of these treatments depends on the type of the seed, size of the seed and type of dormancy etc. However, a few treatments are quite effective for some seeds but not for others. Some of the treatments, which help in improving the seed germination process, are described hereunder :

Soaking Seeds in Water

It is a common horticultural practice to soak seeds in the water before sowing. It promotes germination by softening the hard seed coat, activating enzymes and diluting the effects of inhibitors. Time of soaking varies from species-to-species and nature of seeds. Hot water treatment to the seeds is the most useful method in breaking seed dormancy. In this method, seeds are placed in hot water (80-100°C) for 12-24 hrs and then sown in the nursery immediately after the treatment. Sometimes, the seeds are put in a cloth bag and then dipped into boiling water for 2 to 5 minutes. Not only the hot water but sometimes cold water is also used for soaking of the seeds prior to sowing. The seeds of celery and spinach (1 day), apple (2 days), asparagus (3 days) and mulberry (4 days) are soaked in cold water to facilitate germination in them.

Stratification

Stratification, also called as 'moist chilling', is a practice of exposing the imbibed seeds to a low temperature (0-5°C) for a considerable period to after-ripen the embryo. It brings about certain physiological changes in embryo, which facilitates germination process in the seeds. In this practice, the imbibed/soaked seeds are placed in wooden boxes containing moist-sand or peat and the temperature is maintained just above freezing. However, stratification can be done under controlled refrigerated conditions and even out-doors also. The time required for stratification varies from one-to-six months, depending upon the plant species. Usually the seeds of temperate fruits like apple, pear and plum etc., require stratification for considerable time for breaking dormancy and facilitating seed germination. The optimum temperature and period of treatment for stratification of the seeds of some fruit plants in given in Table 5.1.

TABLE 5.1 : Stratification requirement of seeds of some temperate fruit crops

Species	*Optimum temperature (°C)*	*Period of treatment (days)*
Apple	3	70 to 80
Pear	3	70 to 80
Plum	3	70 to 80
Peach	5	60 to 100
Grape	5	90
Cherry	3	180

Scarification

It is a process of injuring or breaking or softening the seed coverings to make them permeable to water and to improve the gaseous exchange for accelerating the germination process. Various means are employed to achieve scarification. It can be done mechanically or by hot water treatment or by acid treatment as discussed in the chapter on seed dormancy.

Hormonal treatment

GA_3 and cytokinins are commonly used hormones for breaking seed dormancy and enchancing seed germination. The concentration, however, varies from species-to-species. Soaking seeds in GA_3 (200-500 ppm) stimulates germination and partially replaces the chilling requirement of many species. Cytokinins have effects on seed germination similar to the gibberellins. Usually, 10 to 100 ppm concentrations are highly effective, but it varies from species-to-species.

Chemical treatments

Dormancy of seeds can also be broken by treating seeds with different chemicals. Soaking the seeds in potassium nitrate (0.1-2.0 per cent) and thiourea (0.5-3 per cent) or hydrogen peroxide (H_2O_2), hastens and stimulates the germination process and subsequent seedling growth. Thiourea is the most effective in seeds that require light and moisture for breaking dormancy. However, the seeds must not be soaked in thiourea solution for more than 24 hours. Similarly, soaking seeds in a strong solution of H_2O_2 (5 per cent) stimulates germination but prolonged exposure is harmful.

Dry storage

Freshly harvested seeds of some species fail to germinate until and unless they are stored for some period. To facilitate germination in seeds of such species, the seeds should be dried at 40°C for 3 days or 37°C for 5 days and then stored for sometime before sowing.

Light exposure

Light is necessary for germination to take place in 'positively photoblastic' seeds. The quantum of light for such seeds varies from species-to-species but it has been demonstrated that exposing the imbibed seeds to 800 to 1,000 lux light intensity for 8-10 hours is sufficient to break dormancy of 'positively photoblastic' seeds, which facilitates their germination.

Protection of seeds against pathogens

Various types of pathogens may infest seeds and thereby affect their germination. Hence, seeds must be protected from all possible attack of pathogens. Three types of seed treatments are used to control pathogens in seeds, disinfestations, disinfection, and seed protection. Disinfestants are used to eliminate organisms present on seed surface (*e.g.* sodium hypochlorite). Disinfectants eliminate organisms within the seed itself (*e.g.* hot water treatment). Protectants are applied to seeds to protects them from pathogenic fungi present in the soil. These can be applied as a soil drench or applied to seed.

• • •

CHAPTER 6

Seed Viability

The term '*seed viability*' refers to the ability of a seed to germinate and produce an acceptable seedling under favourable conditions. However, simply emergence of a radicle does not mean that the seed is viable. Similarly, a non-viable seed is always fully dead. So, the term non-viable does not simply applied to the dead or diseased seeds but also to those that produce abnormal seedlings, which are not acceptable to the growers. Seed viability varies greatly in different kinds of seeds. For example, the seeds of mango, citrus, litchi, jackfruit and walnut are short-lived and remain viable only for a few days, whereas the seeds are long-lived in leguminous crops. The viability of seeds depends upon the storage conditions also. Some seeds are long-lived if stored at low temperature (0-4°C) and low humidity, whereas, seeds of some species (*e.g.,* litchi, mango, citrus, cocoa, walnut etc.) desiccate and their embryos die, if stored under dry conditions. On the basis of storability of seeds in relation to moisture, seeds of different fruits have been classified into two groups *viz*. orthodox seeds and recalcitrant seeds.

(*a*) **Orthodox seeds :** The seeds, which can be dried to a low moisture level and loose viability with the increase in moisture, are called as *orthodox seeds*. The seeds of *ber*, custard apple, apple, date palm, fig, grape, guava, mulberry, papaya, passion fruit, peach, pineapple, plum, *phalsa,* pomegranate are orthodox seeds.

(*b*) **Recalcitrant seeds :** The seeds, which remain viable at a relatively higher moisture content, and if dried below certain moisture level, these loose viability, are called as '*recalcitrant seeds.*' The seeds of avocado, Barbados cherry, carambola, breadfruit, durian, jackfruit, litchi, mango, mangosteen, rambutan, walnut, and citrus are recalcitrant seeds.

In general, the seeds belonging to second group are short-lived and long-term storage of this type of seeds is difficult. The longevity and storage conditions of seeds of some fruit plants and vegetables are given in Table 6.1 and Table 6.2, respectively.

TABLE 6.1 : The longevity and storage conditions for seeds of different fruits and plantation crops

Common name	*Scientific name*	*Longevity*	*Ideal environment for germination*	*Remarks*
FRUIT CROPS				
Arecanut	*Areca catechu*	Very less	Can't be dried	—
Avocado	*Persea americana*	15 months	5°C, 90% R.H.	—
Barbados cherry	*Malpighia glabra*	4-5 months	Can't be dried	—
Bread fruit	*Artocarpus incisa*	Very less	Can't be dried	—
Carambola	*Averrhoa carambola*	Very less	Can't be dried	—
Coconut	*Cocos nucifera*	16 months		—
Grapefruit	*Citrus paradisi*	6 months	2°C, 88% R.H.	—
Grapes	*Vitis vinifera*	12 months	5°C, 88% R.H.	Ample moisture is required
Jackfruit	*Artocarpus heterophyllus*	Very less	Can't be dried	
Lemon	*Citrus limon*	8 months	2°C, 88% R.H.	Treatment with fungicide is required
Lime	*Citrus aurantifolia*	6 months	2°C, high R.H	Treatment with fungicide is required
Litchi	*Lichi chinensis*	3 weeks (85% R.H.)	3°C, high R.H.	Proper moisture is required
Loquat	*Eriobotrya japonica*	6 months (92%)	5°C, high R.H.	Semi-sealed conditions
Macadamia nut	*Macadamia integrifolia*	6 months (50%)	Ambient temperature and R.H.	—
Mandarin	*Citrus reticulata*	4 year (88%)	7°C, 58% water	Ample moisture is required
Mango	*Mangifera indica*	80 days	Room temperature, 50% R.H.	Temperature below 3°C is lethal
Mangosteen	*Garcinia mangostana*	8 weeks	21-29°C,	Ample moisture, proper ventilation is required
Persimmon	*Diospyros virginiana*	Can't be dried	—	—
Pummelo	*Citrus grandis*	8 months	2°C, 88% R.H.	Treatment with fungicide is required
Rough lemon	*C. limonia*	16 months (46%)	5°C, 46% water	Ample moisture is required
Sapota	*Manilkara achras*	4-5 months	Can't be dried	—
Sour orange	*Citrus ourantium*	11 months	5°C, 88% R.H	—

PLANTATION CROPS				
Cinnamon	*Cinnamomum zeylencium*	1 month	Ambient temp.	—
Cocoa	*Theobroma cacao*	4 months (52%)	10°C, 70-80% R.H.	24% water in seeds is lethal
Coffee	*Coffea arabica*	47 weeks (22%)	Warm, humid	0°C pr 10% water is lethal
Tea	*Cammelia chinensis*	10 months (50%) 0°C, 100% R.H.	—	—

TABLE 6.2 : The longevity and storage conditions for seeds of some vegetables and annual flowers

Common name	*Scientific name*	*Dry storage*		
		Growing condition	*Storage life (years)*	*Germination (%)*
VEGETABLES				
Beet	*Beta vulgaris*	Lab	3	58
Cabbage	*Brassica olracea var. capitata*	Lab	19	19
Carrot	*Dacus carota*	Lab	31	7
Chilli	*Capsicum frutescence*	Lab	28	69
Cucumber	*Cucumis sativus*	Lab	30	77
Lady's finger	*Abelmoschus esculentus*	Lab	10	20
Muskmelon	*Cucumis melo*	Lab	30	96
Onion	*Allium cepa*	Dry, lab Storage	22	33
Pea	*Piscum sativum*	Dry, lab storage	31	78
Radish	*Raphanus sativus*	Lab, 4% water	16	No significant loss in germination
Spinach	*Spinacea oleracea*	Lab, 7% water	11	-do-
Tomato	*Lycopersicon esculentum*	Lab	28	30
Turnip	*Brassica napus*	Lab	12	3
Watermelon	*Citrullus lanatus*	Lab	30	97
FLOWERING ANNUALS				
Calendula	*Calendula officinalis*	Lab	10	28
Carnation	*Dianthus caryophyllus*	Lab	10	47
Chrysanthemum	*Chrysanthemum cinerifolium*	Lab	10	28
Larkspur	*Delphinium chinensis*	8°C, 6% water	16	43
Latus	*Delphinium nucifera*	In herbarium, Storage	237	.
Petunia	*Petunia hybrida*	Storage	11	80
Zinnia	*Zinnia elegans*	Storage, sealed	11	80

Seed viability can be expressed by germination percentage, which indicates the number of seedlings produced by a given number of seeds.

FACTORS AFFECTING SEED VIABILITY

There are several factors, which influence seed viability but genetical factors, environmental conditions under which a seed is produced, mechanical injury to the seed during harvesting, and the storage conditions of the seeds are usually reported to affect seed viability greatly. These factors are described briefly hereunder :

(*a*) Mechanical Injuries

Mechanical injuries may damage certain seed structures, resulting in loss of seed viability. Such injuries usually occur during harvesting and threshing of seeds. The most intensive injuries reduce seed viability immediately, whereas, the small injuries may serve as an infection centre for the microorganisms. Generally, the large seeds are more susceptible to mechanical injuries than the smaller ones. Similarly, the spherical seeds are better protected against such injuries than the elongated and flat seeds. For example, the seeds of cole crops (*e.g.*, cabbage and cauliflower etc.) are spherical in shape and the folded cotyledons provide good protection for internal structures of the seeds. Thus, one must be careful to protect seeds from mechanical injuries during harvesting and threshing and if at all seeds receive small injuries, these should be properly treated during planting to protect their viability.

(*b*) Environmental Conditions

Seed germination and viability vary widely from year-to-year and species-to-species. This variation in seed germination and viability may be due to the direct or indirect variation in the weather conditions prevailing at the time of maturation, ripening and harvesting of the seeds. A warm and dry weather at the time of seed maturity usually favours the production of quality seeds, whereas, adverse weather conditions like occurrence of frost, excessive rainfall or drought at the time of harvesting of seeds may affect their viability adversely. Similarly, the temperature during ripening of seeds can also influence seed viability significantly.

(*c*) Genetical Factors

Some investigators feel that seed viability is also affected by genetic make up the plant. For example, seeds (stones) of mango have poor viability but it is not a problem in cashewnut, through both the fruits belong to the same family *i.e.* Anacardiaceae. Indian lotus has been reported to remain viable for 100 years. Even among short-lived species, there is a greater variability among different species to the external environmental conditions. For example, mango stones can not be dried and stored at a low temperature. For such types of seeds, neither low temperature nor dry storage is useful for prolonging their viability. However, some seeds although can't be dried, but can be stored for long time under low temperature as in citrus. Even differences in seed longevity has been encountered within cultivars of different fruits, vegetables and ornamental plants.

(*d*) Storage Conditions

Proper storage conditions are important for keeping the seeds in good condition for longer time. The environmental conditions prevailing in the storage also influence the seed viability invariably. If temperature, relative humidity, aeration and other conditions are not adequate in storage, it would surely affect the viability of seeds. As discussed earlier, the recalcitrant seeds loose their viability under dry storage but the orthodox seeds require low moisture and cooler storage conditions for maintaining their viability. In general, the humidity and temperature in the storage room or structure should be optimum for maintaining the viability of seeds for longer time. In general, high relative humidity shortens the life of the seeds. On the basis of moisture and temperature requirements of seeds, the following two rules should be kept in mind :

1. For every one per cent rise in seed moisture, the seed viability is reduced by 50 per cent. However, it is true only if the seed moisture is between 5-14 per cent.

2. For every 5°C increase in temperature, the life of the seed is reduced to 50 per cent. However, this range is applicable for temperature range between 0 and 50°C.

In addition to temperature and humidity, aeration is another portent factor, which determines the longevity of seeds. It means that there should be proper concentration of O_2 and CO_2 inside the storage house, room or storage structure. If all these conditions are inadequate, these may favour the development of pathogens and storage insect-pests, which further affect the seed viability adversely.

METHODS OF TESTING SEED VIABILITY

Before sowing, it is necessary to test the viability of seeds. To test the viability, a representative sample of seeds is taken from a lot and one of the following methods may be used for testing the viability.

Germination Test

It is a useful method for testing the viability of seeds having no dormancy problems as in vegetables and some flowering annuals. However, dormant seeds can also be tested through germination test, provided treatment is given to break the dormancy. Different techniques are used for the germination test. However, the most commonly used method is to place the seeds in blotting papers, which are then kept in petri dishes or germination trays. Besides blotting paper, absorbent cotton, filter paper, paper towel, sand or vermiculite can also be used as a germination substrate. The main function of the substrate is to supply moisture to the seeds. Therefore, moisture supply should be regulated throughout the germination test because seeds will not germinate if the moisture supply is inadequate. However, excessive supply of water should also be avoided as it impedes the process of germination by encouraging microbial growth and restricting O_2 supply to the seeds. Similarly, all the material used in the germination test should be clean and sterilized to discourage the growth of the microorganism. The temperature and light inside the germinator is kept as per the

International Rules of the seed testing. In general, it is adjusted between 10-35°C. Air humidity is maintained at 90 per cent or so and usually uniform light (180 to 1300 lux) should be maintained throughout the germination test, depending on the seed type.

Dormancy of seeds often creates difficulty in germination test because it is not possible to distinguish the dormant and non-viable (non-dormant) seeds with eyes. In general, dormant seeds are firm and swollen, whereas, the non-viable seeds are soft and watery. To overcome dormancy, different methods are used in laboratories before germination test. The seeds may either be pre-treated at a low temperature or with KNO_3 (0.2 per cent solution) at the beginning of the germination test to hasten the germination of dormant seeds. The petri dishes along with seeds are kept in a germinator in which all necessary environmental conditions *viz.* temperature, light and humidity are controlled. The duration for which the seeds are to be kept in germinator depends upon different species, which varies from 5 (radish) to 20 (sorghum) days or even more.

The definition of germination for seed testing differs from the botanical definition of germination. Under laboratory conditions, emergence and development of normal seedlings from seeds, is generally taken as the definition of germination. After the prescribed period, the final count in terms of normal seedlings is called as germination per cent or seed viability per cent.

Tetrazolium Test

This test is highly useful for early prediction of seed viability. This is a biochemical test in which seeds are subjected to an indicator dye, known as 2,3,5-triphenyl tetrazolium chloride. This dye is colourless when dissolved in water, but changes to red (formazan) when comes in contact with living tissues. In living tissues, dehydrogenase enzymes are present, which change tetrazolium salt in to red formazan. These enzymes are absent in dead seeds so they do not show any colouration. In this test, the seeds are first soaked in water and are cut lengthwise into two halves to expose their embryos. The cut seeds are then placed in 1 per cent aqueous solution of 2,3,5-triphenyl tetrazolium chloride (popularly known as tetrazolium salt) at room temperature for few hours to few days, depending upon the species. Generally, the pH of this solution is adjusted between 6-7 as the reaction goes smoothly at neutral pH. Further, this chemical is highly sensitive to light and therefore its exposure to light should be avoided. The exposed embryos of seeds will stain with dye and degree of staining of the embryos will give the seed viability results. For example, if the embryos are completely stained, it indicates full viability, if completely unstained, indicates complete non-viability of the seeds but if the embryos are partially stained, it is very difficult to draw any conclusion.

Excised Embryo Test

In some plants, the dormant embryo requires a long period for after-ripening. Usually seeds of such species do not respond to germination test. In such cases, excised embryo test gives fruitful and quick results. In this method, the embryos are excised from the seeds carefully and then placed in petri dishes, containing moist papers kept in a room maintained at 18-20°C, normal relative humidity and light intensity. The embryos showing any growth activity such as development of cotyledons, radicle and plumule etc., are considered as viable otherwise non-viable. The non-viable embryos become soft, brown and are infested by microorganisms very soon.

X-ray Tests

This is an advanced technology for testing the seed viability. The X-ray photographs predict the emptiness of seeds, shrunken embryos, number of embryos in a seed and damaged portion of an embryo. In X-ray test, the seeds are first soaked in water for 14 to 16 hours and then placed in concentrated solution of barium chloride for 1-2 hours. Later, the excess barium chloride is removed by washing the seeds in running water. The treated seeds are then photographed by soft X-ray films. The X-rays are examined and if the embryo is fully developed and undamaged, the seed is said to be viable, otherwise it is dead or non-viable.

• • •

CHAPTER 7

Seed Production

A 'good quality' seed fetches good price in the market. Any organization supplying good quality seed has a reputation in the market. Hence, production of quality seed is always praised and promoted. Seed production is a highly specialized job. A person dealing with seed production should have an in-depth knowledge about soil, climatic requirements, cultural and manurial requirements of crops for which one is producing the seeds. Similarly, one should have thorough knowledge about its flowering habit, pollination, isolation requirements and all pre-requisites required for the production of a good quality seed.

Seed is different from the grain. 'Seed' is used to raise the crop, whereas 'grain' is used for consumption as a food, feed and industrial raw material.

ATTRIBUTES OF A GOOD QUALITY SEED

A good quality seed should be of defined pedigree; genetically pure; physically pure; free from seeds of obnoxious weeds; free from disease and pest infection; reasonably free from other crop seeds; should be viable; duly packed in good containers; properly labelled and sealed. The seeds sold in the market may be verified for the above factors with the label/tag attached to the container.

For the production of high quality and genetically pure seed, the following measures should be followed :

1. Adopt proper cultural practices while raising a crop for seed.
2. Provide adequate isolation distance for cross-pollinated crops.
3. Follow all remedial measures to control insect-pest and diseases.
4. Periodic inspections should be made to rogue out all 'off' type plants in the seed production unit/field.
5. Harvest crop at a right maturity.

6. Threshing and extraction of seeds should be done carefully and in an isolated place.
7. Curing and drying of seeds should be done in the sun.
8. Seeds should be dried to the recommended moisture level.
9. Treat seeds with suitable chemicals (fungicides/insecticides) before packing.
10. Seeds should be packed in airtight and attractive containers.

HYBRID SEED PRODUCTION

Hybrid seed is F_1 generation seed, developed by crossing two different individuals having desirable traits. The major advantage of the hybrid seed is that it gives outstanding yield in comparison to its parents due to hybrid vigour. The hybrid vigour may be manifested with higher yields, early maturity, and uniformity in size, better quality and resistance to various biotic and abiotic stresses. The concept of exploitation of hybrid vigour in vegetables like tomato, brinjal, chillies, radish, onion and flowering annuals like petunia, zinnia, pansy, marigold and gladiolus is gaining popularity day-by-day in India also and many hybrids have been developed, which are far superior to their parents in yield, quality and other desirable attributes. However, the production of hybrid seed is a difficult, labourious and cumbersome job, which results in higher production cost because of the following reasons :

1. Selection of desirable parents is difficult and time-consuming process.
2. Technical manpower is required for emasculation and crossing work.
3. Strict control on isolation distance is required.
4. High and specific knowledge is required on flowering behaviour, anthesis, pollination etc., of the parents selected for hybridization.
5. Hybrid seeds are very costly and are not in the reach of common farmers.

In general, selection of desirable parents is the first and foremost requirement for the production of hybrid seeds. Once the parents are selected (after testing the combining ability), the crossing is made between male and female parents under strict control. In general, it is easy to produce hybrid seeds in dioecious vegetables like spinach and asparagus; monoecious vegetables (cucurbits) and self incompatible vegetables like cauliflower, radish, turnip etc., than in self-pollinated vegetables like tomato, chillies, brinjal etc. In the process of hybrid seed production, the flowers of the male parent are made ineffective by removing the stamens (emasculation). Usually stamens are removed by hand (hand emasculation) in crops having hermaphrodite (bisexual) flowers as in tomato, chillies etc., which is a very tedious task. However, the process of emasculation may be avoided in crops having male sterility (marigold). After removing the male parts of a flower or male plants (cucurbits) crossing is done by taking the pollen from the desired male flower/plants. The crossing can done by using the following methods :

1. Hand emasculation and hand pollination : In this method, both emasculation as well as pollination is done by hand as in tomato, brinjal, chillies and okra. This method is usually followed in self-pollinated vegetable crops.

2. Hand pollination without emasculation : In this method, no emasculation is done but pollination is done with hands. It is useful in crops producing unisexual flowers (most cucurbits) or when stamens of a hermaphrodite flower are not functional due to one or the other reason (tomato, brinjal, capsicum etc.).

3. Hand emasculation and pollination by wind : In this method, emasculation in the male plants or male parts of a plant is done by hand and pollination takes place by wind. It is useful in dioecious plant species like spinach, asparagus etc.

4. Hand emasculation and pollination by insects : In this method, emasculation is done by hands and pollination by insects. The male flowers are removed by hand before they open and insects are used for pollination. It is useful in cucurbitacious vegetables like cucumber, melons, pumpkin, bitter gourd, squashes etc.

5. No emasculation and pollination by insects and wind : This method is used in plants having the problem self-incompatibility as in cabbage, cauliflower, broccoli, Brussels Sprouts, turnip, radish etc., and also in species in which male sterility is present as in onion and marigold etc. In such species, there is no need of emasculation and pollination may takes place by insects or wind.

After crossing, the plants are labelled with a tag, containing the information on the parents, date of emasculation, date of crossings, number of flowers crossed etc. The plants are raised on all the standard cultural requirements. The seeds are harvested, threshed, sun-dried up to standard moisture level and packed.

Hybrid seeds are widely used in advanced countries like Japan, USA and Germany etc. However, now-a-days, there is a great inclination of Indian farmers for hybrid seed production and its use.

Some F_1 hybrids have been developed in different vegetable crops in India by various public and private institutions. A list of different hybrids along with institution is given in Table 7.1.

TABLE 7.1 : A list of some F_1 hybrids of vegetable crops released in India

Crop	*Name of hybrid*	*Institution*
Brinjal	Arka Navneet	IIHR, Bangalore
	Pusa Hybrid-6	IARI, New Delhi
	Pusa Hybrid-5	IARI, New Delhi
	ARBH-201	Ankur Seeds
	NDBH-1	NDUAT, Faizabad
	MHB-10	MAHYCO Seed Co. Jalna
	MHB-39	MAHYCO Seed Co. Jalna
	NDBH-6	NDUAT, Faizabad
	ABH-1	Anand Seeds
	ABH-2	Anand Seeds
	ABG-2	Anand Seeds
	Pusa Hybrid-9	IARI, New Delhi
	Suphal	Indo American Hybrid Seed Co.
	Phule Hybrid-5	MPKVV, Rahuri
	Nisha, Vardhan, Shiva	Century Seeds, New Delhi

Contd....

Tomato	ARTH-4 ARTH-3 MTH-1,2,6,15 & Sonali Arka Vardhan Pusa Hybrid-1,2,4 KT-4 NA-601 Arka Vardhan NA-501 DTH-4 BSS-20	Ankur Seed Co. -do- MAHYCO Seed Co., Jalna IIHR, Bangalore IARI, New Delhi IARI, Reg. St, Katrain Nath Seeds IIHR, Bangalore Nath Seeds IARI, New Delhi Beejo Sheetal Seed Co.
Capsicum	KT-1 Bharat Early Bonty Indira, Lario Heera	IARI, Regional Res. St., Katrain Indo American Hybrid Seeds Sutton and Sons Sandoz India Ltd. Nath Seeds
Cabbage	Pusa Synthetic Sri Ganesh Gol NATH-401 BSS-32 Questo Vishesh Regalia, Stone Head, Green Boy	IARI, Regional Res. St., Katrain MAHYCO Seed Co. Jalna Nath Seeds Beejo Sheetal Sandoz India Ltd. Hindustan Lever Ltd. Sakata Seeds
Carrot	Hybrid-1	MAHYCO Seed Co. Jalna
Chillies	ARCH-236 Agni HOE-888 Mangla, Naveen, Rupali	Ankur Seed Co. Sandoz India Ltd. Hoechest India Ltd. IAHS, Bangalore
Cauliflower	Pusa Hybrid-2 Early Himlata Himani	IARI, New Delhi Century Seeds, Delhi IAHS, Bangalore
Muskmelon	Pusa Rasraj MHC-5 MHC-6	IARI, New Delhi MAHYCO, Seed Co. Jalna
Watermelon	Arka Jyoti Madhu, Milan, Mohini Nath-101, Nath-102	IIHR, Bangalore Indo-American Hybrid Nath Seeds
Lady's finger	Varsha, Vijay Adhunik, Panchali	IAHS, Bangalore Century Seeds, New Delhi
Bottle gourd	Pusa Meghdoot, Pusa Manjri Ward	IARI, New Delhi Mayco, Jalna
Bitter gourd	Vivek	Sungro, Delhi
Cucumber	Pusa Sanyog Priya	IARI, New Delhi IAHS, Bangalore

• • •

Seed Testing and Certification

Most of the flowering annuals, vegetables and some fruit plants are commercially propagated through seeds and thus their production chiefly depends on the quality of seeds. To maintain high seed quality, The Government of India in 1966 has enforced Seed Act. The basic objective of Seed Act is to protect the farmers from purchasing seeds of poor quality and low purity. For the implementation of this Act, seed testing and seed certification programs have been initiated in India also.

SEED TESTING

In the recent years, seed testing has emerged as a multi-dimensional technology of Seed Science. For seed testing, mainly purity test, test for obnoxious weed seeds and germination tests are done. Seed testing is done with the following objectives :

- To determine the seed quality, *i.e.,* their suitability for planting.
- To identify seed quality problems and their probable causes.
- To determine the need for drying and processing.
- To determine whether seed meets the established quality standards.

Sample collection for seed testing : For seed testing, proper sampling is very important because it is the basis for analysis to determine the purity, seeds of obnoxious weeds, germination, variety and many other quality attributes. If the sample is not genuine, the whole exercise putforth to know the purity, germination, presence of seeds of obnoxious seed etc., will go futile. Therefore, the sample should be a true representative of the entire seed lot to be tested. Generally, a large sample should be taken because it is more representative of a seed lot than a small sample.

For seed testing, samples of seeds are collected and sent to the laboratory for testing. Effective seed sampling can be done by hand. The hand should be closed when thrust into the seed, then opened to collect the sample, and closed tightly again in order to prevent any seed from escaping. However, some triers or probes are manufactured, which can be used for sampling (Figure 8.1).

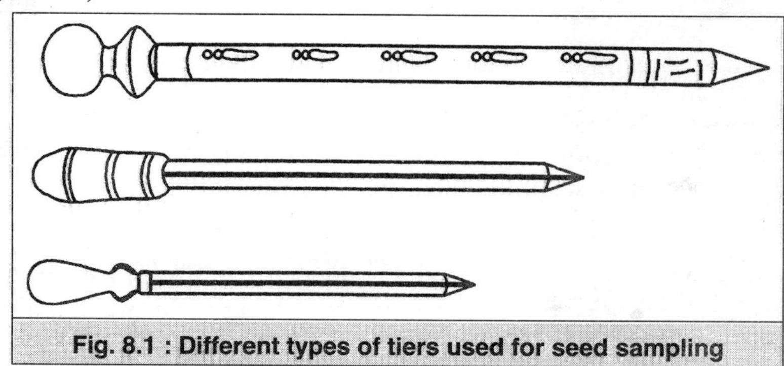

Fig. 8.1 : Different types of tiers used for seed sampling

SIZE OF A SEED SAMPLE

The size of sample varies with the crop-to-crop. In general, the size of the sample to be followed in different horticultural crops is given in Table 8.1.

TABLE 8.1 : The size of sample to be submitted for testing to seed testing laboratory

Sample size (g)	*Name of the crop*
900 to 1000	French bean, cluster bean, bittergourd, squash, ridge gourd, snake gourd, sponge gourd, summer squash, okra, asparagus, globe artichoke, Jerusalem artichoke, broad bean, field bean, lima bean etc.
750	Moth bean
700	Ash gourd, bottle gourd, pointed gourd etc.
500	Velvet bean, sugar beet, garden beet, spinach beet etc.
400	Coriander
350	Pumpkins
300	Rat tail radish, radish etc.
250	Chow chow, multiplier onion, spinach, sweet potato, tapioca, etc.
160	Rapeseed, mustard etc.
150	Cucumber, long melon, musk melon, snap melon, brinjal, chilli, sweet pepper etc.
100	Parshnip, cabbage, cauliflower, broccoli, Knol Khol etc.
80	Onion
70	Tomato (varieties), amaranth, turnip etc.
40	Fenugreek, Parsley, Chinese cabbage etc.
30	Lettuce, carrot etc.
25	Birdwood grass, buffel grass, doob grass, guinea grass, setaria grass, poppy, celery etc.
7	Tomato and brinjal hybrids.

The seed sample should carry the information like date, sender's name, address, name of the crops, name of the variety, origin or class of seed, lot number, quantity of seed in the lot and kind of test required (purity, germination percentage, moisture contents etc.). This information should be placed inside the bag and submitted with the sample. For seed testing, the following tests are done :

(*a*) Purity test : This test is done to identify the different kinds of seeds and other component present in a representative sample of a seed lot. It is of first and foremost importance in seed testing because the subsequent germination test is usually based on it.

Testing of seeds for purity is a difficult job, as it requires constant use of vigilant eyes. For this, one may use hand lens and/or stereoscopic microscope. In this test, the seed sample is first divided in four components *viz*. pure seeds, seeds of other crops, weed seeds and inert material. Then each component is weighed and its percentage is determined. Seeds are sometimes tested to determine the trueness of a variety also. This is a highly complicated job because environmental triggers may influence each and every character of a variety invariably at different locations. For which, laboratory, greenhouse, growth chamber and field tests are done.

The shape, size and colour of the seed can give some indications for the identification of a variety. Similarly, some chemical tests have also been standardized for identification of varieties of some cereals. Though, plant produces clear cut diagnostic characteristics of a variety under controlled conditions (greenhouse conditions) than natural field conditions, however, test of trueness of a variety can only be done by conducting a field trial. Although, field trials are more expensive than other methods of testing the trueness of a variety.

(*b*) Testing for seeds of obnoxious weeds : In this test, the weed seeds are identified and their percentage is determined in a sample. It is a part of the purity analysis because the methods used for conducting these two tests are similar.

(*c*) Germination test : Different methods for laboratory germination of seeds of most horticultural crops are used to test the germination of seeds. In this test, the imbibed seeds are placed on filter paper or blotting paper or paper towels. In addition, medium such as sand, peat moss or vermiculite can also be used. The substrate medium used in the seed testing should be non-toxic to the germinating seedling, relatively sterile and able to supply adequate aeration and moisture to the germinating seedlings. Excessive moisture should, however, be avoided as it may restrict aeration and favour the growth of the microbes. Further, proper spacing of seeds should not be less than 3-5 times the diameter of the seed to be tested. Proper temperature and light conditions should be maintained during the entire period of germination. Automatically controlled germinators are more convenient for this purpose. Similarly, a light of 750 to 1250 klux should be maintained in the germinator for proper germination of the seeds. The germination counts are made after every 2-3 days for quick germinating seeds and once a week for slow growing or germinating seeds. The details on different seed germinating tests are given in the chapter on 'seed viability'.

SEED CERTIFICATION

Seed certification is a legally sanctioned system for quality control of seed multiplication and production. The objective of seed certification is to ensure genuineness and quality of a seed to the grower. A well-organized seed certification programme, in theory and in reality does in fact, become the guardian of pure and good quality seed supply. For seed certification, different states have established State Seed Certification agencies. The broad principles for organizing a seed certification agency are as follow :

1. A seed certifying agency should not be involved in production and marketing of seeds.
2. It should have autonomy.
3. It should adopt uniform seed certification standards throughout the state/country.
4. It should have close links with research institutions.
5. Its long-term objective should be to work on no-profit-no loss basis.
6. Its staff should be adequately trained.
7. It should have adequate facilities for ensuring timely and thorough inspection.
8. It should serve the interests of seed producers and buyers.

Seed certification programmes in India recognize the following three classes of seeds :

(*a*) Breeder's seed : Breeder's seed is a seed or vegetative propagating material, directly controlled or produced by the breeder of an institute or organization. It is produced only in small quantity and its production is done under strict control and supervision of a breeder. From this seed, the foundation seed is prepared. It is considered as the most pure seed. After proper packing, it should be properly labelled.

(*b*) Foundation seed : Foundation seed is multiplied from the breeder seed. Its production is done under strict control and supervision of a Seed Certifying Agency. Genetic identity and purity of the seed is strictly maintained as per seed certification standards. Like breeder's seed, foundation seed should also be labelled with a suitable tag.

(*c*) Certified seed : This type of seed is produced from foundation seed and its production is done under strict control and supervision to maintain the genetic purity and identity of the seed as per the standards fixed by the State Seed Certifying Agency. It is usually done in fields of progressive farmers or government farms or recognized private farms.

STAGES OF SEED CERTIFICATION

Seed certification is a very important step for the production of foundation and certified seeds. It involves the following different stages :

• Verification

The first and foremost step in seed certification is the verification of the seed source to which a farmer is using for seed production. For this, farmers growing different crops for seed production are asked to submit a documentary proof like certification tag, records etc., to the Seed Certification Agency for the verification of the seed source. Verification of the seed source helps in maintaining the strict enforcement of Seed Certification laws. If a farmer does not submit these documents in time, the Seed Certification Agency may demand for the same.

Field Inspection

Different cultural practices are followed by the growers for raising the seed crops. Whether the grower has followed or is following all the standard cultural practices for ensuring the genetic and physical purity of the seed crop or not, the field is inspected time and again by the technical experts of the State Seed Certifying Agency. Certain seed standards have been ascribed by the Seed Certifying Agencies (Table 8.2) and these agencies are responsible for enforcement of these standards through repetitive checks in the farmer's field.

Seed Analysis

After harvesting and threshing, seed is approved by the seed agency only after testing its genetic and physical purity. If the seed meets all the requirement of the prescribed standards (both field and laboratory) only then it is approved otherwise not. In the laboratory, the seed is tested for seed borne diseases, moisture contents etc., and lastly certified under the standards fixed by the Seed Agency.

TABLE 8.2 : Minimum seed certification standards for some vegetable crops

Crops	Isolation distance (m)	Maximum off types (%)	Seed purity (%)	Maximum insert matter (%)	Minimum germination (%)	Maximum moisture contents (%)
Amaranthus	400	0.10	95.0	5.0	70.0	8.0
Beans	50	0.10	98.0	2.0	75.0	9.0
Brinjal	200	0.10	98.0	2.0	70.0	8.0
Carrot	1,000	0.10	95.0	60.0	80.0	8.0
Cole crops	1,600	0.10	98.0	2.0	65.0	7.0
Cucurbits	800	0.10	99.0	1.0	60.0	7.0
Fenugreek	50	0.10	98.0	2.0	70.0	8.0
Garden beet	1,600	0.10	96.0	4.0	60.0	9.0
Lettuce	50	0.10	95.0	5.0	70.0	8.0
Okra	400	0.10	99.0	1.0	65.0	10.0
Onion	1,000	0.20	98.0	2.0	60.0	8.0
Pea	20	0.10	98.0	2.0	75.0	9.0
Pepper	400	0.10	98.0	2.0	60.0	8.0
Radish & turnip	1,600	0.10	98.0	2.0	70.0	6.0
Tomato	50	0.10	99.0	1.0	65.0	8.0

Tagging and Sealing

After the seed is certified, it is packed in bags of different sizes. The certified seed should have an official tag, containing. name and address of Seed Certify Agency, name of crop, name of variety, genetic purity (%), germination percentage, weight of seeds, lot number, name of the producer, an appropriate sign to designate the seed, date of issue of certificate by the Seed Agency, expiry date of seed etc.

SEED TESTING LABORATORIES

Seed testing laboratories are important in the sense that they are the main source to start or to enforce any good seed production programme. For this, the role of a well-equipped laboratory becomes more pertinent. However, the basic functions of a seed testing laboratory are to test the seed produced by a certified producer whether it meets all the requirements of the standards fixed by a Seed Certifying Agency or not, to determine the genetic purity of the seed, to test the seeds of the cultivators and to test the seeds samples submitted by seed inspectors for enforcement of strict seed laws. In India, it is estimated that there are about two hundred laboratories in government and private sectors. Different states governments have established seed testing laboratories at different places.

• • •

CHAPTER 9

Seed Vigour

In general, seed vigour is considered as the best seed quality parameter by the Seed Technologists. As per International Seed testing Association, seed vigour is the sum total of those properties of seed, which determines the potential level of activity and performance of seed or seed lot during germination and seedling emergence. The performance of seed include, (*i*) biochemical process and reactions during germination, (*ii*) rate of uniformity of seed germination and seedling growth, (*iii*) rate or uniformity of seedling emergence and growth in the field, and (*iv*) emerging ability of seedlings under unfavorable environmental conditions. Seed vigour is influenced by several factors, but it is mainly affected by genotype of the seed, environmental factors like light, temperature, wind, rainfall, humidity etc., nutrition of the mother plant, stage of maturity and harvest, seed size, seed weight, ageing and pathogens etc.

VIGOUR TESTING

Sequence of many event takes place for seed deterioration, which ultimately affects the seed germination. The loss of germination, which is the final consequence of seed deterioration, is preceded by a several biochemical and physiological processes. Thus, the changes preceeding loss of germination could serve as vigour test. Ideally, the ultimate vigour test would be that event, which is farthest removed from loss of germinability, *e.g.*, membrane degradation.

MAJOR CONSTRAINTS IN VIGOUR TESTING

McDonald, a famous Seed Technologist had pointed out several limitations of seed vigour test. An ideal seed vigour test must be inexpensive, rapid, uncomplicated, reproducible and it should correlate with filed performance of the seed. Due to all these collective requirements of an ideal seed vigour test, it is very difficult to select one vigour test. However, the following aspects of vigour test should be kept in mind, which are pertinent to the development and the standardization of a vigour test.

1. Bulk seed testing evaluation should be used/followed, which is rapid and less expensive than individual seed evaluation.
2. Only that vigour test should be employed, which has proved its worthiness in many crops.
3. The results of vigour test should be accompanied with date and explanations.
4. Instead of using many vigour tests, try only one because every test will prove to be successful as all tests are well developed and refined.
5. All the vigour tests rely upon some quantitative assessment of the seedling growth to monitor seed growth. Thus, dormant seeds or seeds treated with fungicides should not be considered low in vigour. The interpretation of vigour tests can account such artificially altered performance capability of the seeds.

TESTS FOR MEASUREMENT OF SEED VIGOUR

The constraints imposed on vigour testing have limited the types of vigour tests, which can be employed for testing vigour of the seeds. Despite these limitations, several vigour tests have been developed, which has been described briefly hereunder :

1. Accelerated Ageing Test (AAT)

This method was initially proposed to evaluate the seed storability. In this method, the unimbibed seeds are subjected to high temperature (41°C) and to cent per cent relative humidity for 3-4 days. After removing the seeds are allowed to germinate under optimum conditions. The relationship between initial moisture content and the seed germination is established to interpret the results. It is a rapid, simple and inexpensive method, which can be employed for all kinds of seeds. Further, this test does not require any additional training for correct evaluation of seedlings.

2. Cold Test

In this method, seeds are placed in soil or paper towel lined with soil and exposed to cold for a specific period to provide stress from imbibition, cold temperature and microorganisms. Afterwards, the seeds are placed under favourable growth conditions. As the seed vigour ranking by cold test remains consistent, it is considered as the most useful 'in house' seed test. However, the soil used in the test has no standard, hence it may affect interpretation of the seed vigour test. As we know, soils differ in moisture content, pH, texture, structure, population of microorganisms etc., which may contribute to divergence in the results.

3. Conductivity Test (CT)

It is based on the known fact that seeds with low vigour have poor membrane integrity because of mechanical injury. Thus, when such seeds are imbibed, cells with poor membrane integrity release solutes of cytoplasm into imbibing medium (principally water), which can be quantified by measuring the electrical properties by conductivity. This method is very useful for measuring seed vigour of bulk samples. It is considered as a rapid, simple and inexpensive method, although; the initial seed moisture content and seed size should be standardized, which may affect the results.

4. Cool Germination Test (CGT)

This method of testing seed vigour is different to cold test in that CGT is conducted under standard laboratory conditions at a lower temperature (18.6°C) and does not rely upon microorganisms for creating stress conditions in the germinated seeds. It is almost similar to germination test (GT) as its procedures are almost similar to the germination test. It is mainly used for testing the vigour of cotton seed.

5. Seed Growth Rate Test (SGRT)

This test is based on the concept that vigorous seeds have the ability to synthesize and transfer new materials to emerging embryonic axis more efficiently and rapidly, which results in accumulation of higher dry weight in the seedlings. Thus, the vigour results are expressed as dry weight/germineable seedling. In this method, the growing segments of the embryo of normal seedlings are excised from the storage organs such as cotyledons or endosperm, placed into breakers and dried at 80ºC for 24 hours and then increase in dry weight is measured. This test is also similar to the routine germination test. However, the results of this method are mostly precise, because the seedling growth rate is correlated with vegetative development of the seedling in the field. But, the differences in moisture and light intensity can influence the results, which is considered as the major limitation of this test.

6. Seedling Vigour Classification Test (SVCT)

This is also similar to routine germination test except that normal seedlings are further classified into 'strong' and 'weak' categories. In general, it is not considered more reliable test because the evaluation of the seedlings is done subjectively.

7. Tetrazolium (TZ) Test

It is one of the most reliable tools for seed analysis, which has been described in detail in the Chapter 6 (Seed Viability).

• • •

CHAPTER 10

Seed Storage

Seeds are required to be conserved through proper storage for a short or longer period. Seed storage is necessary because of the following reasons :

1. It is not possible to produce seeds of each and every stock every year.
2. Seed yield can't be estimated easily.
3. The demands for seeds may not always be uniform.
4. Valuable seed stock can be difficult or costly to produce.
5. Conservation is always desirable than consumption.
6. Seeds can be supplied for a longer period.

A small quantity of seed may be stored on the farm using traditional methods, but for conservation of large quantities of certified seed, proper storage conditions are required to avoid the deterioration of seeds. Hence, good storage is a basic requirement in seed production.

FACTORS AFFECTING SEED VIABILITY DURING STORAGE

Type of the Seed

The storage life of the seed in influenced invariably by the kind or type of the seed. For example, some seeds are naturally short-lived and germinate immediately after harvest, and if such seeds are allowed to dry even for a short period, they loose viability, as in mango, litchi, citrus, rubber etc. It is really difficult to preserve such type of seeds. However, a few researchers have advocated the use of ash to store such seeds, as in citrus.

Moisture Content of Seed

The moisture content of seeds is perhaps the most important factor, which influences the seed viability considerably. In general, low moisture contents (4-6%) of the seeds are desirable for their better and long term storage. However, moisture content below 4% may be equally harmful because it may cause desiccation of the seed, affecting its viability considerably. Further, fluctuation in seed moisture content during storage also reduces the seed longevity. Thus, seeds should be dried to the recommended and safe limits of moisture content before storing them. Similarly, seeds must be sealed in moisture resistant containers to avoid moisture fluctuation and thereby the viability and longevity of seeds. The maximum safe seed moisture content of some vegetable seeds during storage at different mean temperatures may be different for different vegetables (Table 10.1).

TABLE 10.1 : Maximum safe seed moisture contents of some vegetable seeds during storage at different temperatures

Sl. No.	*Seed crop*	*Mean temperature of storage (°C)*		
		4.5-10	*21*	*26.5*
1.	Beet	14	11	9
2.	Cabbage	9	7	5
3.	Cauliflower	9	7	5
4.	Carrot	13	9	7
5.	Celery	13	9	7
6.	Cucumber	11	9	8
7.	Kidney bean	15	11	8
8.	Lima bean	15	11	8
9.	Lettuce	10	7	5
10.	Lady's finger	14	12	10
11.	Onion	11	8	6
12.	Garden pea	15	13	9
13.	Pepper	10	9	7
14.	Spinach	13	11	9
15.	Palak	13	11	9
16.	Tomato	13	11	9
17.	Turnip	10	8	6
18.	Watermelon	10	8	7
19.	Muskmelon	10	8	7
20.	Chillies	10	9	7

Storage Temperature

Storage temperature is equally important for regulating the seed viability during storage. In general, storage at low temperature increases the life of the seeds. Harrington, a world famous Seed Technologist has proposed a thumb rule that for each 1 per cent decrease in seed moisture content or for each drop of 10°F in storage temperature, storage life of the seeds in almost doubled, Thus, a combination of moisture resistant sealed container with low moisture content of seeds plus low storage temperature provide excellent conditions for short-lived seeds for a longer period.

Storage Atmosphere

Storage atmosphere has a significant influence on seed viability. Thus, attempts have been made to modify storage atmosphere, such as, increasing CO_2 concentration, creation of vacuum and replacing oxygen with nitrogen. Relative humidity (RH) during storage has a great bearing on seed viability. Below 74% RH, there is a little activity of fungi, which is completely suppressed below 62% RH. Similarly, bacteria require a minimum of 90% RH for their growth and development.

IMPORTANT POINTS FOR SAFE STORAGE OF SEEDS

Keep all the following factors in mind while storing the seeds :

1. Only seeds of valuable stock should be stored, which is healthy and has good germination percentage.
2. Before storage, seeds should be dried to the acceptable or recommended moisture level.
3. Seeds must be treated with chemicals to protect them from the incidence of insect-pests and diseases.
4. Storage conditions must be good enough to provide adequate atmosphere to the seed.
5. Store where seeds are to be stored, should be properly maintained.
6. There should be sufficient provision for the cleaniness and sanitation of the store.
7. Proper measures should be followed to store seed properly during transit, at retailer's shop and at user's farm.

• • •

CHAPTER 11

Nursery Establishment

A nursery is a place, where seedling, saplings, trees, shrubs and other planting materials are grown and maintained until planted/transplanted in a permanent place. Unlike most agricultural crops, establishment of an orchard is a permanent venture. Any mistake made initially can't be rectified in the later stages. Hence, the availability of a quality planting material is the foundation on which an orchard is established. If the plants are reliable and of guaranteed performance, they will be high yielding and will produce quality fruits. Unfortunately, most of the present day orchards in our country have been established from inferior stock. Thus, for any successful horticultural production, availability of true-to-type plants is the first and foremost pre-requisite. Nursery is, therefore, a place where one can produce true-to-type plants by gaining technical skill, maintaining plants properly and by careful planning. Like orchard establishment, establishment of a nursery is also a permanent venture and any mistake made initially, can't be rectified easily in the later stages. Therefore, for establishing a nursery, due care in selection of site, plants to be raised and transportation facilitaties etc., must be given.

PRE-REQUISITES FOR ESTABLISHMENT OF A NURSERY

Following point should be kept in mind while establishing a nursery :

1. As far as possible, the nursery should be located in the important production areas.
2. The locality where fruit, vegetable or ornamental plants are grown on a commercial scale provides an ideal place for nursery establishment.
3. The soil should be deep, fertile, well drained and free from soil borne pathogens.
4. The locality should have an adequate supply of sweet water.
5. Climatic conditions should be favourable for the plants to be propagated.

6. The site should be well connected by different means of communication (road, rail etc.) and must be easily accessible.
7. Sufficient labour, budder and grafters should be available to handle different operations in the nursery.
8. The materials like fertilizers, pesticides, growth regulators, grafting wax, lanolin paste and equipments etc., should be readily available.
9. There should be sufficient provision for different propagation structures like glasshouse, lath house etc.
10. The nursery should have its own resources to provide parent material (mother plants) for propagation.

IMPORTANCE OF NURSERY

Due to the following reasons, a nursery is required :

1. The young plants can be easily maintained in the nursery. Similarly, it is easy to look after the young and tender seedlings in the nursery rather than in the field.
2. Propagation of plants by asexual means require special skill and care of plants before transplanting in the field, which can be easily done in nursery.
3. Many plants do not respond to direct sowing (*e.g.*, cabbage, tomato, and papaya) in the field as compared to transplanting of seedlings raised in the nursery.
4. Cuttings of different horticultural plants for rooting are first planted in the nursery for better care and management.
5. For hardening of seedlings/layers, nursery is a pre-treatment place.
6. Seasoning of the seedlings against natural calamities is only possible in the nursery.

CLASSIFICATION OF A NURSERY

The nursery can be broadly classified into two groups on the basis of its size *viz.* (*i*) Home nursery (*ii*) Commercial nursery.

Home Nursery

It is a small area in the garden in which the plants are grown to meet the demands of grower's own garden. The main objective of such nursery is to provide the quality planting material. Usually, costly methods of nursery practices are followed in this type of nursery for raising planting material of high quality.

Commercial Nursery

The main aim of such nursery is to earn money on the investment. Such nurseries have a large area and costly nursery practices are usually avoided. Moreover, the control on quality of planting material is also reduced. Commercial nursery may be located in cities and towns (urban nursery) or in villages (rural nursery).

Rural nursery : This type of nursery is located in a village near some highway or near a railway station. Usually, the size of a rural nursery is large because land and labour is not a problem in rural areas. The planting material in such nurseries is sold out at a cheaper rates because the cost of raising the planting material as well as the buying power of the identors is less.

Urban nursery : This type of nursery is located in a town or a city. The size of the nursery is usually small because the land is costly and not easily available. The planting material is also costly in these nurseries because the cost of raising the planting material is high due to costly labour and other management practices.

Based on business, the nursery can be of different types, like wholesale nursery, the retail nursery, the landscape nursery, the agency nursery and the mail order nursery.

(*i*) **Whole sale nursery :** In wholesale nursery, the plants are produced in large quantity for sale to retail outlets. These nurseries are usually located in rural areas, where the land and labour are available at cheaper rates. Due to these reasons, rural nurseries can afford to expand their business without incurring much additional expenses.

(*ii*) **Retail nursery** : The retail sellers purchase plants from wholesale nursery. Because the retail nursery is largely dependent on house owners for its trade, it must be located near a town or a city. These nurseries also keep goods like fertilizers, seeds and tools etc., required for raising the garden plants.

(*iii*) **Landscape nursery :** The landscape nursery should be located near a populous town or city because urban people require the landscape plants for beautifying their houses. Such nurseries should be located on the outskirts of a city or town as it facilitates the clientele to obtain landscape services.

(*iv*) **Mail order nursery :** It is a specialized wholesale nursery. It depends primarily on a catalogue display of the stock it offers for sale. Customer order from the catalogue and receive the plants through mail or parcel service. These nurseries are also located in a locality, where land is relatively inexpensive and sufficient labour, water and transportation facilities are available.

(*v*) **Agency nursery :** The agency nursery sells its stock through agents or sales representatives. Such nurseries are highly specialized and are usually few in numbers.

FACTORS AFFECTING THE ESTABLISHMENT OF A NURSERY

Location and Site

The land of nursery should be flat with little slop in one direction for proper drainage. The area selected for nursery establishment should be reputed for one or the other kind of business. It should be a known locality. The skilled labour should be available in the locality.

The area should be well connected with road or by a rail track. It should be easily approachable by the customers.

Soil and Climate

It is of prime importance to select a locality in which a wide range of species and varieties are grown and are in great demand in the market. The soil should be physically good with proper drainage. Usually, for fruit plants, friable loamy soils of high fertility with a pH range of 5.5-6.5 are preferred. A clay loam soil is advantageous for evergreen plants. However, soils with hard and compact sub-soils layers should be avoided. Usually, a depth of 70-80 mm is sufficient for growing plants in nursery. Soils with impregmented layers of calcium carbonate may be avoided as it affects permeability and aeration, resulting in poor root development.

Type of Sub-soil

For establishment of a successful nursery, both soil and sub-soil should be suitable and the pH must be in an optimum range. The soil should be physically good and should have proper drainage. Usually, loamy soils of high fertility with pH range of 5.5-6.5 are preferred. A clay loam soil is advantageous, as in this type of soil, the lifting of evergreen plants with earth-ball intact is easy. However, soils with hard and compact sub-soil layers should be avoided. Soils with impregmented layers of calcium carbonate may be avoided as it affects permeability and aeration, resulting in poor root development. Highly acidic, alkaline and saline soils are not preferred for nursery purpose.

Water Supply

Frequent light irrigations are required for nursery plants. So, availability of regular soft water in area is of prime importance. For this purpose, a surface well or a tube well may be dug out. However, a water reservoir may be constructed in areas with short supply of water for meeting the requirements during water shortage.

Demand of Plants and Varieties

While making choice of kind of plants and their varieties, one has to take into consideration a number of factors. Demand for fruit plants of different kinds and varieties is the most important factor. In the nursery, only those plants and varieties should be produced in bulk, which are in high demand in that particular locality.

Availability of Chemicals, Manures and Fertilizers

In the nursery, many chemicals, manures, fertilizers and plant protection measures are required for timely application and management of seedlings and plants in sound manner. Organic manures like farm yard manure, leaf mould, compost, fertilizers like urea, single super phosphate, potassium sulphate and several fungicides, insecticide, weedicides are required and these must be timely available in the locality in sufficient quantities.

Availability of Trained Gardeners and Skilled Labourers

The orchardists who want to take up nursery production work, should be well conversant with different propagation techniques and other nursery operations. They should have with them trained budder/grafter, who could perform these operations with a good success.

PARTS OF A NURSERY

A modern nursery should have the following parts :

Building Structures

An ideal nursery should have a office, sale counter, packing shed, store house, implement shed, bullock shed and residential quarters etc.

The office is the most important amongst all the building structures because all the business transactions are to be made from the office. It should look beautiful and attract customers. Its size depends on the size of the nursery.

The store and office can be in the same building. The store can be utilized for keeping packing material such as baskets, labels, nails, ropes, equipments, fertilizers and planting material etc. The owner of the nursery can use the office. The office should be near to the gate so that visitors have no difficulty in locating it.

Progeny Block

In progeny block, true-to-type mother plants are maintained in the nursery. According to the choice of customers, suitable plant types with existing superior varieties should be collected and maintained in the progeny block. The plants in the progeny block should be healthy, disease-free, genetically pure and free from the attack of insect-pests and diseases. The plants of each variety should be labelled correctly. Proper care regarding control of insect-pests and diseases should be taken in respect to mother plants in the progeny block.

Propagation Structures

Many times, the out door conditions are not favourable to raise plants successfully. Low or high temperature or the hot desiccating winds may be harmful for the seedlings and their subsequent growth. Hence, there should be a provision for modern propagation structures like greenhouse, hot bed, cold frame, lath house, net house and mist chamber etc. These structures provide optimum conditions (light, temperature, moisture and humidity) for seed germination, rooting of cutting and for hardening of young seedling before transplanting them in the field.

Seed Beds

Seeds can be accommodated in a comparatively smaller area. They should be nearer to a source of water supply and to the office so that they can be kept under vigilant control. The beds should be at a slightly higher level than the surface of the ground so as to provide a good drainage. Seedbeds should be located in an open situation because many seeds do not germinate properly and seedlings are infested by 'damping off' disease, if kept in shade.

Nursery Beds

Nursery beds should be located in an open area near to a water source. Seedlings from the seedbeds are removed and transplanted in these nursery beds. A nursery bed should be prepared by repeated ploughing and adding appropriate quantity of organic manures. Some fertilizers should also be mixed during their preparation. There must be an adequate provision for proper drainage. Nursery beds should be a little raised, with 3 m length and 1 m width. It is better if the nursery beds are divided into sections. For example, one section can be a mango seedbed area, a second section can be set apart for rooting of cuttings, another for fruit plants and still others for citrus plants etc. This order should be changed next season to have a rotation and thereby keeping the fertility of the soil intact. The nursery beds should be laid out in such a way that there is an access to all the beds through roads or paths so that workmen may not have to walk over the bunds or in water channels.

Pot Yard

The pot yard should be in shade because the tender plants require shade as compared to the hardy plants. This section should be near to a water source because the plants will have to be watered at frequent intervals. Trenches can be provided for keeping the potted plants closely packed together. This facilitates easy watering once or twice a week.

Packing Yard and Working Shed

The packing yard is used for packing the plants before sale or dispatch to outstations. The yard can be combined with the working shed. In packing yard, there should be a plenty of space to enable a number of workers for sorting out and packing the plants with ease.

Water Source

The timely availability of good quality water in sufficient amount is an integral part of nursery production. In nursery, several plants of different stages of their growth are raised, therefore, timely supply of quality water is essential. Thus, proper arrangements should be made to meet the water requirement of a nursery. For this purpose, provisions for water storage tank, motor or diesel engine operated tubewell should be made. Not only the source of water but the water channels to distribute the water evenly and orderly is also important. These channels should be connected to every part of the nursery for timely use.

Compost Pit

Nursery production of horticultural and forestry plants require huge amount of organic manures, like F.Y.M., compost, leaf mould etc., for different purposes. Therefore, arrangements should be made at nursery level to produce enough quantity of compost etc., for own purpose. One compost pit of permissible size should be located at any corner of nursery layout. At the same time, several waste products of nursery can also be utilized for the same and extra effort to dispose off waste material is not required.

A model of a well-planned nursery has been given in Figure 11.1.

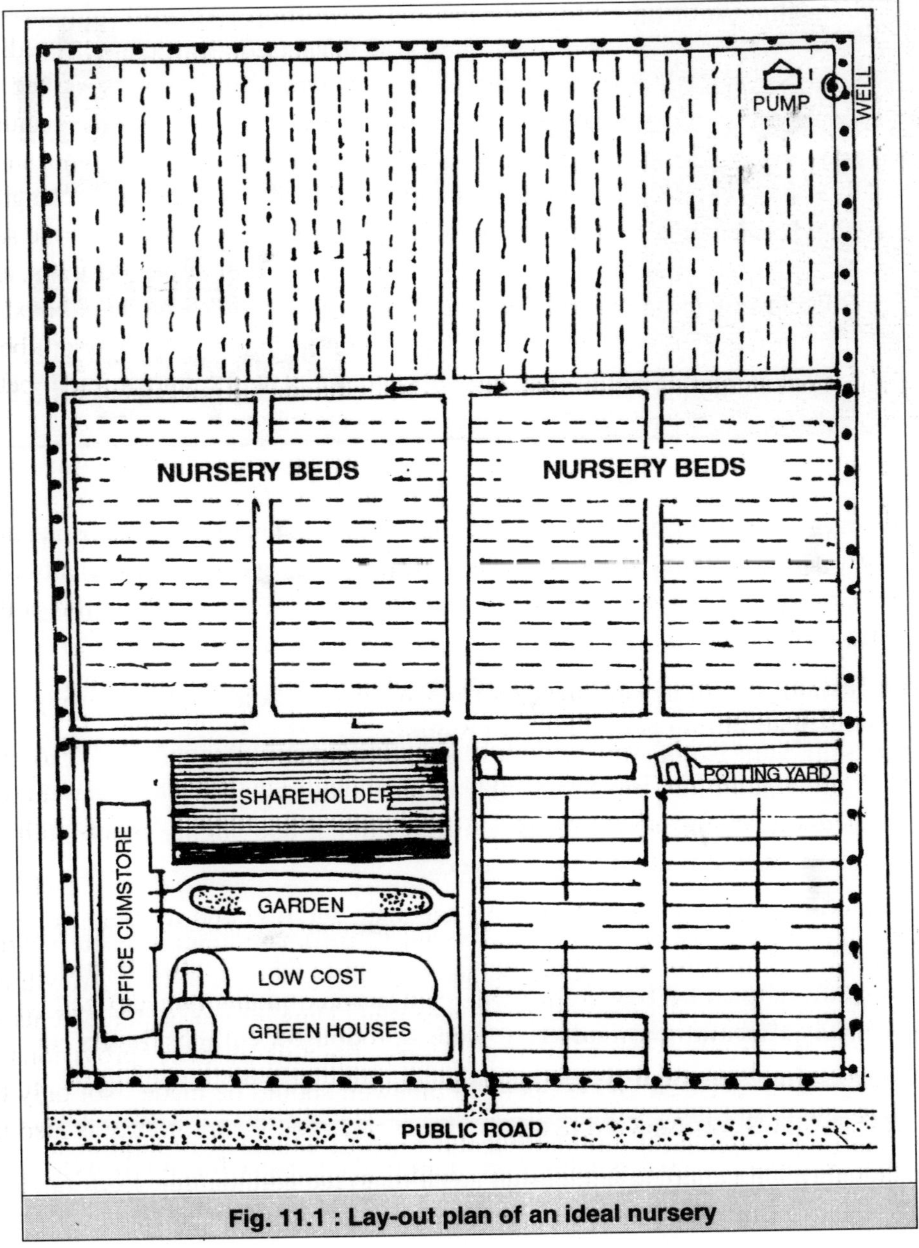

Fig. 11.1 : Lay-out plan of an ideal nursery

• • •

CHAPTER 12

Media for Propagation

The nurserymen and propagators use different kinds of media for the propagation of different horticultural plants through seeds, cuttings or other methods of vegetative propagation. One may use a medium of his choice but an ideal medium should have the following characteristics :

1. The medium should be sufficiently firm and dense to hold seeds or cuttings or layers firmly in place during germination or rooting.
2. The volume of the medium should remain constant upon drying or wetting.
3. The medium should have enough ability to retain and supply required amount of moisture to the seeds or cuttings or layers.
4. It should be porous enough to permit adequate aeration and to drain out excessive moisture.
5. The medium must be free from seeds of unwanted plants (weeds) and pathogens, which may affect the germination of seeds or rooting of cuttings, layers etc.
6. It must be free from high levels of salts.
7. Its pH should be between 5.5 to 6.5.
8. It should be able to provide adequate nutrition to the germinating seeds or cuttings or other plant parts.
9. It must be capable of being pasteurized with steam or chemicals without harmful effects.
10. It should be cheap and easily available.

CRITERIA FOR THE SELECTION OF A PROPAGATION MEDIUM

Before using any propagating medium, the following criteria should be kept in mind, while selecting a medium :

1. As far as possible, we should always select a locally available material as propagation medium.
2. Check the quality of the medium before buying it.
3. Take guidance of a technical expert or a person dealing in nursery business.
4. The selected medium should have sufficient porosity and ideal pH (5.5-6.5).
5. The selected medium should easily mix with other growing medium.

COMMONLY USED PROPAGATION MEDIA

The characteristics of commonly used media by propagators, nurserymen and scientists are described briefly hereunder :

Soil

The portion of earth's crust, which is formed by the decomposition of rocks and minerals by physical, chemical and biotic forces is called as soil. The soil is mainly composed of solid, liquid and gaseous states. For proper growth of the plants, all these contents should be in a proper proportion. Solid portion of soil has both organic and inorganic matters, which determines the structure and texture of the soil. The inorganic part is mostly the residue, which is the resultant of decomposition of parent rocks by weathering process. These vary in size, ranging from large sized gravel to extremely minute particles of clay. The coarse particles serve as supporting framework of the soil and the clay fraction as the store-house of the nutrients, which are absorbed by the plants. The organic portion of the soil constitutes the living and dead organisms (fungi, bacteria, and insects). The residue of decay of dead organisms is called as *humus,* which is highly colloidal in nature and is mainly responsible for holding of water and plant nutrients. The liquid part of the soil is mainly water, which contains dissolved mineral elements and dissolved O_2 and CO_2 gases. The gaseous portion of soil is very important for plant growth as in water-logged (poorly drained) soil, water replaces the air, thus, plant roots are deprived of essential microorganism, which are required for proper growth and development of a plant.

Soil texture is an important physical property of the soil, which determines the success or failure of seed or cutting or other propagule. It depends upon relative proportion of the sand, silt and clay. The main textural classes of soil are sand, sandy-loam, silt-loam, clay-loam and clay etc. A soil having 40 per cent sand, 40 per cent silt and 20 per cent clay is considered as the best for seed germination of many plant species. Sandy-loam soils are excellent for preparation of soil mixtures for container growing plants. The pH of soil also determines the success or failure of seed germination or rooting of cuttings of layers. In general, a soil having neutral pH (5.5 to 6.5) is always preferred for better germination of seeds and rooting in cuttings or layers. In addition to texture, the structure of soil, which refers to the arrangement of soil particles in a soil mass, has great bearing on the germination of seed and rooting of cuttings, or layers. Water holding capacity of soil is usually determined by its structure. Thus, a soil having good texture with water holding capacity should be preferred for using as a propagating medium. In general, heavy soils should be avoided, as working with these soils is very difficult. Moreover, these soils form heavy clouds and have very high water holding water capacity, which is injurious, both for seed germination and rooting of cuttings, layers or other propagules.

Sand

Sand is formed as a result of weathering of rocks. Its composition depends upon the type of the rock from which it has formed. The size of sand varies from 0.05 to 2 mm in diameter. Sand contains silica with almost no mineral. Quartz sand, which consists chiefly of silica compounds, is mainly used for propagation by the nurserymen. Sand is the heaviest of all the rooting media and therefore it should be used in combination with some other organic material. Commercially, sand is available in many grades and sizes. However, sand should preferably be washed, fumigated or heat-treated before use to kill harmful pathogens present in it.

Vermiculite

Vermiculite is a micaceous mineral, which is widely used as a propagating medium in advanced countries. Chemically, it is a hydrated magnesium-aluminium iron silicate, which expands markedly upon heating. When expanded, it becomes very light, insoluble in water but absorbs very high quantity of water. It has a high cation exchange capacity and can hold nutrients for longer time and releases them later. It contains sufficient quantity of Mg and K for supply to plants. Horticultural vermiculite has four grades, depending upon the size of the particles. No. 1 grade has particles from 5-8 mm in diameter, No. 2, from 2-3 mm, No. 3 from 1-2 mm and No. 4 from 0.75 – 1 mm. In general, No. 2 grade vermiculite is a regular horticultural grade and is widely used. No. 4 grade vermiculite is most useful as a seed germinating medium.

Peat

Peat contains mainly the partially decomposed parts of aquatic vegetation. Its composition depends upon the vegetation from which it originated, state of decomposition, mineral content and acidity. Peat is mainly used as a propagation medium in the USA and there, it has been classified into different types *viz.* moss peat, reed sedge peat and peat humus.

Moss peat, which is commonly called as ‘peat moss’ is the least decomposed, which is derived from sphagnum or hypnum or other mosses. It varies in colour, ranging from light-tan to dark-brown. It is highly acidic (pH 3.2-4.5) but has high moisture holding capacity. It contains small amount of nitrogen but little or no P and K.

The reed sedge peat mainly consists of the remains of grasses, reeds, sedges and other swamp plants. It varies considerably in colour and composition, ranging from reddish-brown to brown to blackish-colour. Its pH range is between 4.0 to 7.5 and water-holding capacity is also high.

The peat humus is highly decomposed material, derived either from hypnum moss or reed sedge peat and is brown-to-black in colour. It has very poor water holding capacity. Peat is not a uniform product and can be a source of weed seeds, insect-pests and pathogens. Thus, it must be pasteurized before using as a propagation medium.

Among different classes of peat, moss peat is very popular propagating medium among nurserymen of most of the advanced a countries, because :

1. It is an uniform product.
2. It is free from insect-pests, diseases and weed seeds.
3. It can be easily mixed with other propagating media like sand, saw-dust, perlite or vermiculite etc.

Sphagnum Moss

The sphagnum moss is commonly found in hilly tracts of India. The dehydrated moss is either a residue or a living part of *Sphagnum papillosum, S. capillaceum* and *S. palustre*. Sphagnum moss is relatively sterile, light in weight, having very high moisture holding capacity. It has a pH of about 3.5 to 4.0. It is widely used for keeping the live material moist during distant transport and in air layering (goottee) of plants. It can also be used as a rooting medium for cuttings. However, moss should be shredded either by hand or mechanically before use. Similarly, it should be moistened with water before use.

Pumice

It is the most widely used medium either alone or in combination with peat moss for seed propagation in the USA and some other advanced countries. It contains chiefly silicon oxide and aluminum oxide with varying amount of iron, calcium, magnesium, and sodium in oxide form. Pumice is screened to different grades but not heat-treated. It increases aeration and drainage in rooting mixtures if used in combination with other propagating media.

Perlite

Perlite is a gray white silicaceous material of volcanic origin. It is prepared by heating glassy volcanic lava in furnace at 760°C. Perlite is a sterile medium with pH of 6-8, high cation exchange capacity and water holding capacity. It contains no mineral nutrient and is useful in increasing aeration in a mixture, when used in combination with other propagating media. Perlite is available in different grades with particle size ranging from 3-8 mm but in general, a particle size of 1.6 to 3.0 mm in diameter is used mostly by the nurserymen. It is considered as an ideal rooting medium for rooting of cuttings if used in combination with peat moss.

Leaf Mould

Leaf mould is prepared by decomposing fallen leaves of the plants. Sometimes, chemical fertilizers, FYM (farm yard manure) or cow dung are also added to it. Its colour and composition depends mainly on the original material. It contains almost all the essential nutrient elements in it. Leaf mould should be properly treated with fungicides and insecticides before use to kill the pathogens.

Compost

Compost is prepared by decomposing bulk organic wastes biologically under controlled conditions in piles or bins. Compost is a rich source of mineral nutrients with good water holding capacity. It can be used as a medium for propagation but by mixing with soil. The various microorganisms like bacteria, fungi and nematodes and insects like beetles, cut-worms, snails and slugs can often be found in compost piles. Therefore, compost should be properly pasteurized before use. The whole process of compost making occurs in three steps. In the first step (for a few days), decomposition of easily degradable material takes place. In the second step (several months), cellulose compounds are degraded at a high temperature and in 3rd stage, the microorganisms recolonise the material.

Soil Mixtures

Different soil mixtures are used as medium for propagation. Usually, soil mixtures are prepared by mixing sand, loam soil and leaf mould in different proportions. However, an ideal soil mixture should be porous and should have good water holding capacity. Soil mixtures are not only used as propagating medium for seed germination or rooting of cuttings but also for filling of containers.

Plastic Aggregates

Synthetic plastic aggregates are also used in many countries, especially in Europe, as a substitute for sand or perlite. The commonly used synthetic plastic aggregates are expanded polystyrene fractions and urea-formaldehyde foam. The polystyrene fractions are usually neutral, don't absorb water and don't decay. These usually improve the drainage and aeration. Urea foams are usually sponge like structures having enough N content and high water holding capacity.

Shredded Bark and Saw Dust

Shredded bark, saw dust and wood shavings of different plants can also be used in mixtures with various propagating medium. Due to low cost, light-weight and easy availability, these are mainly used in soil mixes for container grown plants. These mixes usually contain lower amount of nutrients and hence additional amount of nutrients may be added to the mixes, while using as propagation medium. However, the shredded bark and wood shavings may contain toxic materials like phenols, resin and tannin etc., when fresh. Therefore, these should be composted for 10 to 15 weeks before use.

Cocopeat

It is also called as '*cocodust*'. It is a byproduct of cuttings and shifting of coconuts for fibre production. It is becoming very popular propagating and growing medium now-a-days because it has an excellent pore space (25-30 per sent) and fine structure, which are absolutely necessary for proper growth and development of the roots of seedlings. Moreover, it is a rich source of nutrients and can easily be mixed with other growing media.

Rock Wool

It is used as rooting and growing medium in many advance countries of the world. It is derived from basalt rock, melted at a temperature 1600°C, then cooled and spun into fibres. Rock wool is available in several forms, such as shredded, pellets, slabs, blocks, cubes etc. Rook wool has good water holding capacity and retains good level of O_2. However, it is advisable to conduct small-scale experiments before using it as growing medium at a large scale.

Calcined Clay

When mineral like clay is calcined (heated) at a high temperature, it produces some stable aggregates. Such aggregates have no fertilizer value, but are porous, and resistant to break down and water absorption. Calcined clay and other aggregates are mainly used to mix with other growing media. Leca, Terragreon, Turfice, Haydate and Hortag are commercially used calcined aggregates as propagation and growing media in many advance countries like the USA, Japan, Germany etc.

• • •

Propagation Structures

For proper growth and development, different plants have specific requirements for light, water, temperature, gases and mineral nutrients. In the same way, for successful propagation of plants, these basic necessities should be met and arranged properly to get optimum results. In addition, outdoor conditions like freedom from pathogens, pests and salinity of media must be available for the nursery plants. In order to get better results, different propagation structures are used for better multiplication of plants, the important being greenhouse, hot bed, cold frame and lathhouse. Greenhouses and hot beds are primarily used for providing the controlled environmental conditions for seed germination or rooting in cuttings or layers, whereas, the lathhouses and cold frames are used for hardening of young plants before transplanting. A brief description of different propagation structures is given hereunder :

Greenhouses

A greenhouse greatly extends the variety and scope of propagation. Many kinds of greenhouses are used for propagation but the most suitable type is the one that admits the maximum amount of light. This is important, particularly in areas where most of the plant propagation work is done in late winter and early spring. Good light conditions are essential for the sturdy growth of the seedlings. Thus, its panes should be large, preferably not less than 24 inches wide. The framework supports should be as slender as the necessity for a certain degree of strength and stability in the structure for light obstruction. Steel and aluminum fulfil these conditions best, but these are usually more expensive than timber. Experiments have shown that a greenhouse that runs from east-to-west direction is the best for better light penetration in winter and early-spring, spring, and consequently preferable for raising seedlings at this time of the year. Moreover, it is important that the greenhouse should be well away from any kind of shade, such as, a tree or buildings and other greenhouse. However, some protection from north-to-northeast winds is desirable. These glasshouses may be polyethylene covered or made from fiberglass. Modern greenhouses are well equipped with elaborate structures and have precise control on temperature, light intensity and humidity. The size and type of greenhouse, primarily depends upon the need of the plant propagator. Glass covered greenhouses are expensive but they have long life. However, for short-term benefits, plastic covered greenhouses can also be made. Two types of plastics are available for their construction, polyethylene and fiberglass. Both are light-weight and inexpensive as compared to glass. Plastic covered greenhouses are lighter than the glass covered ones but there is very high increase in humidity in such houses, especially in winters, which results in undesirable water drops on the plants. It can, however be, overcome by making adequate ventilation. Polyvinyl chloride (PVC) has also been used in construction of greenhouses but it tends to darken per-maturely in sunlight.

In India, construction of temporarily low-cost polyhouses is in fashion for raising nursery of fruit plants in off-season. Such low cost greenhouses are constructed either on wood or metal framework and are covered with polyethylene sheet of 0.10 to 0.15 mm thickness, which is resistant to ultra-violet rays. These houses are equipped with thermostat, cooler or an air conditioner or humidifier etc., for rigid control on temperature and humidity. In greenhouses, two types of beds are used for growing seedlings. The beds may either be prepared on ground itself or raised beds or bench type beds are prepared.

Hot Beds

The hot beds are small low structures, which are used for propagation of nursery plants under controlled conditions. Usually, hot bed consists of a frame, the cover and a heating system. The frame is made up of wood, steel or aluminium, and covered with a plastic or glass, while heat is provided artificially below the propagating medium by hot air or hot water. A thermostat automatically controls the temperature. Hot beds are widely used for raising plants in western countries but in India, these are rarely used as the cost involved in their construction is very high. As in greenhouse, close attention must be paid on shading, ventilation, temperature and humidity control in hot beds. For better results, hot beds should be placed in sunny, protected and well-drained location. Hot beds can be used throughout the year, except in areas, which experience severe winter. In such areas, use of hot beds may be restricted to spring, summer and fall.

Cold Frames

Cold frames are more advantageous when plants are raised in open. Soft-wood cutting from herbaceous plants and shrubs can be raised in them during warmer months. In autumn, semi-mature or hardwood cuttings can be planted in cold frames. Frames are also useful for starting early vegetable crops and some other from seeds. Cold frames can be used for the conditioning or hardening of the rooted cuttings, young seedlings or container plantings and starting new plantings in late spring, summer or fall, when external heat supply is required. Sowing can be done under the continuous type of cloche to give seedling an early start. Cold frames are similar to hot beds except that there is no provision for supply of bottom heat. In cold frame, heat is provided through sun only and hence these are kept in sunny places only. For their construction, a standard glass (0.9 × 1.8 m) is often used as a covering material for the frame but less expensive polyethylene or fiberglass covering can also be used. The covered frames should fit tightly to retain heat and humidity. Cold frames should be placed in sunny locations, protected from winds and should have covering sloping down from north-to-south. To get desired results, a careful attention should be paid on ventilation, shading and winter protection in cold frames.

Lathhouses

Lathhouse provides outdoor shade and protect container-grown plants from high summer temperature and high light intensity. They may reduce moisture stress and decrease the water requirement of the plants and thus these are useful for protecting young seedlings or rooted cuttings or container—grown plants from high summer temperature and light intensities. Lathhouse serve as an intermediate between cold frame, greenhouse or a hot bed. The construction of a lathhouse varies as per the need of the propagator. These can be made from pre-fabricated aluminium or wood but aluminium made structures are costly. Mostly, wood or pipe supports are used, which are set in concrete with necessary supports. For shading, woven plastic material is widely used to allow proper light inside the house. In summers, however, overhead sprinklers are more useful than shade for cooling down the temperature. Water requirements of plants grown under lathhouse is usually low because of low rates of evaporation and transpiration of the plants grown in it.

Propagation Frames

Sometimes, in greenhouses, the humidity is not enough to allow satisfactory rooting in the leaf cuttings. In such cases, enclosed frames covered with glass or plastic material may be used for rooting of cuttings (Figure 13.1 and 13.2). These frames are useful only in grafted plants as these retain high humidity during the process of healing.

Large inverted glass can also be kept over a container having cuttings. Although, high humidity is required in such frames but proper ventilation and shading is necessary after the rooting process has started in the cuttings. Warm and humid conditions inside these structures provide excellent environment for growth of pathogenic organisms, which may infect the propagating material. It is, therefore, necessary to maintain cleanliness and proper sanitation in these structures.

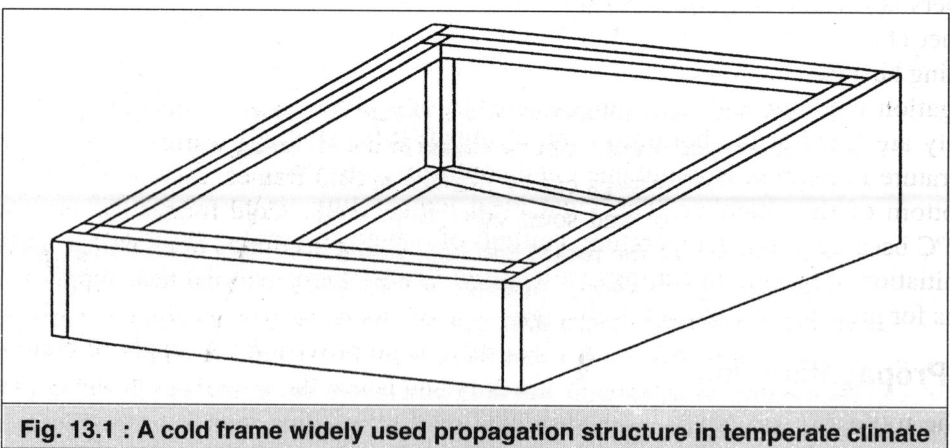

Fig. 13.1 : A cold frame widely used propagation structure in temperate climate

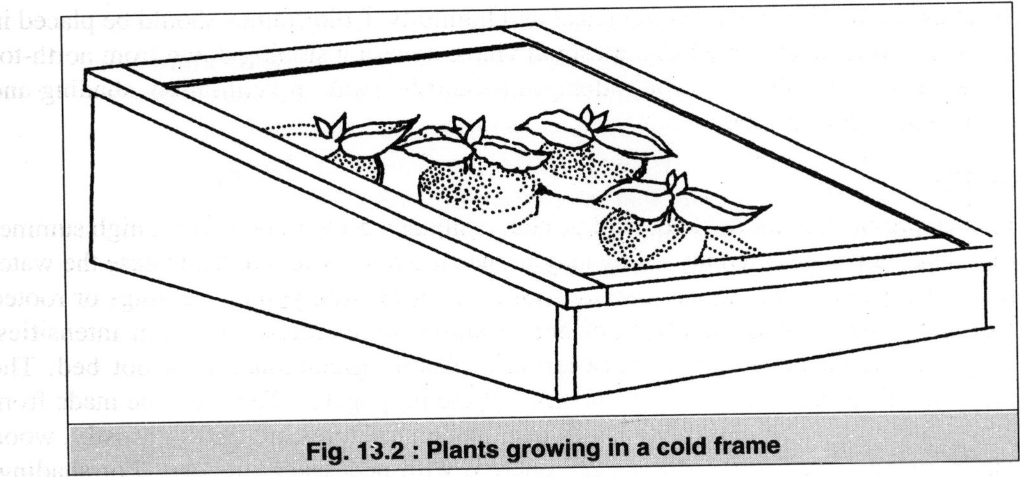

Fig. 13.2 : Plants growing in a cold frame

Net Houses

Net house is widely used as propagation structure in tropical areas, where artificial heating is not required and artificial cooling is expensive. In these areas, a net house may be constructed with roofs covered with glass or plastic film and its sides are covered with wire net. It provides necessary ventilation and maintains an ideal temperature for germination of seeds and subsequent growth of the seedlings. The roof of net house may be covered with gunny cloth or even with live plant creeper to cut-off the solar radiant energy and to keep the house cool. Net house can be constructed as per need of the propagator and therefore its size varies with the requirement of the nurserymen.

Bottom Heat Box

It is a simple device for promoting rooting of cuttings in difficult-to-root fruit plants like mango and guava. It was devised at IARI, New Delhi-12. It consists of two chambers made from galvanized iron sheets. The outer chamber has a height of 70 cm with 46 cm width and the inner chamber has a height of 68 cm and width of 44 cm. The space between the two chambers is filled with glass wool for heat insulation. The another chamber is fitted inside the inner chamber with the height of 35 cm, fitted with two electric bulbs at the bottom for providing heat to the cuttings. The innermost chamber is filled with soil mixture or any other propagation medium and the cuttings are inserted in it. Two electric bulbs of 100-watt capacity are fitted at the bottom of the chamber to provide heat light to the cuttings. The temperature in the box is maintained and controlled automatically by a thermostat fitted at the bottom of the chamber. The most ideal temperature to be maintained in the box is 30 ± 2°C because at this temperature, cuttings of mango and guava root easily and profusely. The initiation of rooting in cuttings varies from species-to-species but in general, it takes 1-2 months for proper development of the roots.

Mist Propagation Unit

The rooting of softwood leafy cuttings under spray or mist is a technique now widely used by nurserymen and other plant propagators throughout the world. The aim of misting is to maintain a film of water continuously on the leaves, thus reducing transpiration and keeping the cuttings turgid until rooting takes place. In this way, leafy cuttings can be fully exposed to light and air because humidity remains high and prevents damage even from bright sunshine. Mist also prevents disease infection in cuttings by way of washing off fungus spores they attack the tissues. While the leaves in this process must be kept continuously moist, it is important that only minimum water should be used. This is because excessive water leaches out nutrients from the compost, which may cause starvation in the growing plants. Moreover, a directly injurious effect on the cuttings may occur from over watering. Hence, it is necessary to utilize nozzles capable of producing a very fine mist. A small mist propagating units are mostly used by small farmers (Figure 13.3), whereas, highly advanced intermittent units are used by the commercial nurserymen in advanced countries (Figure 13.3).

Mist propagation units are used for propagation of 'difficult-to-root' cuttings in most advanced countries. Mist beds are constructed within a greenhouse. A fine mist is sprayed intermittently over the cuttings at regular intervals during day and night. The mist unit is controlled by a time clock, operating a magnetic solenoid valve and is set in a way to, turn on, the mist for 3-5 seconds to wet the leaves and, turn off, for some time and when the leaves are dry, the mist is again turned on.

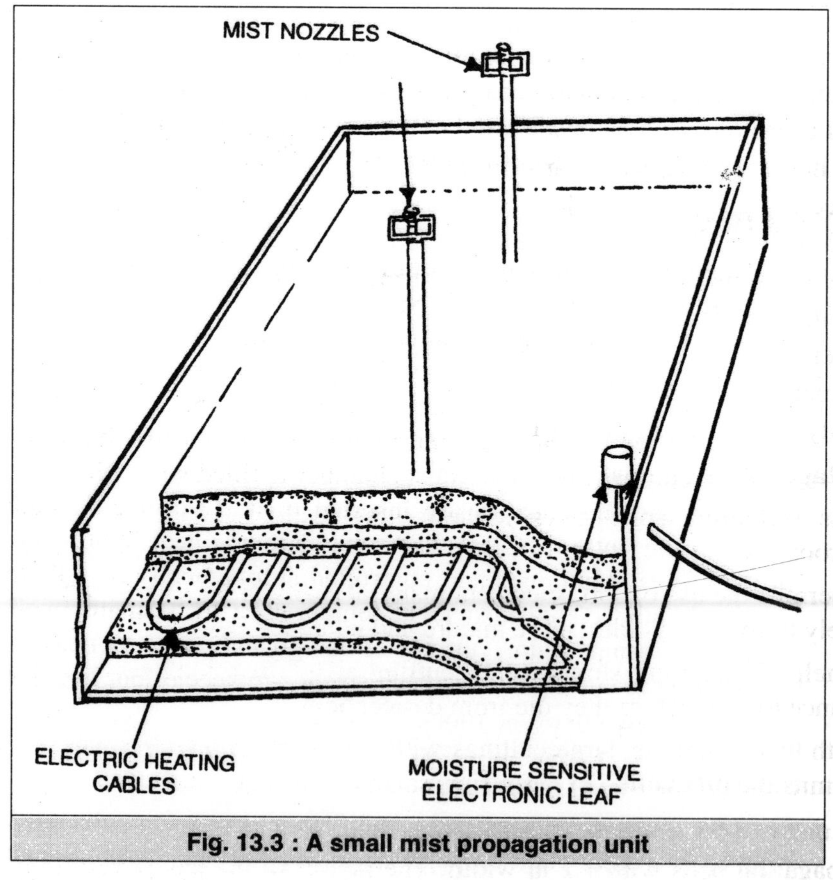

Fig. 13.3 : A small mist propagation unit

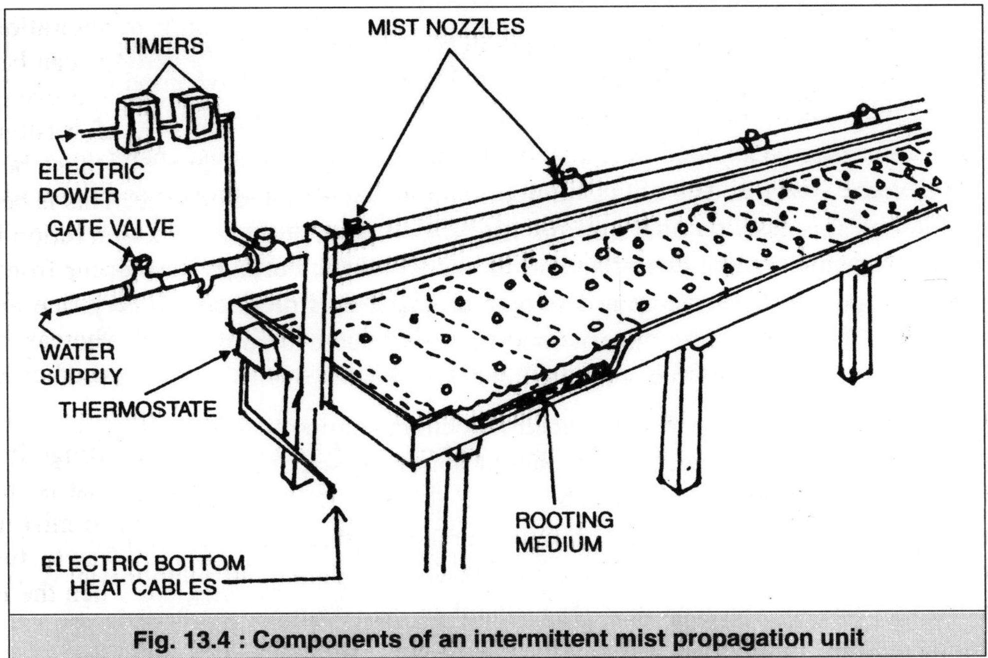

Fig. 13.4 : Components of an intermittent mist propagation unit

In general, the mist has 5 control mechanisms (Fig. 13.4). Timer, electronic leaf, thermostat and timer, screen balance and photoelectric cell. The two types of timers are used in a mist unit, one turns 'on' in the morning and gets 'off' at night and the second operates during day hours to produce an intermittent mist, usually 6 seconds 'on' and 90 seconds 'off'. In electronic leaf, a plastic with two terminals is placed under the mist alongwith cuttings, the alternate drying and wetting of the terminals breaks off the current, which inturn controls the solenoid valve. A thermostat controls the temperature of the mist. In screen balance control mechanism, a stainless steel screen is attached to a lever with mercury switch. When mist is on, water is collected on the screen and when weight of water is more, it trips the mercury switch. The photoelectric controls are based on the relationship between light intensity and the transpiration rate.

Advantages of using mist units

- This technique makes possible the rooting of cuttings of plants which are difficult-to-root.
- It permits the use of soft, succulent material easily in the season, which is much more likely to root than older, more mature and hardened wood.
- It helps in keeping slow-rooting cutting alive for a long period, thus, provides a chance to root before they die from desiccation.
- With this technique, large cuttings with considerable leaf area can be rooted, which permits the production of large-size, salable plants in a short time.

A mist unit can be set up in a glasshouse or in a polyethylene tunnel. Usually, it is set up on the propagation beds with 1.2 m width. The layout of the jets is very important. While installing mist propagation unit, all the jets should be at a equal height. For proper functioning of the mist, there must be proper supply of water. The water should have good pressure and it must be free from salts. The optimum pH of water to be used in mist unit is 5.5 to 6.5. Hard or alkaline water may be avoided as it may block the nozzles of the mist chamber and it may accumulate on the young leaves of the cuttings and thus inhibit the growth and development of roots. Further, it is essential that a well-drained rooting medium is used and there should be a proper provision for adequate removal of excess water. Further, development of blue and green algal growth is very common in mist propagation structures, which is considered very harmful to the propagating material and thus every care should be taken to keep mist propagating unit free from any type of algae.

After rooting of cuttings in the mist, their hardening is important for better success in the field. When cuttings are rooted, misting should not be ceased abruptly as this may help in drying out of the young plants followed by scorching. Hence, weaving off process should be adopted in which misting is continued but the number of sprays/day are gradually reduced. Lessening the 'on' periods and increasing 'off' period can do it. The another way is to shift the rooted cuttings to a greenhouse, fog chamber, and frames, maintained at a higher temperature and low relative humidity. After phase-wise hardening only, the rooted cuttings are planted at the permanent locations.

Plastic Tunnels

A simple but effective method of protecting cuttings has been developed by Greenhouse Crops Research Institute, Littlehampton, England, which is being used by many nurserymen throughout the world. It consists of a plastic tunnel similar to the tunnels made for strawberry protection. The three feet wide wire hoops, which support the polyethylene are made from 0.2-inch diameter wire and are erected at 30-inch intervals. A white translucent polyethylene sheet 6 feet wide is then stretched over the hoops and is held in position by fastening two lines of polyethylene bailer twin to alternate hoops on either side of the tunnel (Figure 13.5). The seeds or cuttings or other propagating material is sown/planted inside the tunnels during winter. Nowadays, perforated tunnels are used because in non-perforated tunnels, temperature and relative humidity rise sharply, which may affect rooting of cuttings or further growth of the propagating material.

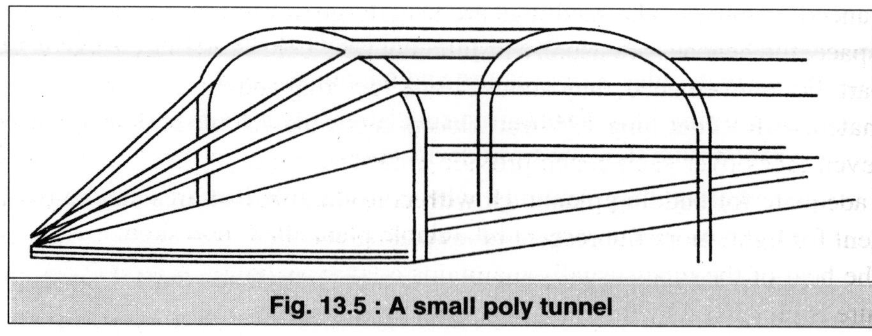

Fig. 13.5 : A small poly tunnel

Propagating Units

There is now wide range of self-contained propagators, heated and unheated, in the market. The simplest type consists of nothing more than a standard plastic seed tray with a transparent cover known as a 'propagator top', which is made to fit over the tray. The propagator tops are made from clear polystyrene and are fitted with one or two adjustable ventilators (Figure 13.6).

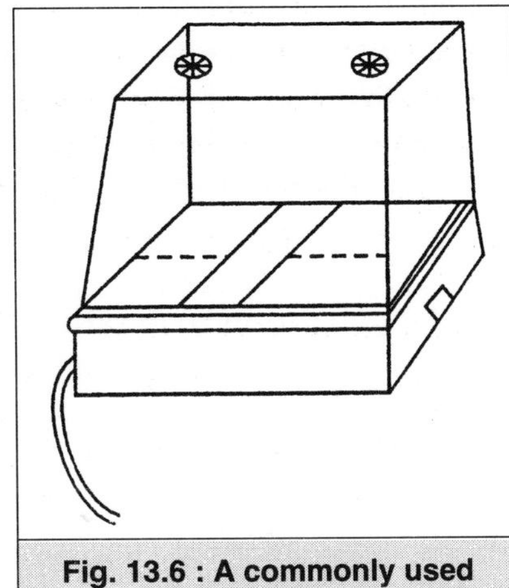

Fig. 13.6 : A commonly used electric propagator

Larger and more sophisticated propagators are also available. One of there is rather like a glass cage with a strong metal framework. It is nearly 3 feet long, 2 feet wide and 2 feet high. This propagator is fitted with sliding glass panels both in the front and rear making for easy access. Similar models are available in small sizes and each has a damp-proof base about 6 inch deep.

Another excellent type of propagator is rather like a miniature greenhouse with sliding panes on both the sides and the roof to provide access and ventilation. A model of this type has an aluminium framework and a fiberglass base. Its dimensions are 4 feet 2 inch long, 2 feet 2 inch wide and 2 feet 9 inch high at the ridge.

Most propagators are provided with heating equipment either with bottom heat or fitted with an air heating system or both air and bottom heat but electrical heating is considered as best. However, a gardener could easily make a simple type of propagator for himself. To do this, make a sand bed at the base of a deep box with bottom heat and provide a layer of about 3-inch of rooting compost. Over the bed, erect a light framework to carry a plastic sheet. A deeper box could have simply a sheet of glass laid over as a cover.

Growing Rooms

A growing room is an insulted building from which natural light is usually excluded. In it, illumination is provided by artificial means. Growing rooms are now widely used commercially for the production of seedlings of bedding plants, tomatoes and cucumbers in most advanced countries. The seedlings are usually grown in trays or pots kept on benches. To save space, the benches are usually installed in tiers being vertically about 2 feet wide and 6 inch apart. For convenience, they are usually 8 feet long and 3 feet 6 inch wide. Each bench is illuminated with 8 feet long 125-watt fluorescent tubes mounted 1 feet 6 inch above the bench. Seven tubes over each bench provide a light intensity of 500 lumens per square feet, which is adequate for bedding plants. However, if plants grown in such houses have higher requirement for light, more fluorescent tubes can be installed over the benches containing the plants. The heat of the tubes usually maintains a temperature of at least 70°F and the rise in temperature is prevented by the use of fans installed in the building.

The Automatic Greenhouses

Today, the modern greenhouse can be almost completely automated thus assisting propagation. For instance, by the use of thermostats, the air and bed temperatures can be maintained as per the requirement. Similarly, automatic ventilation allows the ventilators to open and close in relation to temperature. Even automatic systems of irrigation are installed in the modern greenhouses and water is supplied to the plants through drip or trickle system to each pot or plant by individual nozzle or tube. This system is controlled by automatic 'on' 'off' valves or by means of a time switch.

Most advancement in raising of plants under highly controlled environmental conditions has been the use of phytotron facility. National Phytotron Facility is available at IARI, New Delhi, which can be used by scientists for raising plants by fixing temperature and relative humidity as per requirements.

• • •

CHAPTER 14

Raising Plants in the Nursery

To raise plants in the nursery may be easy but to maintain them in good state is very difficult. Raising of seedlings and maintaining the young plants in a nursery is a continuous process. One must have full idea about the sowing time, sowing depth of a seed, cutting or layers to raise them effectively. Similarly, knowledge about transplanting of seedling, their handling and further management in the nursery is equally important for raising quality material.

SEED SOWING

In many plants, the seeds are first sown in the seedbeds and after sometimes, the seedlings are transplanted in the actual field. Similarly, rootstocks of fruit plants are also first raised in the nursery. Growing of plants in the nursery seems to be easy in words but in reality, it is a difficult job and can be done effectively and efficiently only after attaining some experience. Thus, it is considered as a specialized job because one has to acquire fair knowledge about the plants to be raised. Similarly, one should have sufficient knowledge about the cultural requirements of the plants to be raised in the nursery.

Preparation of seedbeds : The most important steps for raising seedlings in the nursery are to select a proper site for seedbeds, seed sowing, care of seedbeds, care of seedlings and then transplanting of the seedlings. Seedbeds should be made in a locality where the soil is fertile, free of clods, rich in nutrients and has adequate drainage and aeration. The beds should be able to provide good growth to the seedlings. For the preparation of beds, the soil should be prepared thoroughly by repeated ploughings. During land preparation, suitable chemicals should be added to protect the seeds and seedlings from the attack of insect-pests, diseases and weeds. The soil should be labelled properly and beds of desired size are made. Usually, the beds of 10′ × 3′ × 6′ are preferred for raising vegetable seedlings.

Sowing time : Seeds of most of vegetables (*e.g.*, tomato, chillies, brinjal, cabbage, cauliflower etc.) and flowering annuals (*e.g.*, daisy, calendula, zinnia, marigold, larkspur etc.) are usually sown about a month before planting. The seeds of tropical and sub-tropical fruits are sown during monsoon (June-July) or in the beginning of spring (February-March). Generally, the seeds of mango. Kagzi lime and jackfruit are sown during June-July while that of guava, *ber* and *aonla* are sown during February-March. Seeds of temperate fruits are generally available during June-to-October and their sowing should be done after the dormancy period is over. Seeds of some fruits (*e.g.*, citrus, mango, loquat, litchi, jackfruit, etc.) lose viability quickly, so these should be sown immediately after their extraction from the fruit. Under North Indian conditions, the seeds of different citrus types are available during winter months and their germination is usually low due to the prevailing low temperature. Hence, use of alkathene tent on seedbeds during December-January has been found useful for increasing the germination percentage and faster growth of the seedlings.

Sowing depth : The seeds are usually sown at a depth 3-4 times of their size. Sowing may be done a little deeper in light soils than in heavy soils. The seed rate depends on many factors, particularly on the species, purpose and spacing. After sowing, a thin layer of grass may be spread over the bed.

After sowing, one should follow the following precautions to get good seedlings :

1. Avoid moisture stress.
2. Avoid over heating.
3. Avoid over watering.
4. Avoid excessive shading.

Sowing of seeds is now commonly done in polyethylene bags, earthen pots and in pans. Papaya seeds are usually sown in polyethylene bags or in earthen pots. For epicotyl grafting, mango stones are sown in polyethylene bags. While raising the plants in polyethylene bags, care should be taken to develop straight seedlings and if these are coiled, these should be uprooted in the beginning. Similarly, seedlings having twisted roots are difficult-to-establish in the field due to poor anchorage and must be discarded.

Planting density and care : It is essential to maintain the proper plant density in the bed. Crowding of plants may result in weak and lanky growth of the seedlings. A great care must be taken for the provision of optimum light, temperature, moisture and aeration to the seedbeds for quick seed germination and subsequent seedling growth. Over watering, shading, over sun exposure, frost, or disease incidence, all may affect seed germination and subsequent seedling growth. So, proper measures should be taken to avoid all these problems. The seedlings may remain in seedbeds for 1-2 month to 1-3 years, depending upon the kind of plant. Vegetable and ornamental seedlings are kept in seedbeds for 1-2 months, whereas the rootstock seedlings are usually kept in seedbeds for 1-2 years.

***In situ* sowing :** In some fruit crops like walnut, pecannut, jackfruit and *ber*, the tap root system is very vigorous. So, during the process of transplanting, root system is disturbed, which ultimately affects their establishment in the field. Therefore, for these fruits, sowing of seeds *in situ* is recommended. In rocky soils of Odisha and Gujarat, *in situ* sowing of mango stones and their grafting late in the season has been recommended. Similarly, for high density planting of Amrapali mango, *in situ* orchard establishment is recommended.

Care and handling of seedlings : After germination, the seedlings must be gradually exposed to the prevailing environmental conditions particularly to light and temperature. Over watering should be avoided as it may invite 'damping off' of seedlings at an early date. Similarly, over-shading should also be avoided as the seedlings may have lanky growth in shade. The frequency and amount of irrigation should be increased with the growth and development of seedlings. Protection of seedling from frost during winter and from hot desiccating winds during summer is necessary. Similarly, some insect-pests and diseases (especially damping off) may infest the seedlings at younger stage and thus care should be taken to spray seedlings with suitable insecticides and fungicides as and when required.

Hardening of seedlings : The seedling plants in the nursery are very succulent and tender. Seedlings usually do not withstand transplanting well unless they are properly hardened before transplanting in the permanent locations. It can be achieved by withholding water supply and reducing temperature or by combination of both. Thus, the reduced growth rate enables the plants to withstand desiccation which will help in proper development of roots. However, over hardening may be avoided at every cost as it may result in extremely stunted plants. Thus, hardening of the seedlings should be done gradually for streamlined growth reduction. The practice of hardening results in (1) reduced growth, (2) thickening of cuticle, (3) increased waxy covering on leaves, (4) increased dry matter, (5) decreased transpiration, and (6) increasing sugars in the leaves. All these conditions are essential for better establishment of seedlings in the field.

Transplanting : Plants raised in the beds are transplanted in the permanent locations *i.e.*, in the field. Sometimes seedlings are transplanted once or twice before final setting in the field. Usually seeds are sown thickly in the seedbeds and a soon as the seedlings become crowded, they are transplanted to the another bed or pot. This practice in common in flowering annual plants and vegetables and is commonly known as *pricking*. When the seedlings attain good size, these are transplanted in the field. During transplanting, maximum possible soil should be retained with the seedlings. After transplanting, plants are firmly pressed and watered immediately. To avoid wilting after transplanting, some provision of shade should be made in the nursery, which should be removed towards the evening. The transplants should be irrigated frequently. However, water stagnation should be avoided at every cost. A starter or booster solution containing mineral nutrients can also be given to the nursery plants for better growth of growing seedlings. Usually, there is a high rate of mortality of seedlings after transplanting because of the following reasons :

1. If the plants do not have well-developed roots.
2. If the roots are too soft and succulent.

3. If the plants are not watered adequately in the nursery.
4. If plants are not given proper shade initially.
5. If the seedlings are not watered immediately after transplanting.
6. If the plants are not protected from scorching sun and wind.
7. If the environmental conditions existing in the nursery bed are different than those in the field.
8. If the seedlings are not properly covered with earth ball before transplanting.
9. If the plants are not pressed firmly (to avoid air space) and watered immediately after transplanting.

• • •

CHAPTER 15

Nursery Management

The young nursery plants are very tender, easily destroyed by very low or high temperature and arc highly susceptible to the attack of insect-pests and diseases. Thus, proper upkeep and management of the plants in nursery stage is highly essential. Utmost attention is required to protect the young and succulent plants by following proper management practices as under :

Irrigation

The young seedling are very exact in their water requirement. The seedlings require frequent irrigation and may desiccate in short water supply. However, over watering is as harmful as is under watering. In order to maintain proper turgidity in the developing seedlings, irrigation should be given at weekly intervals in the summer months and at bi-weekly intervals during winter. One must be very careful in maintaining the sub-surface drainage, otherwise it is likely that whole nursery may be destroyed by one or the other diseases. Water in the nursery can be applied by traditional methods (flood, check basin or furrow) or modern methods (micro-sprinkler or drip).

Nutrition

Like field grown plants, the nursery plants also require adequate amount of nutrients in the optimum concentration for their growth and development. Like other plants, nursery plants also require all major (N, P, K, Ca, Mg, S) and minor nutrients (Zn, Fe, B, Mn Mo, Cl etc.) in ample quantity. If there is deficiency of any of the nutrients, it is likely to affect the plant performance and ultimately the quality, which is the major aim of a nurseryman. Thus, all the nutrients must be present or supplemented artificially in the growing media.

The nurserymen and propagators are more often encountered with some deficiency symptoms of these nutrients but because these symptoms are so confusing that it is difficult for them to identify as which of the nutrients is actually deficient. Further, a particular deficient nutrients doesn't exhibit the same deficiency symptoms on the other plants, which makes the problem much complex. However, the peculiar deficiency symptoms of some important major and minor nutrients are given below so that a nurseryman may be able to identify the deficient nutrient and correct the deficiency of the same (Table 15.1).

TABLE 15.1 : Deficiency symptoms of some important nutrients in the nursery plants

Deficiency symptoms	*Deficient nutrient*
Leaves are light-green to light-yellow (chlorosis) with short and slender stalks.	Nitrogen
Leaves are dark green, often developing red and purple in colour. The lower leaves may be yellow, which may turn greenish, brown and lastly black.	Phosphorus
The lower leaves may be mottled or chlorotic, without dead spots with margins curved upward. The stalks are short and slender.	Magnesium
Development of dead spots on lower leaves, usually at the tips and between the veins, internodes are short and leaves have slender stalks.	Potassium
Development of dead spots on lower leaves, involving larger areas. The leaves are thick and stalks with shorter internodes, showing rosette appearance.	Zinc
Initially, terminal buds of the young leaves become typically hooked, which dies soon at tips and margins.	Calcium
The terminal buds of young leaves are light green at base. Later, the leaves become twisted and die at the tips.	Boron
The young leaves may be permanent wilted or show chlorosis. The twig of stalk just below the tip is unable to stand erect in later stages. Cracking of bark may also take place.	Copper
Spots of dead tissue are found scattered over the young leaves. The smaller leaves tend to remain green.	Manganese
Chlorosis of young leaves with veins and tissues between veins remaining light green in colour. The stem may be thin and erect.	Sulphur
Chlorosis in young leaves with major veins remaining green. The stalks of leaves are short and slender.	Iron

CORRECTION MEASURES FOR DEFICIENT NUTRIENTS

The deficiency of nutrients can be corrected by applying the deficient nutrient to the soil or by foliar application or by combination of both the methods as discussed below :

Nitrogen : The deficiency of nitrogen can be corrected by the use of any nitrogenous fertilizer. N can easily be applied through foliar application of urea (1 per cent) during active growth period of the plant.

Phosphorus : Follow recommended doses of P. Application of DAP (2 per cent) or single super phosphate (1 per cent) is equally useful.

Potassium : Use recommended doses of K or foliar application of sulphate of potash (1 per cent). Foliar application of muriate of potash (KCl) is equally useful but it should be avoided in alkaline/saline soils.

Calcium : Apply lime to the soil or growing media, depending on the pH of the soil or media. Foliar application of calcium nitrate (1 per cent) is equally useful.

Magnesium : Soil application of dolomite or gypsum corrects Mg deficiency effectively. Foliar application of magnesium sulphate (0.5 per cent) is helpful.

Zinc : Soil application of zinc sulphate @ 20-25 Kg/ha or its foliar application (0.5 per cent) helps to correct Zn deficiency effectively.

Iron : Application of ferrous sulphate (40-50 Kg/ha) to the soil or growing medium or its foliar application (0.5 per cent) is recommended to correct Fe deficiency.

Manganese : Soil application of manganese sulphate (20-25 Kg/ha) or its foliar application (0.2—0.4 per cent) is useful for correcting Mn deficiency in nursery plants.

Copper : Soil application of copper sulphate (10-12 Kg/ha) or its foliar application (0.5 per cent) is recommended dose to correct Cu deficiency in nursery plants.

Boron : Borax (15-20 Kg/ha) applied to soil or growing medium is quite effective to check B deficiency in the nursery plants. Foliar application of borax (0.2-0.5 per cent) is equally advisable.

Molybdenum : Foliar application of ammonium molybdate (0.1 per cent) or its soil application (1-2 Kg/ha) effectively checks Mo deficiency in the nursery plants.

In general, nutrients are applied immediately after the symptoms have appeared. In general, fertilizers in the nursery are applied either as a surface application, sub-surface application or as foliar sprays. However, for nursery plants foliar, application of the nutrients is very effective as compared to soil application because nutrients are absorbed at a faster rate. Moreover, wastage of nutrients is also very less. For foliar application, the nutrients should be applied when there is sufficient foliage on the plant. The nutrient should either be applied in the morning or late evening hours. Similarly, it is better to add some wetting agent like Tween-20 or Triton-X-100 etc. in the spray solutions as it helps to absorb the nutrients more effectively.

Weed Control

Weeds are unwanted plants, which create serious problems to the very existence of young seedlings in the nursery. Weeds compete for nutrition and moisture with the nursery plants and suppress the growth of the seedlings. Weeds may harbour many harmful insects and pathogens also. Once the seedlings are covered by weeds or infected by any pathogen or insect, it will be difficult to get healthy and quality seedlings from the nursery. Therefore, the nursery beds should be kept free from weeds by shallow hoeing. The hoeing also creates better aeration in the soil. Further, the irrigation channels and bunds of the beds should be kept free from the weeds. Weedicides can also be used, if necessary.

Protection from Adverse Effects of Weather and Climate

The plants in the nursery are very delicate and sensitive and thus should be protected from adverse weather and climatic conditions. The high temperature and desiccating winds in summer and low temperature and chilly winds in the winter may kill the nursery plants. Therefore, the seedlings should be protected from scorching sun by providing some shade during the daytime. In winter, a thatch house or a polyethylene tent house may be constructed over the seedling plants or beds. Raising of windbreaks on the northern and the western side of the nursery is also useful. Similarly, the frequency of irrigation should be increased during summer months to avoid the damage caused by the water stress. The young nursery plants are also highly susceptible to frost. Occurrence of frost even for a day can destroy the whole nursery. Smoking, irrigation at frequent intervals, providing a thatch covers to beds, mulching with grass or polythene and provision of wind-breaks etc., can protect the nursery plants from being damaged by the frost.

Plant Protection from Insect Pests and Diseases

In addition to the protection from adverse weather and climatic conditions, protection from insect-pests and diseases is equally necessary. The seedlings in the nursery being tender are infested by insect-pests or diseases easily and thereby cause considerable damage to them. It is therefore necessary to keep them under control by following some suitable management several insect-pests and diseases attack the nursery plants (Table 15.2). In general, powdery mildew, some leaf spots, blights and some other diseases infest nursery plants but 'damping off' in the major disease in the nursery. Therefore, a programme should be initiated to control it in the nursery itself. Damping off is reported to be caused by many fungi but *Pythium, Phytophthora, Rhizoctonia* and *Botrytis cinerea* are chiefly associated with it. It symptom of this disease is the girdling of the seedling stem at the soil surface. The affected seedlings topple down immediately. It usually affects young and succulent seedlings. Similarly, it is more pronounced in nurseries having poor sanitation. The prevention of damping off and other diseases in the nursery can be done by treatment of seeds, soil and by control on environmental factors.

TABLE 15.2 : Common insect-pests and diseases of nursery and their management

A. Insect-pests	*Management Practices*
Scales	Single spray of rogor or metasystox (0.05 per cent) is quite effective. Repeat it after 15 days interval. Crude oil spray is also useful for controlling the scales.
Thrips	Two sprays either of metasystox or rogor (0.05 per cent) at fortnightly interval on new growth reduce the population of thrips considerably.
Mealy bugs	1. Collect and destroy the adults. 2. Soil raking in December-January to kill the pupae. 3. Spray dimathoate (0.05 per cent) on the crawling insects during February-March. 4. Use ostico sticky bands on the trunks of the mother plants.

White flies	Spray phosphamidon (0.02 per cent) or other systemic insecticide in the early stages of infestation.
Mites	Give two sprays of acaricides like dicofol (0.05 per cent) and wettable sulphur (0.2 per cent) at fortnightly intervals.
Leaf eating caterpillars	Spray sevin/carbaryl (0.1 per cent) at the first appearance of the caterpillars. Neem oil (1 per cent) is also effective.
Cut worms	1. Use light traps for catching the adults. 2. Use poison baits consisting of malathion (0.1 per cent), wheat bran and jaggery. 3. Spray malathion or quinalphos (0.05 per cent) before the expected attack of the worms.
Leaf miners	Two sprays of metasystox or rogor (0.05 per cent) at fortnightly interval on newly emerged growth flushes are sufficient for controlling leaf miners in the nursery.
Snails and slugs	1. Hand picking and destruction is the most effective method. 2. Spray of common salt (2 per cent) is very useful in controlling snails and slugs. 3. Use metaladdehyde pallets.
B. Diseases	
Damping off	1. Treat seeds before sowing with ceresan, thiram or agrosan @ 2 g per kg seeds. 2. Spray captan or bavistin (0.2 per cent) in the nursery.
Powdery mildew	Give one prophylactic spray of karathane or calixin or wettable sulphur (0.2 per cent). Repeat it again if cloudy weather persists for a longer period.
Leaf spots	Give a single spray of Dithane Z-78 or bavistin (0.2 per cent).
Blights	A single spray of Dithane Z-78 or bavistin (0.2 per cent) is very effective to check blight disease in the nursery plants.
Dieback	1. Prune the dead portions of the plant. 2. Apply blitox or bordeaux paste to the cut portion. 3. Spray benlate (0.2 per cent) as soon as the symptoms become visible.

To avoid the attack of diseases in the nursery, seed treatment with chemicals, soil treatment and environmental control is must, which is briefly described below :

A. Seed Treatment

The seeds carry out inoculum of many diseases. If infected seeds are sown, they do not germinate properly, and if germinate, their growth is lanky and such seedlings may even die soon. Thus, seeds should be treated properly to kill the source of inoculum. There are four types of seed treatments to control seed borne diseases. It includes disinfestation, disinfection, protection and prevention.

(*a*) Disinfestation : It is the process of eliminating organisms from the seed surface. Many chemicals are used to disinfest seeds before sowing to avoid the incidence of diseases in the seedbeds but ethyl alcohol, calcium hypochlorite and mercuric chloride are most commonly used. The concentration and the duration of treatment chiefly depend on the size and kind of the seeds. In general, 50 per cent ethyl alcohol, 2 per cent calcium hypochlorite and 0.01 per cent mercuric chloride concentrations are used. The duration of treatment varies from 5 to 30 minutes, depending on the size and type of the seed. However, prolonged exposure of the seeds to disinfectants may be avoided as it may injure the seed and thereby affect germination. After treating the seed with chemicals, these are rinsed with tap water and immediately sown in the nursery.

(*b*) Disinfection : It is the process of treating seeds in a way to eliminate seed borne pathogens *i.e.*, the pathogens, which are present within the seeds. For disinfection, hot water treatment is the most effective. The seeds can also be treated with formaldehyde and mercuric chloride for this purpose. However, nowadays, aerated steam is being used commercially. Hot water treatment is very effective in eliminating pathogens from within the seeds, which normally looks healthy. In this treatment, dry seeds are kept in not water (48 to 55°C) for 10-30 minutes, depending upon the kind of seeds and the pathogen. During the process of treatment, the water should be stirred continuously. Similarly, there should be precise control on temperature and duration of treatment otherwise it may result in many indirect effects on the seeds. The harmful affects include delay in germination, stunted growth of seedlings, injury to seed embryo or even death of the seed. Generally, injured seeds are more prone to injury than the healthy ones. It is advisable to spread the seeds in thin layers after the treatment. Similarly, the treated seeds should be kept in sun or under fan for drying and sown afterwards. Although, hot water treatment is advantageous in many cases, it has, however certain disadvantages also, which are enumerated below :

1. It is a cumbersome and labourious process.
2. It is effective only if there is rigid control on time and duration of the treatment.
3. Seed viability may be lost if not done properly.
4. There may be injury to the seed coats or even embryos, which affects growth of the seedlings after germination.
5. The seeds may absorb plenty of water and therefore these should be dried properly before storage or sowing.
6. During the process, some seeds may exude resinous material, as result of which, the seeds may stick together, which makes their handling difficult.

For these reasons, aerated steam is being increasingly used nowadays for disinfection. In this method, there is precise control on temperature and vapour flow rate, as a result of which, injury to the seeds is very less.

(*c*) Protection : To protect the seeds from the infection of pathogens, these should be treated with different chemicals. Many chemicals are used as protactants, but two methods are in vogue for protection of seeds from the diseases *viz.* dry seed treatment and wet seed seed treatment methods. Dry seed treatment method gives better protection against soil borne diseases, whereas, wet method gives protection both from soil and seed borne pathogens.

In dry seed treatment method, the seeds are thoroughly shaken with chemicals in a rotating seed duster for sometime so that the chemical sticks well to the seed surface. The most commonly used chemicals for this treatment are agrosan-G, ceresan, brassicol, captain, thiram and dithane-M-45. In wet treatment method, the seeds are kept immersed for a certain period in the liquid suspension of the chemical. The seeds should be stirred continuously during immersion for proper contact of the seeds with chemical and to avoid excessive accumulation of chemical at one portion of seed. After treatment, the seeds are dried in sun or shade quickly. The most commonly used chemicals in wet treatment method are agallal and thiram. However, certain antibiotics can also be used for wet treatment method.

(*d*) Prevention : A propagator must not always stick to curative measures but should also follow certain measures to avoid or prevent the pathogen attack. Preventive measures include the proper selection of seed production field, use of pathogen free seeds, control of diseases in the seed field, avoiding irrigation with sprinklers in seed fields, harvesting of seed crop at a proper time and maturity and following clean practices for threshing etc. By this way, many diseases can be prevented in the seeds.

B. Soil Treatment

Many times, soil contains certain harmful pathogens, which interfere with growth and development of the seedlings. These harmful pathogens can be eliminated by pasteurization, changing the site of nursery or by chemical treatment. However, soil pasteurization is not desirable and possible because it is highly difficult and cumbersome job. Moreover, it may kill several useful organisms also. Usually, pasteurization is done at 60°C for 30 minutes to kill harmful organism in the soil and soil mixes. Propagators should use steam pasteurization because it kills harmful pathogens and not the beneficial ones, though, it requires expensive instruments, which is sometimes beyond the reach of a common man. In addition to pasteurization, certain chemicals can also be used for killing the pathogens. Commonly used chemicals are formaldehyde, methyl bromide, chloropicrin, vapan etc.

Nowadays, soil solarization is becoming quite common in nursery. In this practice, nursery beds are covered with alkathene sheets for some specific period. Due to this treatment, the temperature of nursery bed is raised several degrees, which kills all the pathogens.

C. Environmental Control

To avoid certain diseases, rigid control on environmental conditions is necessary in the nursery. For example, damping off is a serious problem in nursery, if drainage is a problem. However, it can be avoided to some extent by preparing raised seeds in localities of high rainfall areas. Sometimes, drenching of soil is done with certain fungicides like captan, thiram to keep harmful fungi under control. In addition, one has to keep control on growth environment of the seedling to avoid the incidence of insect-pests and diseases.

• • •

CHAPTER 16

Water Management in Nursery

Water is one of the most important inputs essential for the production of nursery plants. Water influences photosynthesis, respiration, absorption, translocation and utilization of mineral nutrients and cell division besides some other metabolic processes. Its shortage as well as excess affect the growth and development of seedlings adversely. The young seedlings are very exact in their water requirement. The seedlings require frequent irrigation and may desiccate in short water supply. In order to maintain proper turgidity in the developing seedlings, irrigation should be given at weekly intervals in the summer months and at fortnightly intervals during winter. One must be very careful enough in maintaining the sub-surface drainage, otherwise, whole nursery may be destroyed by one or the other diseases. The scientific utilization of water for nursery production involves the consideration of the suitability of water for irrigation and then the planning and management practices.

QUALITY OF IRRIGATION WATER

Irrigation water contains impurities in varying concentrations. The suitability of irrigation water mainly depends on the amount and type of salt present in the water. The main soluble constituents in the water are calcium, magnesium and sodium as cations and chloride, sulphates and bicarbonates as anions. The other ions present in minute quantities are boron, selenium, molybdenum and fluorine. The presence or absence these cations canions determines the quality of water, which also has an impact on seedings. Thus, the growth of nursery plants irrigated with poor quality water decreases due to increased osmotic stress and poor physical conditions of highly dispersed soil. The degree of harmful effect varies with crop, variety and stage of growth of the nursery plants. The quality of irrigation water is judged with the three parameters (*i*) total salt concentration, (*ii*) sodium adsorption ratio, and (*iii*) bicarbonate and boron content.

Water containing total dissolved salt to the extent of more than 1.5 m mhos/cm is considered as 'saline'. Salt content of irrigation water is measured as electrical conductivity

(E.C.). In addition to E.C., which has been used as a main criterion to determine the quality of irrigation water, sodium adsorption ratio (SAR), residual sodium carbonate (RSC) and boron content are also used to find out suitability of irrigation water. Irrigation water, which contain more than 3 ppm boron is harmful to the nursery plants. Hence, before establishment of any nursery in an area, the availability of good quality water must be ensured.

METHODS OF IRRIGATION IN NURSERY

Water application methods in the nursery may be broadly classified in to following two groups. Conventional methods, modern methods, as described briefly here under :

Conventional Method

It is also known as surface irrigation or gravity irrigation in which water applied is distributed by means of open surface flow. This may further divided into following groups such as flood irrigation, check basin irrigation, and furrow irrigation.

These methods are not suitable for irrigation in nursery due to their several ill effects. Water in all these methods is applied freely and after a considerable interval of time, thus, causing flood and moisture stress alternatively over the tender plants. Therefore, these methods are not suitable for irrigating plants in the nursery.

Modern Methods

To avoid flooding and severe moisture stress in the nursery, modern methods of irrigation have been developed. These are : micro-sprinkler irrigation, drip irrigation, and capillary irrigation.

Sprinklers are suitable for dense growing seedlings. It may be used in field or controlled environment such as polyhouse or in poly tunnels. It is difficult to irrigate seedlings, which are being grown in contains with this method of irrigation. But almost all types of vegetable crops may be successfully grown under micorsprinkler irrigation. It may also be useful for seedlings grown in trays and frames. This method is very well-suited for wide-spaced high-valued crops. Precise amount of water is applied to replenish the depleted soil moisture at frequent intervals. For optimum plants growth, this system enables the application of water and fertilizers at an optimum rate and at regular intervals to the plants. The amount of water applied to the soil could be adjusted to equate the daily consumptive use.

Drip irrigation is the best suited for container grown crops but is not suitable for dense, trays and bed-grown seedlings due to its inherent properties of oozing water at a particular position. Horticultural crops, which are being propagated by layering, grafting and budding, can also be irrigated by drip irrigation.

Capillary irrigation is the most suitable method for densely-grown crops nursery. Capillary irrigation can be used for both outdoor and indoor plants. Capillary irrigation system can be constructed in several ways. A simple capillary irrigation system consists of a 5 cm deep bed of sand, the depth mainly depends on the root depth of crop to be grown. The dimension can be conveniently based on the size of bed, hydraulic conductivity of soil and root depth of crop.

• • •

CHAPTER 17

Nursery Registration and Certification

Major cause for low productivity of horticultural, plantation and other perennial crops in the country is mainly due to supply of poor quality planting material. In other words, it can be said that unavailability of genuine planting material to the growers is the cause of low production or productivity of major horticultural and plantation crops. Central Government had constituted model nursery regulation act long back in the year 1954. But the implementation of the rules and regulations are still awaited. Due to several undefined reasons, the implementation of regulation act could not be achieved. Although registration and certification of nurseries by the competent government authorities is of utmost importance to equip it with the rules and regulations for its proper functioning but local nurseries are mushrooming in every area, which merely follow the rules and regulations. Poor or inadequate inspection by the agencies also favours them. Inefficiency of the competent authorities responsible for inspection at the time of issue of the license and random inspection and visits to the nursery to check and regulate the supply of quality planting materials, synergizing the monopoly of local nurseries. However, a nursery should be registered for the purpose of providing quality seed, rootstock and scion sources. A nursery stock should be inspected, tested for truness-to-type, freedom from viruses, insect-pests and diseases.

LICENSING REQUIREMENTS FOR THE NURSERY

(*a*) Any persons(s) engaged directly with the distribution of plants, products, plant material, nursery stock is required to inform the department of the existence of their operation. He/she should obtain a nursery industry license prior to initiating such business.

(*b*) The aforementioned person(s) are required to renew their nursery industry licenses each year that they remain in operation. Nursery industry licenses issued by the department shall be valid for only one year.

INSPECTION OF NURSERIES AND NURSERY-STOCK CERTIFICATION

After receiving the nursery industry license application, the department or its representative shall examine or inspect all plants, plant material, or nursery stock located or grown on the business location or any other applicable location. The department shall conduct the aforementioned inspection or examination yearly or at such time as it deems the best, with or without notice. The department reserves the right to conduct unannounced inspections as frequently as it deems necessary. Upon the successful completion of an inspection, the department shall issue a nursery stock certificate to the licensed nursery.

A. Responsibility of the Applicant

The applicant nurseryman shall be responsible, subject to approval by the department, for the selection of the location and the proper maintenance of registered plantings being grown. The applicant shall be responsible or maintaining the identity of nursery stock in a manner approved by the department.

B. Location of Plantings

Each planting location shall be subject to approval by the department and shall be in an area having minimal risks for spread of infectious pests by drainage, flooding, irrigation or other means. Fruit tree nursery stock being grown shall be on rootstocks approved by the department and shall originate from the registered seed trees or from registered stool beds. Certified nursery stock should have been produced on soils that received a preplant nematicide or that were sampled before planting and found to be free from virus, vectors and nematodes.

Nursery stock meeting the requirements, when sold, shall have the variety, interstock and rootstock designated, where applicable.

C. Refusal, Suspension or Cancellation

Registration or certification may be refused, suspended or cancelled for any plants in part or all the plantings under any of the following conditions :

1. If nursery is not working as per the standards prescribed by the rules and regulations of Government of India.
2. If the plants are found to be virus-infected or there are 'off-types' in the nursery.
3. If a registered tree is found upon indexing to be virus-infected.
4. If the pest control requirements are not followed properly upon inspection.
5. If the plants have not been properly maintained.
6. If the registration number is misused.

D. Civil Penalties

(*a*) The civil penalty for violation of any section of this chapter not already stated will be the assessment of a civil penalty, the confiscation or destruction of any and all plants, plant materials or nursery stock found on the premises or contained in the shipment in question and/or suspension or revocation of the current nursery industry license or any future operation privileges granted under this chapter.

(*b*) Any person who violates any rule or regulation or interferes in the enforcement or rules, as determined in an administrative hearing, shall be assessed for a civil penalty.

• • •

CHAPTER 18

Marketing of Nursery Plants

The main aim of establishing a nursery is to supply plants either to one's garden or to out-station indentors to fetch good income out of it. For getting adequate returns on the investments made by a propagator, one should have fair idea about lifting, packing and storage of the plants.

LIFTING OF NURSERY MATERIAL

The plants raised in the nursery are to be lifted for their permanent place of establishment or for sale. The careless lifting may affect may affect the establishment of nursery plants adversely. To achieve high rate of success in the field, it is always advisable to lift the saplings with a ball of earth. It is particularly true for evergreen fruit plants like mango, guava, citrus and litchi etc. However, the deciduous plants like peach, grape, apple, kiwifruit, almond, walnut etc., can be lifted bare-rooted also. To avoid the damage to the roots, it is advisable to irrigate nursery, 3-4 days before the actual date of lifting the plants. The lifting of sapling should, however be avoided in case there is heavy rain in the rainy season because earth ball does not form properly if soil is too wet. One of the oldest suggestions is to cut the tap root system of the evergreen plants especially mango alteast one month before the actual date of lifting. This practice helps the sapling to tolerate the shock of lifting. Moreover, there is development of secondary roots in the sapling, which helps in better survival rate of the saplings in the field.

After lifting, the seedlings/saplings should be properly packed. The bare-rooted saplings (*e.g.,* grape, peach, apple pear etc.) should be packed in moist sphagnum moss grass and wrapped with hessain cloth to avoid drying of the roots. The sapling of evergreen plants along with earth ball should be wrapped with moist grass or rice straw before putting into polyethylene bags and then kept in the basket. The strawberry runners are first packed in moistened moss grass and then kept in the hessain cloth bags or in baskets for long distance transportation.

STORAGE OF SAPLINGS

The high or low temperature and humidity may favour desiccation of the plants lifted from the nursery. Before sending them to distant places, these should be stored in a store house. The store house should have proper ventilation and it should be maintained at a low temperature (2-3°C) and high relative humidity (85 to 90 per cent). These environmental conditions reduce the transpiration and respiration rate, avoids withering or desiccation of the seedlings/saplings, thereby enhances the chances of their survival rate in the field.

PACKING OF NURSERY MATERIAL

Packing of nursery material has to done from time-to-time. Usually, much emphasis should be given while sending plants to a distant place but it does not mean that the plants to be sent to nearby places should no be packed properly. To get a better price of the plants, a nurseryman should pay high attention to the packing of the planting material. A packing material should have the following characteristics :

- It should be cheap and easily available.
- It should be highly suitable for the material to be packed.
- It should deliver the plants conveniently without affecting the quality.
- It should look attractive to the indentors.
- It should ensure protection against drying out of the material and mechanical injury.
- It should be well adapted for convenient and economical handling.
- It should be well adapted to transport, loading with security and economy at a space.
- It should not make the material bulky, rather it should add minimum weight to the material.

Packing either of seeds or plants or bulbs etc., should be done carefully so as to supply the planting material in safe and sound condition. It is particularly true when planting material is to be sent to distant places. For distant places, majority of material is sent by rail and trucks, where it is handled very roughly and therefore proper packing becomes more pertinent.

PACKING MATERIALS

The different packing materials used for packing of the nursery plants are hessain cloth, sacking cloth, plastics, paddy straw, sphagnum moss grass, dried grass etc. Hessain and sacking cloth is traditionally used by almost all the nurserymen. It is used in balling and burolopping of the plants. The hessain cloth is made from the best jute fibres, whereas sacking cloth is made from the raw grade jute fibre. Plastics and its derivatives are used for packing nursery material by the modern nurserymen. Different plastic materials like low-density polyethylene (LDPE), high-density polyethylene (HDPE), polypropylene (PP) and

Nylons etc., are used for packing purpose. Paddy and wheat straw and hay, dry grasses are used for wrapping the earthen ball of the saplings. Similarly, moistened moss grass is used to wrap up the delicate planting material like vegetable and ornamental flower seedlings, cuttings and strawberry runners before packing them in hessain cloth.

SALE OF THE NURSERY PLANTS

Marketing of the commodities has been a big problems in India because there is no proper marketing system in our country. When one has to produce the plants, these are to be sold. Nurserymen sometimes get good price in the market but not always because the price depends upon the supply and demand. If the supply of the commodity is high, it is likely to give fewer prices to the grower in the market. However, a nurseryman should give emphasis on the following points to get higher price for his plant material :

1. Major emphasis should be given on the production of a quality material.
2. Plants offered for sale should be of uniform size.
3. Proper grading of the plant material should be done before packing.
4. Material should be packed in an appealing packing material.
5. Plants should have attractive foliage, it should be free from insect-pests and diseases.
6. Plants should be labelled and displayed attractively at the sale spot.
7. Proper handling of the material during transit avoids damage to the material and helps to fetch better price in the market.
8. There should be some catalogue and pamphlets containing package of practices to be followed for handling, planting and management of the material offered for sale.

RAISING PLANTS IN CONTAINERS

Every day new techniques are developed, and if appealing, these are taken up by the users. In the recent years, there had been great inclination of the plant lovers towards container grown plants. Accordingly, nurserymen are now widely adopting the technique of growing plants in containers. The container-grown plants have the following advantages :

1. Such plants are not frequently damaged during packing and long distance transportation.
2. Fruit plants and ornamental foliage plants, which require greater care and management, are highly suitable for growing in containers.
3. Containers used for raising for plants can be reused repeatedly.
4. Most of the containers have higher durability.
5. The mortality rate of the container-grown plants is very less.
6. Plants in containers look attractive and help to fetch good price in the market.

Containers of different types, shapes and sizes are used by the nurserymen to raise plants. The chief characteristics of major types of containers used by the plant propagators are given below :

Clay Pots

Clay pots are mainly used for raising some fruit and ornamental plants. These are available in different shapes and sizes but round type pots are avoided for raising plants because they cover more area in the nursery. The clay pots are porous in nature and thus permits frequent aeration and water movement. However, if such pots are continuously used, it is likely that pores of the walls are accumulated with salts, which may be harmful in later stages. These pots are mostly used in India.

Rigid Plastics Pots

Rigid plastic pots are also available in different shapes and sizes. They are non-porous but occupy less space in the nursery. However, feasibility of their repeated use and light weight make their handling very easy during transportation.

Fibre Pots

Fibre pots are manufactured from peat and wood fibre and are available in different shapes and sizes. These are biodegradable in nature and can be put into soil along with plant. These are widely used in advanced countries but rarely in India.

Polyethylene Bags

Polybags have replaced other containers (clay and rigid plastic pots) and these are now widely used by the plant propagators mainly because these are cheaper and can be used time and again. Moreover, handling of plants in poly bags is very easy. These are light in weight and are easily available. Polybags of different sizes and thickness are available in the market, but in general, polybags of 30 × 20 cm size are used for raising fruit plants and 10 × 6 cm sized for raising seedlings of ornamentals and vegetables. However, before planting, these bags should be perforated with needle for proper drainage and aeration.

Coated Paper Containers

These containers are also used widely in advanced countries but rarely in India. These are light in weight and easy to handle during storage and transportation. These are manufactured from paper, cardboard fibre, impregmented with paraffin wax or bitumen or other water-repellent material of non-poisonous nature. These are cheaper than other containers.

• • •

Tools and Accessories Required in Nursery

Several operations are carried out in nurseries every day. For various operations, many tools and accessories are required from time-to-time. Some tools are simple and are used for simple operations such as budding and grafting. However, sometimes special equipments are required for particular method of budding, grafting or other operations. Some of the tools (Figure 19.1) and accessories required for such different operations in the nursery are described below :

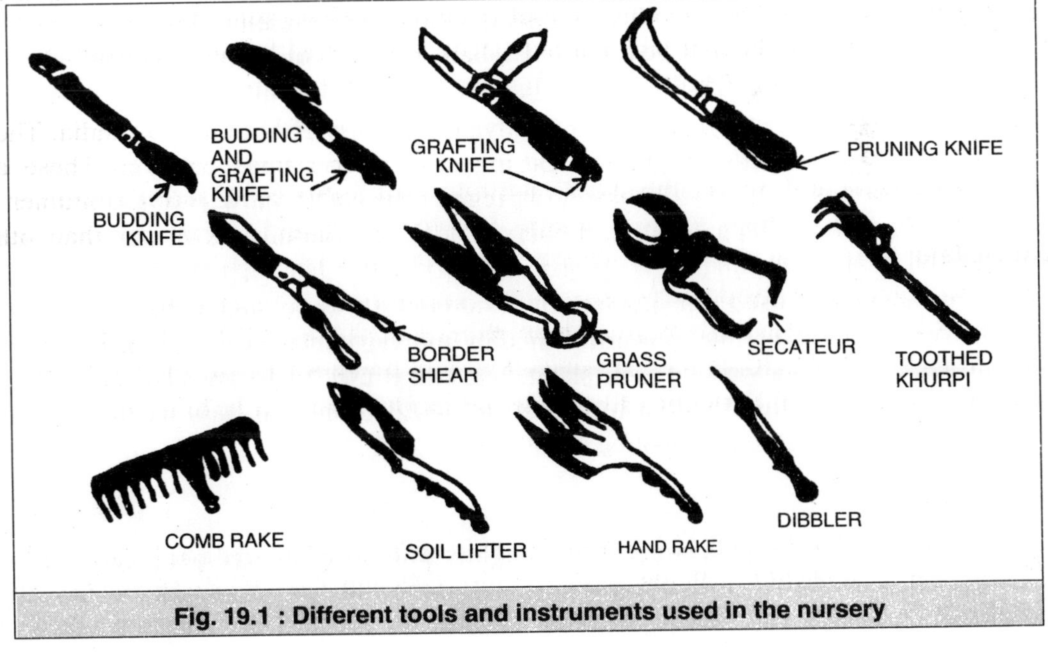

Fig. 19.1 : Different tools and instruments used in the nursery

Knives

For propagation work. two types of knives *viz.,* budding and grafting knives are required. However, very often a budding knife can be used for grafting purpose also. These knives may have either a folding or fixed blade. The blades of the knives should be made from high carbon steel and should always be very sharp. A grafting knife in general has a straight 7.5 cm long blade and a strong long handle. The budding knife on the other hand may have a straight or a bit curved blade of shorter length. It has, however, a spatula at the end of a handle, which is used for lifting the bark during the budding operation. Sometimes, a knife having two parallel blades (double bladed knife) is also necessary. It is specially required for lifting or removing the patch of a bud from the bud-wood. Similarly, a patch budder tool, having two pairs of blade is specially required for patch budding.

Kudali

Kudali is an important tool for digging of pits or digging of soil for other operations. Kudali has a metal blade attached to a wooden handle.

Garden Fork

Garden fork is mainly used for breaking the soil clods during preparation of nursery beds or fields.

Weeding Fork

It loosens the soil and is equally useful in weeding. It consists of a long handle with a blade having teeth.

Khurpi

Khurpi is widely used for various cultural operations in the nursery but mainly for the removal of weeds. It has a sharp edged triangular blade fitted with a wooden handle.

Crow Bar

It is an iron rod with one end pointed and other as wedge shaped. Crow bar is mainly used for digging of pits.

Water Cans

Watering cans are used for irrigation of annual flowers, vegetables and newly transplanted seedlings of various flowering annuals and vegetables or young plants in pots. These are made up of galvanized iron sheet, but sometimes these are also made up of plastics. These cans are fitted with a rose over the nozzle, which is very useful for equal distribution of water over the germinating seedlings.

Garden Rakes

Garden rake is used for collecting stones and bits of bricks from the bed, scarifying the grass surface and gathering the fallen leaves. A toothed rake is always better for such operations.

Secateurs

Secateur is considered as the most important tool for a propagator or a nurseryman. Secateur is used for excising scions, lopping off the rootstock, removing the undesirable sprouts/shoots from the stock, preparation of the scion sticks and for pruning operation. The blade of the secateur should be of a good quality because poor quality blades may not give smooth cuts to the stock and scion.

Grafting Machines

Many machines have been developed and commercially used for the preparation of scion sticks and bud-wood for budding and grafting operations in many developed countries of the world. These machines are, however, not used in India.

Pruning Saws

Several types of saws are required for different operations involved in propagation of horticultural plants. The commonly used are crescent saws, tapered saws and straight saws. All of them have long and widely set teeth to facilitate pruning or cutting of green wood. In general, a small saw is required for cutting the hard branches of the stock plant and sometimes the same saw can be used for pruning also. Its blade should be narrow so that it can pass through the narrow or closely spaced branches.

Pruning Shears

The different types of shears like hand shears, lopping shears, tree trimmers etc., are needed in a garden. Pruning shears should not be very expensive but these should be made up of good steel. Similarly, these should make a smooth and clean cut with least injury to the plant.

Ladder

In propagation work, ladder is required for operations like cutting of bud-wood, pruning of vigorous plants and top working of old and declining plants. In general, step ladder or straight ladders or hook ladders are used for such operations in the propagation field.

Tying and Wrapping Material

In most of the grafting and budding methods, it is essential to hold scion and stock firmly to have successful graft/bud union. For this, a suitable tying or wrapping material is required. Accordingly, nurserymen and other propagators use different materials to achieve desired results from budding and grafting. Commonly used tying wrapping materials are as under :

(*a*) **Waxed string and cloth :** It is the most commonly used tying material by the nurserymen. It can be easily prepared by dipping narrow strips (25-30 cm wide) of long cloth in the melted bee-wax. These waxed cloth strips can be used for tying the stock and scion after budding or grafting. The added advantage of using these strips is that these eliminate the need of further waxing.

***(b)* Waxed string :** Next to waxed cloth, waxed strings are the most commonly used trying material. It can be prepared by soaking a ball of cotton twine in a mixture of melted resin (1 Kg), bee-wax (500 g), linseed oil (250 g) and paraffin (120 g) for 10-15 minutes. This cotton ball should be rotated repeatedly while kept in the above mixture. For large scale grafting operation, waxed string is the most convenient tying material because it adheres to plant parts nicely even without tying. However, it should be strong enough to hold the grafted plant parts together and week enough to be broken by hand.

***(c)* Raffia fibre :** It is a strong fibre, which is not easily damaged by heat or rain. For proper working with raffia fibre, it must be moistened by wrapping it in damp cloth overnight.

***(d)* Rubber strips :** Rubber strips provide better grip to the stock and scion than raffia fibre or waxed cloth. These are highly elastic in nature and can be stretched to a great extent. The greatest advantage of these strips is that these don't girdle the plant even if these are not removed after the union has taken place.

***(e)* Polythene strips :** Plastic strips of different thickness are used as a tying material in budding and grafting by the nurserymen. In India, plastic films are more common amongst nurserymen. These are slightly elastic in nature and don't girdle the plant.

In addition above mentioned tying material, adhesive tapes, similar to surgical adhesive tapes but lighter in weight, are also used by commercial nurserymen. Similarly, paraffin tape is also used by some propagators but in advanced countries.

Baskets

A gardener should have plenty of baskets. These baskets are mainly used for carrying the seedlings or plants from one place to other or for distant places.

Labels

A commercial propagator should have different types of labels. These labels are used for proper labeling of plants before sale. The labels may be made up of paper, cardboard, wood, celluloid, aluminium and plastic etc., but usually the plastic or aluminium labels are preferred. However, for temporary use, the paper or cardboard labels can be used. In advanced countries, plastic labels are widely used.

Pots

Pots of different shapes and sizes are used in the nursery. The clay and plastic pots are available in different shapes and sizes. Usually, 4″ (10 cm) pots are suitable for small annual seedlings, six inches (15 cm) for palms and ferns, 8 inches (20 cm) for single specimens. In India, earthen pots are mostly used but now there is a trend towards plastic pots because plastic pots are durable and are easy-to-handle.

Seed Pans

Seed pans are usually smaller pots, which are used for raising seedlings of annual vegetables and ornamental plants. The seed pans are usually flat with size varying from 30 to 50 cm in diameter.

Seedling Trays

Seedling trays are primarily used for hardening of the micropropagated plants and for germination of seeds of different horticultural crops. The size of these trays varies with the number of cavities, ranging from 25 to 250. The propagators may choose the seedling trays of different capacities or as per their need and the crop to be raised.

Planting Supports

Bamboo of different sizes are required for staking the plants during their initial stages of growth. These should be available with the person who is raising nursery. In place of bamboo, iron can also be used.

Grafting Wax

Grafting wax is used by nurserymen atleast with two objectives (*i*) to seal the graft union for preventing the moisture loss and desiccation and death of the cells of the cut surfaces, which are essential for healing of the graft union, and (*ii*) to prevent the decay of wood by way of checking the entry of the pathogens.

• • •

CHAPTER 20

Hormonal Powders, Solutions and Pastes

Hormone can be defined as a substance produced in one part of a plant, transferred to the other part, where it affects a specific physiological function. Thus, hormones are produced at one location and are transported to the actual site of action. These are required in a very minute quantity to affect a specific physiological process. The term 'phytohormone' is applied to those hormones, which are exclusively produced in a living plant. The term growth regulator is given to those hormones, which responsible for growth by cellular enlargement. These hormones are usually synthesized artificially and used for various manipulations in horticultural crops. In fact, horticulture industry is indirectly dependent on plant growth regulators as these are used for various purposes in different horticultural plants. Further, for propagation of horticultural plants, different hormones and growth regulators are also used for the treatment of seeds for breaking dormancy and cuttage and layers for inducing roots in them. These are applied as powders, solutions or as a paste. Although, some ready-made commercial products are also available in the market. However, one should be able to make the solutions or pastes at home. A brief procedure for the preparation of hormonal powders, solutions and pastes is given in this chapter. In general, it should be kept in mind that one milligram pure hormone is required in one kilogram of inert material (talc etc.) or distilled water or lanolin for making one ppm strength of that particular hormone or growth regulator. Similarly, it is necessary to remember that most of the hormones don't dissolve in water. These are therefore first dissolved in a suitable solvent. For this, the required quantity of the hormone is weighed and dissolved in a few drops of alcohol or other solvent, shaken until it dissolves completely. Only then, the final volume is made with distilled water.

PREPARATION OF HORMONAL POWDERS, SOLUTIONS AND PASTES

Hormonal Powders

For preparing hormonal powders, the required quantity of the hormone is weighed precisely in an electronic balance. It is dissolved in $\frac{1}{2}$ litre ethanol, methanol, or acetone in a breaker. This material is poured into one kilogram of talc taken in mortar and mixed thoroughly with a glass rod. After mixing, the mixture is kept open in air for few hours. The alcohol will evaporate soon, after which, the dried talc is ground to a fine powder. This fine powder should be kept in airtight containers and can be used for treating cuttings as and when required.

Hormonal Solutions

For the preparation of hormonal solution, the required quantity of the hormone is weighed accurately in a physical or electronic balance. It is then dissolved in a few drops of acetone or alcohol.

When the contents are fully dissolved, the final volume is made with distilled water. The pH of the solution should be nearly neutral or so. If the pH of the solution is high or less, it may not induce rooting, rather it would adversely affect the rooting process. The pH of the solution can be adjusted with the help of 0.1 N HCl or 0.1 N KOH. Hormonal solutions should preferably be prepared fresh, otherwise, these are likely to become inactive. These can, however, be stored at a low temperature in a refrigerator for sometime.

Hormonal Pastes

For preparing hormonal pastes, the required quantity of the hormone is weighted accurately and dissolved completely in a few drops of alcohol. The required quantity of the lanolin (wool fat, a product similar to grease, but greenish-yellow in colour) in also weighed and heated slightly in a beaker under gentle flame. When, the lanolin is slightly liquidified, the dissolved hormone is poured in it. The contents are dissolved thoroughly by constant stirring with a glass rod. The contents are then allowed to cool down. The paste is thus ready for use. Until use, the paste may be kept in a cool dry place.

Aerosols

Growth regulators are used in aerosol form in greenhouses for rooting in softwood and herbaceous cuttings. Sometimes, mother plants are also sprayed with hormonal solutions prior to excising cuttings from them. Aerosol sprays are performed at a concentration of 25 -100 ppm about 30 to 40 days before expected date of excision of cuttings.

PRECAUTIONS TO BE FOLLOWED WHILE USING PGRS

The use of root promoting hormones for the induction of rooting in the cuttings and layers has now become the backbone of the nurserymen. Such formulations are readily available in the market in various forms but it is always better to prepare them at home. The hormones not only induce rooting but also help in growth and development of the induced roots. Indole butyric acid (IBA) and Naphthalene acetic acid (NAA) are most effective in cuttage and layers and these are used in tissue culture systems also. To have better and desired results from growth regulators, the following precautions should be followed for their preparation and further use :

- First, check for expiry date of the hormonal powder.
- Hormones/growth regulators should be weighed precisely and accurately, preferably on an electronic balance.
- As most of them don't dissolve in water, a proper solvent (acetone/alcohol etc.) should be used.
- Only a minute quantity is used at once. So don't buy more than what is required or nearly so.
- Hormones are liable to deteriorate soon if kept under warm place. Therefore, keep them in a cool and dry place, preferably in a refrigerator.
- Solutions should always be made fresh. If required to store for sometime, use of refrigerator is desirable.
- Use hormonal solutions for treatment of cuttings and lanolin paste for layers.

METHODS OF APPLICATION OF GROWTH REGULATORS

As stated earlier, plant growth regulators are widely used as an effective aid for breaking seed dormancy, and rooting of layers and cuttings. Among different plant growth regulators, auxins are the most effective and widely used as a rooting aid. Among synthetic auxins, IBA and NAA are found to the more effective than the others. However, 2,4-D, 2,4,5-T, 2,4,5-TP etc., are also effective in inducing roots in some plant species if used at low concentrations.

In general, the effectiveness of a plant growth regulator is not only dependent on the concentration, type of the plant species but also on the method of application. Different methods are used for the treatment of cuttings and layers with plant growth regulators. These are described briefly hereunder.

Prolonged Soaking Method

In this method, the basal ends of cuttings are dipped in the dilute solution of the hormone for 24 hours in a cool dry place. After the treatment, the cuttings are planted in the nursery or in a suitable growing medium. The concentration of the hormone or growth regulator usually varies from 20 ppm to 200 ppm, depending on the plant species and type of the cuttings. The concentration is usually low for easy-to-root species and *vice versa*. The nurserymen rarely use this method of treatment because it is a cumbersome and laborious method. However, it is very useful for difficult-to-root species, where some materials like vitamins, sugars and nitrogenous compounds are also used along with the growth regulators for facilitating rooting in such species.

Quick Dip Method

This method is an improved version of the previous method. It is very effective method of treating cuttings with growth regulating chemicals and is usually followed by most of the plant propagators for raising plants through cuttings. In this method, the basal ends of cuttings are dipped in the concentrated solutions of a hormone for a short time, usually for 5 seconds to 2 minutes. The treated cuttings are then planted in the nursery or field for rooting. The concentration of hormone for quick dip method may be 500 ppm to 10,000 ppm, depending on the species and type of the cuttings but usually a concentration of 4,000 to 5,000 ppm is used. In general, the higher concentration should be avoided as it may cause injury to the cuttings.

Powder Dip Method

In this method, the basal ends of freshly prepared cuttings are dipped in the hormonal powder and carrier (talc) for sometime. After treating the cuttings, extra powder adhering to the cuttings should be removed by shaking and cuttings are immediately inserted in the rooting medium. Seradix, a popular formulation is used by the nurserymen by this method. For effective rooting, the cut ends of the cuttings should be moistened before the treatment, but excess of powder should be shaken off, otherwise it may cause adverse effect on the rooting process.

Spray Method

Spraying of growth regulators is sometimes done to the mother plants before taking cuttings from them. Spraying of stock plants with 2,4,5-TP in concentrations ranging from 25 to 100 ppm is done about 30 to 40 days before taking cuttings from them. Cuttings taken from such plants root better as compared to un-treated plants.

Lanolin Paste Method

As described earlier, the growth regulators (especially IBA) are applied to the girdled portion of a layer or stool in lanolin paste for inducing rooting in them.

• • •

CHAPTER 21

Routine Nursery Practices

In nursery, there are certain operations that should be followed judiciously for the production of quality planting material. Most of these operations are of vital importance for the growth and development of the plants grown in the nursery. A brief description of these operations is given below :

Clipping and Cutting

To give proper shape to the hedges and shrubs of the nursery, regular clipping and shearing of hedges is very important. It makes them healthy and attractive. The hedges can be trained into different shapes, but usually square-shaped hedges look beautiful and can be easily maintained. In contrast, coniferous hedges look attractive if trained to pyramidal shape, which is their natural form also. Usually, quick growing hedges need frequent pruning. Shearing of hedges can be done in any season but rainy season is the best. Clipping and shearing can be done by a sharp pair of pruning shears and woody branches can be cut with the help of secateurs or pruning saws.

Defoliation

It is the practice of removing large foliage of a plant. This is mainly done in the flowering plants to reduce the vegetative growth and induce flowering in them. It is equally important in plants specially grown in arid climate, where availability of water to the plants is always a problem, basically to reduce water losses taking place through transpiration. Defoliation is also done while transplanting the seedlings in the field or while packing the plants for distant places. It can be done by hand or with some tools or by the use of certain chemicals. Defoliant chemicals are rarely used in India, however, pentachlorophenol is used widely as defoliant in jasmine and some other horticultural crops.

Deshooting

It involves the removal of shoots of a plant, which are not required. Basically, some flowering plants produce numerous side shoots and if all of them are allowed to flower, the size and quality of the flowers is adversely affected. Thus, initially, only a few shoots are retained and the unwanted ones are removed. Deshooting can be done by hands or with a blade. It is particularly useful and required in container grown plants or if the plants are to be prepared for some exhibition.

Disbudding

It refers to the removal of superfluous buds, flowers and shoots from the plants. The main purpose of disbudding is to get quality flowers. In many horticultural plants, several superfluous buds (vegetative and flower buds) develop in the leaf axils. These buds share the nutrients and water of the main stems, branches and shoots. Therefore, to divert the sap to the main branches, buds or required plant part, the unwanted buds are removed at an early stage of the plant growth. This operation is mainly done in flowering annuals, herbaceous perennials and in many container-grown plants, maintained particularly for exhibition purpose.

Mulching

Mulching is the practice of placing some material in the seedbeds mainly to regulate soil temperature, moisture conservation and suppress weed growth. Seeds in the beds are irrigated frequently and these can be washed off easily. Thus, placing of some mulch material prevents the blowing away of the seeds. Mulch also saves the seeds from being stolen away by the birds. Similarly, covering of seedbeds is equally useful to prevent the young seedlings from frost damage during winters and scorching sun during summers. Several kinds of mulch materials like hay, straw, stone pebbles, dried, leaves, corn cobs, pea nut hulls, cotton seed hulls, plastic sheets etc. are used in different horticultural crops but hay or straw are the best suited materials for covering the seeds beds. These are not only cheaper but are easily available too. These also ensure proper germination of the seedlings. However, construction of tent house during winters over citrus nursery is considered highly suitable for better germination and further growth of the seedlings. Similarly, construction of low-cost poly tunnels is widely used in winters over propagation beds of strawberry. Such tunnels not only saves the plants from low temperature and frost damage during winters but also help in the production of profuse, healthy and higher number of runners. Mulch material in general should be placed over the beds immediately after sowing and removed after the seedlings have germinated and have attained sufficient height or when the danger of frost or high temperature is over.

Pinching

Pinching involves the removal of growing points of shoots of a plant alongwith some leaves. This operation is also known as *stopping*. The main objectives of pinching are to induce bushy growth and to produce higher number of flowering stalks in a plant. Pinching is done mostly in annuals and herbaceous perennials and is hardly required by any flowering tree. It is the most wanted operation in dahlia, carnation, chrysanthemum, brachycome and marigold. In fruit crops, pinching has been found to be highly useful practice in training 'Amrapali' mango plants. In flowering plants, pinching is usually done when the plants are young and between 7 to 15 cm in height. Pinching, in general, requires no tool and can easily be done with hands. It is basically important and most wanted if a nurseryman wishes to prepare his plants for some exhibition, where quality but not the quantity is the main criterion for the selection of the best plants.

Pricking

Pricking is the practice of transferring the young seedlings to another pan, tray or place from its original site or location. Pricking helps seedlings to have better growth and development. Pricking is usually done when seedlings have attained sufficient height. The operation of pricking should be done with greater care because fragile root and shoot system of the seedlings may be damaged with a single jerk occurring during careless handling. The seedlings can be pricked when these are sufficiently large with the help of a pointed stick in a pot or tray. After pricking, the seedlings can be transplanted in the field when they have developed 4 to 6 leaves.

Shading

Almost all the plants in a nursery are very delicate and are easily damaged by scorching heat during summer and low temperature during winters. Therefore, shading and protection of the plants is necessary and very important operation in the nursery. The seedlings in the nursery need shading from scorching sun in summers. If not properly protected, the leaves of the delicate seedlings may get scorched or even killed. During winters, frost may be very harmful to the nursery plants in temperate and sub-tropical climates, though frost is not a problem in tropical climate. Therefore, protection of plants is of prime importance. Different materials can be used for shading and protection of plants but usually, paddy straw is widely used. Now-a-days, nurserymen and propagators use polyethylene tents or low-cost poly houses in winters to raise the temperature of beds for facilitating the seed germination process and avoidance of frost. Individual mother plants can be thatched with paddy straw during winters but care should be taken to keep the eastern and southern side open for better sunlight. Similarly, a row of windbreak (sissoo, eucalyptus, acacia or some fruit plant etc.) also provides shade to the nursery plants during summers and also prevents the formation of frost during winters.

Soil Sterilization

It is an important operation to eliminate the soil borne diseases caused by different pathogens. Soil sterilization can be done either by steam or chemicals. The soil meant for seed sowing or planting of mother plants is sterilized to save them from the attack of pathogens. Steam sterilization may be done in a sterilized chamber with steam or heat. The other method for soil sterilization is the use of 2 per cent formalin or formaldehyde. In this method, the formalin or formaldehyde is mixed thoroughly with the soil and soil is covered with tarpaulin or gunny bags for 48 hours. Afterwards, the soil is worked up thoroughly, dried up for sometime and later can be used for seed sowing or transplanting of the seedlings. Some fungicides like captan, benlate, blitox, brassicol etc., are also used for soil sterilization. For this, 0.2 per cent solution of any of the above mentioned fungicides is used for soil drenching *i.e.*, for the soil to be sterilized. Similarly, soil solarization is practiced in advanced countries as a major method of soil sterilization. It is a very effective method of soil sterilization but is very costly. Keeping Indian conditions in mind, steam sterilization is the best-suited methods as it is the cheapest and most effective method for combating soil borne insects and other harmful pathogens present in the soil.

Staking

Some plants show lanky growth in the initial years of growth in pots or even in the field and therefore require some support. Such plants should be provided with a suitable stake because it not only supports the plant but also prevents the plants from rocking about or blowing away by the strong winds. Staking can be done with the locally available materials like bamboo, sticks of pruned wood, dried stems of jute, cotton, maize and pigeon pea etc. Even angle iron rods are also used for staking. For single stemmed plants, only one support is required, while for multi-stemmed plants, 4 to 5 or more sakes may be required, particularly if plants are required to be prepared for exhibition purpose. To give as aesthetic look, the stakes may be painted green. To secure plants tightly with stakes, these may be tied with the help of a green thread but precaution should be taken to avoid injury to the plants. The plants should be checked at regular intervals to see that it does not obstruct the growth and injure the plant.

Training and Pruning

Training and pruning are two interrelated operations required in the nursery and field. Training is primarily done to give proper shape to the nursery grown plants, while pruning is mainly concerned with the productivity or vitality of the plants. Both these practices involve the judicious removal of the undesirable plant parts (branches, limbs, twigs, flowers, shoots or roots etc.) of a plant. These practices require technical knowledge on the flowering and bearing habit of the plants but in a nursery, a plant propagator can learn these techniques with the passage of time, experience, observations and practice. Disbudding, pinching and defoliation practices are also some forms related with training and pruning of the horticultural plants in the nursery.

Watering and Feeding

The most common mistake done by the amateur's is the daily watering of the established seedlings, which is very harmful. Daily watering is needed to the seedlings only for about a week or so but afterwards, flooding of the beds is the best method. Now-a-days, sprinkler and drip methods of irrigation are getting popularity among the nurserymen.

Both these methods not only ensure the regular supply of the water to the plants and seedlings but also save water to a greater extent. It is better if liquid fertilizers or foliar feeding of nursery plants is followed for efficient absorption and better utilization of nutrients by the plants.

Wintering

Wintering is an important cultural practice done in roses and jasmines in North and East India but should never be done in hot areas. It is considered as an alternative to root pruning. In wintering, the supply of water to the plants is withheld for sometime during rest period and the roots are exposed to sun by removing the surface soil around the trunk. The time of exposure of roots to the sun varies from 3 to 10 days, depending on the plant age, type of species and intensity of sun light. After root exposure, the roots are covered with the same soil but mixed with farmyard manure (FYM) and watered immediately. However, Floriculturists feel that wintering is a devitalizing process and should never be practiced.

● ● ●

CHAPTER 22

Chemicals and Plant Growth Regulators in Plant Propagation

Many chemicals and plant growth regulators (PGRs) play prominent role in the propagation of horticultural plants. These may play a vital role by facilitating the seed germination, breaking seed dormancy and by inducing roots in the layers or stools. The use of different chemicals and PGRs in seed and vegetative propagation is described briefly in this chapter.

A. SEED PROPAGATION

Many chemicals and plant growth regulators play one or the other role in facilitating the seed propagation as discussed briefly hereunder :

Treatment of Soil and Soil Mixes

Soil or the propagating medium may contain weed seeds, nematodes and many other pathogens, which may cause severe loss to the seed before germination or even after germination. To avoid the loss caused by these pathogens, many chemicals are used for the treatment of soil or soil mixtures. Formaldehyde (2 per cent) solution, chloropicrin, methyl bromide, vapam, captan and benomyl are the chemicals used for pre-treatment of soil and other propagation media. As a result of fumigation (vapam formaldehyde, methyl bromide) and soil treatment with fungicides (captan, benomyl etc.), the pathogen, insect-pests and nematodes may be killed and heavy losses causes by them can be avoided. In fact, these chemicals are useful for proper sanitation in and around the nursery area, thereby control the further spread of diseases, insect-pest and other pathogens.

Seed Treatment to Control Diseases

Many diseases of the plants are carried out by the seeds. The diseased seed may decay during storage and many act as a source of inoculum for further spread of diseases. The infected seeds do not germinate properly and if germinated, these may be affected by 'damping off' or other diseases, which results in mortality of the seedlings. Thus, disease infected seeds require chemical treatment to kill the infecting organisms. Seed borne diseases are controlled by dis-infestations, dis-infection and by protection. The most important and commonly used disinfestants are ethyl alcohol (50 per cent), calcium hypochlorite (2 per cent) and mercuric chloride (0.1 per cent). The duration of treatment varies in different kinds of seeds, usually from 5-10 minutes. Hydrochloric acid is also sometimes used a disinfestant. Disinfestation is useful treatment to eliminate organisms present on the seed surface from small seed lots. For protection of seeds from seed borne diseases, the dry seeds are treated with organo-mercuricals like agosan-GN, ceresan and non-mercuricals like brassicol, captan, thiram and dithane M-45.

Seed Treatment to Overcome Dormancy

Seed dormancy is a serious problem in the seeds of many vegetables, flowering annuals and fruits of temperate climate. Different chemicals and plant growth regulators are used to overcome this problem. Acid scarification is one of most important processes to overcome seed dormancy in which dilute sulphuric acid is used. Gibberellic acid (500-1,000 ppm) and kinetin (100 ppm) are the most commonly used growth regulators to overcome dormancy in seeds of many plant species. Similarly, thiourea, (3 per cent), sodium hypochlorite and calcium hypochlorite are also very effective chemicals for overcoming seed dormancy. The concentration and time to treatment of these chemicals and growth regulators is different for different horticultural plants. Similarly, one chemical may be highly suitable to break the dormancy in some kinds of seeds, but it may be ineffective for the other lot of seeds.

Testing seed Viability

For testing the seed viability, different methods and chemicals are used. Among different methods, tetrazolium test is the most widely used and accepted method used for testing the seed viability. In this method, seeds are soaked in 2,3,5- triphenyl tetrazolium chloride (TTC) solution for certain time. The living tissues of the seed changes TTC to an insoluble red compound (formazan), whereas the dead or non-living tissues remain unstained.

B. ASEXUAL PROPAGATION

Some growth regulators and chemicals play vital role in asexual (vegetative) propagation and facilitate early and easy rooting in cuttings or layers as under :

Treatment of Cuttings with Growth Regulators

Indole-3-butyric acid (IBA) is universally used rooting hormone for rooting of the cuttings. Its concentration usually depends on different species and type of the cutting but in general, a concentration of 2,000 to 5,000 ppm is commonly used by the nurserymen and plant propagators. In addition, 2,4-D (2,4-dichlorophenoxy acetic acid) and NAA (Napthaleneacetic acid) are also used for inducing rooting in cuttings in some species. For propagation by leaf and root cuttings, production of shoot alongwith roots is essential. For this, treatment of such cuttings with cytokinins such as kinetin or benzyl adenine is helpful for producing adventitious shoots in cuttings.

Treatment of Cuttings with Fungicides

During the process of rooting and after rooting, various microorganisms may attack cuttings. To protect the cuttings from such pathogens, their treatment with fungicides like captan is very useful. For this, captan may be used as a power dip followed by IBA treatment or the IBA in talc may be mixed with captan powder. Benomyl is another effective systemic fungicide, which can be used to control many diseases of cuttings during or after rooting.

Treatment of Layers

Root promoting growth regulators, especially indole-3-indole butyric acid (IBA) has prominent role to play in plants propagated through layersing. In air layering of many plants (*e.g.*, litchi, guava, mango and rubber plants etc.), IBA is applied to the upper cut of the ring. IBA is usually made in lanolin paste. Similarly, lanolin paste is commonly used for inducing rooting in stools of guava, mango and litchi.

Use in Micropropagation

For various techniques of micropropagation, many chemicals, growth regulators, vitamin, salts etc., with concentrations depending on type of medium and explant are used. For example, in callus production, a medium (MS or B_5 or other culture media) is supplemented with higher concentration of auxins compared to the cytokinins. Similarly, for rooting of the plantlets, plant growth regulators like IBA or NAA is used with varying concentrations, depending on the plant to be propagated, under controlled and aseptic conditions.

• • •

CHAPTER 23

Role of Plastics in Plant Propagation

Plastic is widely used not only at the home and industry but has become very popular among the gardeners, nurserymen, and the propagators. The role of the plastics in modernizing the horticulture industry is so great that the present day gardening is almost transforming itself, so to speak, it as plastic gardening.

Among the various types of plastic materials polyethylene, popularly known as *polythene,* has properties, which make it invaluable for the gardeners and the nurserymen. The word polythene has been derived from a Greek word 'Poly' meaning 'many', is an abbreviation of polyethylene. It is manufactured by subjecting ethylene from crude oil to a very high pressure. This was discovered by the scientists in the United Kingdom in 1933. The sheet film, which is invaluable for gardening, was for the first time, manufactured in 1938 in the United Kingdom. Sheet films in different colours and thickness are now available in the market.

As a sheet film, polyethylene has properties, which in some respects make it similar to the outer skin of the plants. It is waterproof, transmits light and allows gaseous exchange of oxygen and carbon dioxide. These are the qualities, which have made it so valuable for the packing and storing of plant material and for plant propagation. Unfortunately, ultra violet radiation weakens polythene film, thereby limiting the length of time to which it can be used under-field conditions.

USES OF PLASTICS

Different types of plastics are available for different horticultural uses. Some of the most important uses of plastics are described below :

Use in Construction of Greenhouses

Polythene lining of glasshouses has become an established practice in the temperate regions of the world. It has seen that a glasshouse completely lined with polythene film is as efficient as a double glazed glasshouse in reducing the heat loss. Plastic films with high transmission and heat conservation properties are inexpensive for raising seasonal flowers and vegetables. The countries like Holland, Denmark, Belgium, United Kingdom, Japan and the U.S.A. are using glasshouses commercially with polythene film to change the environmental conditions inside the glasshouse. In tropical countries, where humidity is high, polythene film is not so widely used as in temperate regions. In a polythene greenhouse, light, heat and moisture are adjusted according to the requirements of the plants. Usually, clear and transparent plastic films are used in greenhouses. Such films transmit as much visible light as does glass and more of infrared and ultra rays, which are found beneficial to the rooting of cuttings. In India, now it has become as common practice to use low-cost polyhouses for raising nursery of vegetables and flowering annuals.

Use in the Construction of Polytunnels

Polytunnels and tents are also used for raising cuttings and seeds of rootstocks. Sometimes bags of polyethylene are placed over the top or tray, which are supported with a loop of wire with its end strung in the soil (Figure 23.1). Polyethene tunnels can easily be made by supporting the plastic sheet with sticks or wires and sealing the ends with some tying material. For tunnels, usually a plastic sheet of 200-300 gauge is used. Polytunnels are being effectively used for raising strawberries in most advanced countries. For this, low-cost polytunnels are constructed and placed temporarily over the beds during winters. It raises the temperature of the beds and thus plants have better growth and produce runners frequently, which is otherwise reduced under natural conditions. The plants produced under plastic tunnels are usually very healthy and have lust green foliage.

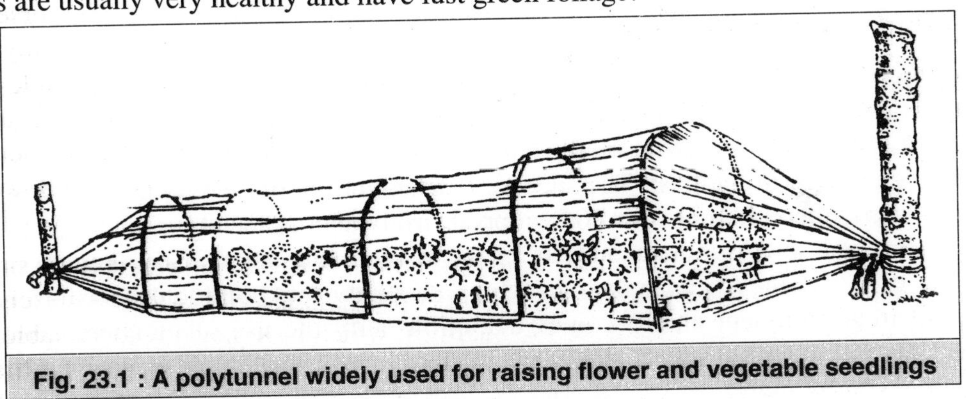

Fig. 23.1 : A polytunnel widely used for raising flower and vegetable seedlings

Use in the Propagation by Cuttings

It is a well-recognized fact that softwood and leafy cuttings must be kept in fresh and turgid conditions for better rooting. Heavy shading, syringing, and the use of propagating frames were the methods employed in the past to keep them fully turgid. However, now automatic methods have been developed, which have led to important changes in plant propagation practices. Humidification and mist techniques are two different practices developed with the aim to reduce transpiration from the cuttings during propagation. In the humidification process, fine particles of water are dispersed into the atmosphere to raise or maintain an optimum relative humidity around the cuttings. The humidity is ensured with the help of a polythene film cover. In the mist technique, which aims at mechanical and controlled syringing, the relative humidity is kept constant and the surface of leaves and stems of cuttings are kept constantly covered with a thin film of moisture. As water evaporates from the surface, the temperature of the leaf tissue is reduced below the surrounding air, consequently, there is very little transpiration from the leaves. Thus, the cooling effect of a continuous film of water on leaves constitutes the the difference between mist and humidification. Both these techniques employ polythene film houses, tents or chambers for ensuring maximum light and for controlling humidity and heat. Mist propagation creates better conditions for survival, a higher rooting percentage of hard-to-root species and faster rooting in easy-to-root species.

Use in Air-layering

In *goottee* or air-layering, moisture in the medium and humidity of the atmosphere in the past were used to be maintained by resorting to the propagation in suitable seasons with frequent watering in fruit crops like litchi, lime, lemon, guava, cashewnut, etc. The use of polythene film, however, has ensured similar conditions in conserving moisture in the rooting medium and humidity around the surface irrespective of seasonal variations. At present, millions of air-layered plants of different horticultural plants are propagated using sand and leaf mould as a rooting medium in a polythene wrap. By this simple technique, propagation of air-layered plants has been speeded up, reducing the labour costs to the minimum. Although, coloured polythene can be used as a wrapping material but transparent sheet should be preferred because one can see the initiation and further development of roots through it easily.

Use in Budding and Grafting

Polythene films of 150 gauge thickness are commercially used by the nurserymen in budding or grafting of citrus, mango, cashew, roses, mulberry etc. The film has replaced raffia fibre, budding cloth, wax cloth, banana fibre and several other locally available materials being used in the past. The success is found to be relatively higher in a bandage made with a polythene film as it can be wrapped firmly and uniformly over the graft or bud union. Further, the photosynthetic activity of the inserted bud-scion is also impaired due to the ability of polythene to transmit light. Huge mortality that commonly occurs in plants during rail trans-shipment also becomes negligible with the use of moisture conserving polythene lay flat tubings. Further, due to the negligible weight of the film, packing costs are also reduced considerably.

Use of Plastic Bags in Plant Propagation

Earthen pots are used universally for grafted or budded plants of mango, sapota, rooted cuttings of grape and pomegranate and many other horticultural crops. The pots are inexpensive as they are made locally. But due to their porous and brittle nature, these pots lose moisture at a very rapid rate and can break after a slightest shock. In addition, the heavy weight of these pots is a serious handicap in transport and their handling during transport is very difficult. Thus, in long distant trans-shipments, the cost of packing and transport claims a greater share of expenditure than the cost of plants. Use of polythene bags is now becoming quit popular for transplanting rooted cuttings of grape, pomegranate, rose, bougainvillea, etc. Polythene flat tubings of 200 to 500 gauge in different colours are now available in the market as a cheap substitute to the earthen pots. Similarly, polythene films of 200 gauge are found quite inexpensive compared to other wraps like gunny bags, wet-cloth, straw mats, etc.

Plastic Multipots for Mass Propagation

Now-a-days, some nurserymen are using plastic pots for propagation of different horticultural crops. These pots are not only light in weight but are quite cheap and durable. Moreover, these can be reused time and again. Their handling and transportation is also very easy. Being impervious, they prevent seepage of water. These pots ensure uniform conditions and thus provide optimum growing conditions for the seedlings. In a limited space, seedlings of vegetables such as tomato, capsicum, cabbage, cauliflower, lettuce, chillies, and fruits plants such as papaya, acid lime, guava, and rootstock materials can be raised without resorting to seedbeds. The growth of seedlings in multipots is quite uniform and much quicker than in open seedbeds.

Miscellaneous Uses of Plastics

Plastics, particularly black polythene is being wedely used as a mulch material in fruit orchards, especially in strawberry plantations. It helps to conserve moisture, regulates temperature, and have rigid control on weed growth, and nutrient loss. Plastic tunnels or cloches are also used for controlling weeds in the nursery or in the field. It is particularly useful in dwarf papaya plants. Plastic sheets are used to cover the entire plantations of low-headed fruit plants for protection against frost.

Nowadays, plastic sheets are commercially used for soil solarization, which kills harmful pathogens and saves a lot of money, which is essentially required to kill the incidence of such pathogens by insecticides, fumigants and fungicides.

Polythene has become a great boon to the plant propagators. It can revolutionize the plant propagation industry, if thoughtfully utilized by the nurserymen.

• • •

CHAPTER 24

Vegetative Propagation

Many plants are propagated commercially through asexual or vegetative means. In asexual propagation, vegetative parts of the plant are used for propagation. It is possible because all the living cells of a plant have a capacity to regenerate into a full plant under favourable environmental conditions. This capacity of plant cell is called as *totipotency*. The term totipotency was first coined by a Physiologist, Haberlandt in 1902. The plants produced through vegetative means are therefore genetically identical and similar to the mother plant, possessing all the characteristics of the mother plant. Vegetative methods of propagation are important because they permit large-scale multiplication of a genetically identical individual plant. Vegetative methods of propagation have certain advantages and disadvantages over the sexual method of propagation.

ADVANTAGES OF VEGETATIVE PROPAGATION

There are many reasons for using vegetative methods of propagation for multiplication of several horticultural plants because they have certain advantages over sexual method of propagation as under :

- The vegetatively propagated plants are true-to-type and are uniform in growth, yield and fruit quality.
- Vegetatively propagated plants come into bearing earlier than the seedling trees.
- Propagation by vegetative means is considered as an ideal method of propagation in plants having seed dormancy (*e.g.*, peach and olive etc.).
- Vegetatively propagated plants are less vigorous and thus can be easily maintained and managed as compared to seedling plants.
- Some fruit plants like banana, pineapple, fig and varieties of grape, guava and lemon

produce seedless fruits and have no viable seeds. These fruit plants can only be perpetuated through vegetative means.

- It is possible to regulate the tree size, fruit quality, and precocity in bearing in vegetative propagation by exploiting the desirable effects of rootstocks.
- It is possible to exploit the desirable abiotic effects of rootstocks on scion cultivar by budding or grafting.
- Self-incompatibility exists in a number of fruit plants such as mango, *aonla*, apple, almond and cherry etc. Provision for adequate cross-pollination in an established fruit orchard can be obtained by top-working the desirable pollinizers on scattered trees throughout the orchard.
- Indicator plants can be used for studying the various viral diseases. For example, Kagzi lime can be used as a rootstock (indicator plant) for detecting the presence of tristeza virus in citrus.
- In various fruit breeding projects, the young seedling selections if left to grow on their own roots, they may take 5 to 10 or even more years to come into bearing. This period can be shortened by grafting it on to a large established tree or on certain dwarfing rootstock for early assessment.
- A ' clonc' can only be perpetuated through vegetative means.
- The benefit of certain intermediate stock or interstock can only be obtained through some vegetative method of propagation.
- Damaged part of a tree trunk or root can be repaired by bridge grafting or inarching.
- A number of varieties can be grown on a single plant, which is a novelty.

DISADVANTAGES OF VEGETATIVE PROPAGATION

There are certain limitations of vegetative propagation also. These are :

- The vegetatively propagated plants, particularly budded and grafted ones, are short-lived than seeding plants.
- No new variety can be evolved by vegetative means of propagation.
- It is a specialized task and requires technical skill.
- Vegetatively propagated plants are usually prone to suckering.
- In many cases, vegetative propagation is more expensive than seed propagation.
- Vegetative propagation is failure in some plants such as papaya, pineapple and phalsa etc.
- Vegetatively propagated plants are usually more prone to insect-pests and diseases.

Various horticultural plants are propagated by one or the other vegetative method. One method may be highly suitable for one crop but may be completely failure for the other. A list of general vegetative methods employed for raising horticultural crops has been given in this chapter.

VEGETATIVE PROPAGATION METHODS

1. *Propagation by apomictic seedlings*
2. *Propagation on its own root system*

   Propagation by cuttings

   Propagation by layering
3. *Propagation on the root system of other plants*

   Propagation by grafting

   Propagation by budding
4. *Propagation by specialized structures or organs*

   Propagation by bulbs

   Propagation by tubers

   Propagation by rhizomes

   Propagation by corms

   Propagation by suckers

   Propagation by runners

   Propagation by offsets

   Propagation by bulbils

   Propagation by pseudobulbs

   Propagation by stolon

   Propagation by other vegetative structures (slips, crown etc.)
5. *Micropropagation*

   Propagation by using different explants (shoot tip, embryos, cell or protoplast etc.)

Vegetative propagation, in general, is very useful for the multiplication of heterozygous plants, which don't breed truly by seeds. The details on principles and methods of the above mentioned vegetative methods have been discussed in the subsequent chapters.

• • •

CHAPTER 25

Clone : Origin, Management and Variability

The main aim of vegetative propagation is to produce the plants identical to the mother plant. The biological process of producing plant similar to source/mother plants is known as '*cloning*' and the resultant population is called as '*clone*'. In nature, the process of cloning is continuously taking place by specialized vegetative structures like bulbs, runners, rhizomes etc., but it is not favoured in nature because it does not provide an opportunity for variability, which results from seed propagation, whereas, vegetative propagation provides a strong opportunity in the selection of clones grown in cultivation. Further, with vegetative propagation, it is possible to select a desirable type to maintain and to propagate it indefinitely. In this chapter a brief discussion has been given on the history of clone, production of a clone, maintenance of a clone, and sources of variability in a clone with measures to exclude pathogens from the propagating material.

HISTORY OF SELECTION OF A CLONE

Since time immemorial, man has domesticated a large number of food plants, such as potato, sweet potato, sugarcane, banana, yam etc., for his survival and discovered various methods of vegetative propagation to perpetuate these desirable plants. Later, many good cultivars in fruit plants like grape, banana, olive, fig etc., of known history were selected, which are being maintained even till today through different methods of vegetative propagation. Similarly, man readily propagated popler and willow trees, which were mainly used as hedges and fences during those days. Tree fruit plants, such as apple, pear, cherry, plum, apricot were not responsive to propagate by cuttings but later grafting techniques were discovered, which led to easy maintenance of these fruit crops. In tropics, where humidity is usually high, different methods of layering were invented, which helped in the reproduction of many tropical fruit plants like mango, litchi, guava, breadfruit, durian and longan etc.

Thus, clone has a great significance in horticulture because it makes feasible to exploit a single superior plant of unique genotype. A clone can, therefore be defined as *'the group of genetically uniform individuals, derived originally from a sexually produced individual or from mutations and maintained exclusively by asexual means.'* Because the entire population of the clone is genetically uniform and the individuals of a clone are almost similar in size, shape, appearance, ripening or blooming time, hence it is possible to standardize the uniform production technology for a clone.

A large number of clones of horticultural interest have been selected and perpetuated by man. The notable examples are Bartlett pear, Delicious apple, Cabernet Sauvignon and Sultana grape and Dashehari mango. It is assumed that Bartlett pear originated during 1770 in England and Delicious apple in Peru during 1870. Dashehari mango originated as a chance seedling in Dashehari village near Lucknow. Cabernet Sauvignon and Sultana grape have been grown and maintained for the last 2,000 years with out any change.

Previously, it was believed that a clone deteriorates with age and its vigour can only be restored by seed propagation. However, it is now established that the decline in the vigour of a clone is not due to age but it my be due to its infestation with some pathogens especially viruses. In addition, if a clone is grown continuously under unfavourable conditions of soil and climate, it is likely that it may add to its progressive deterioration. But, if environmental conditions are favourable and the clone is managed properly, it can be maintained trueness-to- type for several years. In general, a clone is more prone to deterioration as compared to seedling plants because in a clone all the plants are of similar genetic make up as compared to seedling plants, which have dissimilar genetic make up. As a result, seedling plants are better adapted to adverse environmental conditions as compared to a clone.

ORIGIN OF A CLONAL CULTIVAR

Most of the clonal cultivars have originated either through seedlings, mutations or through biotechnology.

Seedling Selection

Most cultivars grown in different parts of the world have originated from a single plant in a seedling population. This original population might have either originated as a chance seedling or through careful planned breeding programme. Clonal multiplication of the selected types is either done by (1) multiplication from the same plant, and/or, (2) consecutive generations of the progeny from the same plant by the breeder. In general, introduction of a seedling selection in a particular area/locality is done by conducting a series of experiments because genotype x environmental interactions have great bearing on its performance in that particular locality. This is the reason that one variety may excel others in a particular area, but may be completely failure in the other.

Many cultivars of horticultural importance have been selected as 'chance seedlings.' For example, most cultivars of mango grown in India like Dashehari, Langra, Chausa, Bombay Green, Baneshan, Neelum, Alphonso etc., have been selected as chance seedlings. The world famous Bartlett pear, Delicious apple, Gros Michel banana, Sultana grape etc., have also been selected as chance seedlings. Similarly, some ornamental shrubs, woody perennials are clonal selections.

Mutations

Chance mutations within established clones are the second major source of new cultivars. A mutation is genetic change involving some part of DNA molecule. It may result in structural changes in DNA molecule or a change in chromosome number. Further, mutation can be brought out artificially with the treatment of mutagenic agents like X rays, gamma rays and certain chemicals (*e.g.*, colchicine). Striking examples of natural mutations have appeared in apple, grape and citrus. In these fruit crops, many cultivars are bud sports, which have suddenly appeared as a change in the branch of a plant, which later proved better than mother plant. Pusa Nanha variety of papaya is a mutant, which has been developed by treating seeds of Pusa Dwarf with gamma rays. Similarly Rubired grapefruit is bud sport.

Biotechnology

It is a modern science in which a number of strategies can be used to alter the genotype of a clone. Recombinant DNA technology in which DNA fragments of one organism are cloned and transferred to another clone. In horticulture and forestry, this technique has great significance because a single and desirable trait can be transferred in to an established cultivar, lacking in that particular desirable trait.

PRODUCTION AND MAINTENANCE OF A CLONE

The production of a clone needs to involve the procedures that maintain the genetic purity, detect modifications and exclude pathogens in a population. It is a known fact that clones may deteriorate owing to these factors. It is likely that a clone does well in a particular locality but may fail in the other owing to different environmental factors or otherwise.

Maintaining Genetic Purity

In the production and maintenance of a clone, it is most essential to maintain its genetic purity *i.e.*, it should be true to name and trueness-to-type. True to name means that the propagation source is the correctly identified plant and trueness-to-type means that the cultivar to be propagated does no have any variation in the phenotype that may affect the performance and production in the later stages. For testing the trueness-to-type usually visual inspection at the time of flowering or fruiting is a right method. However, sometimes such inspections may fail if certain effects are masked by environment. Hence, progeny tests are used, which are cent percent effective in identifying a clone.

Excluding Pathogens

Many pathogens such as fungi, bacteria, insects, mites, viruses or mycoplasma like organisms (MLO's) may be involved in propagation, which may deteriorate the clone. Thus, control of these pathogens must be an essential part of the propagation operations. Sometimes, the pathogens may be present but there may not be any specific symptom and sometimes the symptoms are present. The main aim should be to get pathogen-free stock of the clone for its further multiplication. To exclude pathogens from a clone, the following different steps should be followed :

1. Selection of true-to-type stock.
2. Proper maintenance of the stock.
3. Proper distribution of the stock.

1. Selection of a true-to-type stock : Initial selection of a true-to-type stock is the first and foremost step for excluding pathogens from a clone. For this, only plants of known history should be selected and should preferably be inspected for viral infection or other pathogenic diseases. For such inspections, visual observations/inspections may be sufficient but indexing with indicator plants is the most reliable method. Indexing is of two types, the *culture indexing* and *viral indexing*. Culture indexing is done for detecting the infection of fungi and bacteria, while virus indexing is done to detect viruses in the desired stock. Nowadays, electron microscopy and serological tests are done to detect viruses but ELISA test (enzyme linked immunosorbent assay) is the most reliable technique. However, there are some simple procedures to eliminate the pathogens from the plant parts as discussed hereunder :

***(a) Select healthy plants* :** Some parts of the plant may be infected by a pathogen and others are not. Thus, selecting an unaffected plant part will surely eliminate the chances of pathogens from a stock. For example, apical shoot meristems have actively growing and dividing cells, devoid of any viral infection. Selecting this portion will help in propagation of stock free from viruses. Similarly, *Phytophthora,* a soil borne fungi can be excluded by taking growing tips of vegetative shoots, which are far away from the soil.

***(b) Follow shoot-tip culture* :** As stated above, the growing tips of plants are free from viral infection. Using shoot-tip culture under controlled and aseptic conditions will help in propagating virus-free material. Shoot-tip culture protocols have been standardized for many horticultural crops like apple, strawberry, Chinese gooseberry, grape, citrus, pineapple, banana, cauliflower, garlic, chrysanthemum, carnation, dahlia, orchids, gladiolus, iris, narcissus etc., which can effectively be used for eliminating pathogens from a clone.

***(c) Thermotherapy* :** Thermotherapy has been used widely to exclude pathogens from the stock plants and their parts. In this technique, the plant or its parts are subjected to a high temperature (110-135°F) for 1/2 to 4 hours, depending on the species, plant part and pathogen to be excluded. It can be combined with shoot-tip culture for getting better results. In combined treatment, the explants are given heat treatment at 38° to 40°C and then these are cultured in a desired culture medium under controlled and aseptic conditions for proliferation and plantlet development.

***(d) Chemotherapy* :** Some chemicals can be used to eradicate pathogens from the stock plants. The treatment of plants with chemicals for the eradication of pathogens is called *chemotherapy*. Chemotherapy is done only for excluding the externally carried pathogens. For example, the stock plants can be made free from *Phytophthora* fungus if soaked in a formaldehyde solution (2 per cent) for half-an-hour or so, depending on the plant species. Sodium or calcium hypochlorite (2 per cent), mercuric chloride (0.01 per cent) or ethanol (80-90 per cent) can also be used to eliminate the pathogens from the explants.

(*e*) ***Growing apomictic seedlings*** **:** Production of apomictic seedlings is a characteristic feature of some plants like citrus and apple. Growing such seedlings is an useful mean for preserving the clone and excluding the viruses from a clone.

2. Maintenance of the stock : Once a true-to-type stock has been obtained, it should be multiplied and maintained under strict control so that no deterioration takes place in it. For proper maintenance of the stock, proper isolation, sanitation and periodic inspections are essential. The plants should be established in a locality, which is quite away from infective agents like viruses. Sanitation is essential to eliminate disease agents from the surroundings, tools and equipments. Similarly, periodic inspection will be helpful to detect and eradicate 'off types' if present in the population. Visual inspection can be used in combination with indexing and other cultural methods. Some special procedures have also been standardized for individual crop,

3. Distribution of the stock : Like seed certification, distribution of true-to-type propagation material is the only aim of a distributing agent or an agency. For this, the distribution authorities design different standards. The propagators should have fair knowledge about the quarantine rules before selling the stock. In India, however, no such rules are yet available. However, selling of a true-to-type material has always a bearing on the name of the agent, agency, or an orchard.

METHODS FOR DETECTING PATHOGENS IN THE PROPAGULES

Various methods can be employed to detect pathogens in the propagating materials. Such methods have been described briefly hereunder :

Visual Inspection

Visual inspection is the best method to detect pathogens in the propagating material. On pathogen attack, susceptible plants exhibit some symptoms, on the basis of which, the pathogen is identified. Some pathogens, particularly viruses exhibit clear cut symptoms and the others may not. For example, ring spot virus in papaya, yellows in strawberry, nectrotic ring spot in genus *Prunus*. Viruses usually express their symptons at cool temperature (22°C) and their activity is suppressed at higher temperatures. However, difficulty arises when no symptom is produced by the pathogens. Hence, visual inspection is not considered as a reliable technique of viral detection because it requires technical skill and knowledge about the symptoms.

Culture Indexing

In this method, small sterilized plant parts are cultured on a nutrient medium, which favours the growth of the pathogens. These pathogens are then identified either by morphological features or biochemical tests. It has proved to be an effective tool for detecting fungi and bacteria in some ornamental plants like carnation and chrysanthemum.

Virus Indexing

In this method, scion varieties are budded or grafted on susceptible (indicator) plants, which then develop clear cut deficiency symptoms. It is a powerful technique for detecting various viral diseases in citrus. For example, Kagzi lime and Maxican lime are very good indicator plants for *Tristeza* virus in citrus. In addition, leaf grafting is used in strawberry.

Serological Tests

In serological tests, ELISA (enzyme-linked immuno-sorbent assay) is most commonly used to detect a particular pathogen on the basis of specific association of a particular pathogen with a specific protein. Now-a-days, commercial kits are available to screen the infected samples.

Biochemical Methods

Various biochemical methods like polyacrylamide gel electrophoresis are used to detect viroids and the replicated form (double standard RNA) of RNA plant viruses.

VARIABILITY IN A CLONE

Non-genetic Variations within a Clone

Genetic variations within the clones usually lead to permanent changes in the plants. Such changes are sometimes desirable, and even may lead to the selection of new varieties. However, sometimes these may be undesirable and thus lead to the production of 'off type' plant, which fetches poor price in the markets. In addition to genetic variations, certain non-genetic changes may occur within the clones, which have been discussed briefly hereunder:

(i) Environmental effects on clones

Horticulturists worldover are of the common view that although individuals of a clone are genetically similar but their performance is variable when plants of a single clone are grown at different sites and under different management practices. This is probably due to genotype x environment interaction, which may affect the phenotype (the appearance) or the performance of plants. Perhaps, this is the reason that before releasing a variety, it is tested under multilocational trials. The main objectives of such tests are to (*a*). make observations on desirable characteristics following vegetative propagation, and (*b*). whether its appearance is affected in different environments or not. For example, Bartlett pear fruits produced in Washington and Oregon are longer and more pear shaped than those produced in California, primarily due to climatic variation. Some commercial varieties of mango in south India tend to produce small-sized fruits having poor quality under north Indian conditions.

(ii) Phase variation and change in maturation

Phase variation is an important non-genetic variation. Its application can be beneficial to plant propagators depending upon the species and cultivar. Plants differ in their ability to change in phenotype with age. For example, those plants, which don't show much change in phenotype with age are called as '*homoblastic*' and those, which show distinct variation in specific traits with age, are called as '*heteroblastic*'.

(iii) Topophysis

Topophysis is the phenomenon in which different parts of a plant show phase variation and if propagated through meristems, the different phases are perpetuated in the offspring, (Figure 25.1). This phenomenon is also called as '*determination of the meristems*' Plants of certain species produced by cuttings taken from upright shoots (*orthotropic*) will reproduce plants in which the shoots grow vertically. Similarly, plants produced by cuttings taken from lateral shoots (*plagiotropic*) will grow horizontally. Many examples can be cited in which topophysis is manifested following vegetative propagation. For example, the new shoots from cuttings in coffee grow horizontally or laterally depending on the branch of the mother plant from which cuttings are taken. Some conifers and *Araucaria sp*. show strong tendency to manifest topophysis, when propagated through cuttings. The mechanism involved in the control of phase changes is not well understood, however, epigenetic variation and physiological changes may perhaps be associated.

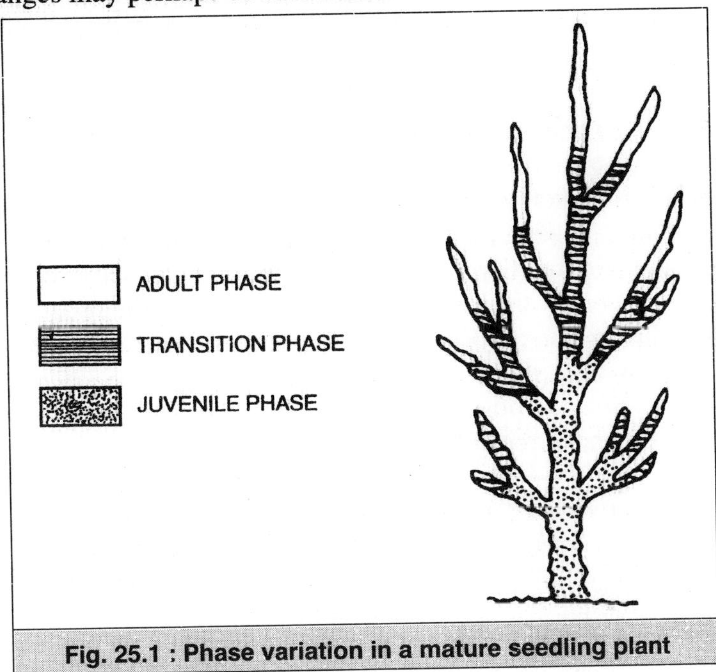

Fig. 25.1 : Phase variation in a mature seedling plant

GENETIC VARIATIONS IN ASEXUALLY PROPAGATED PLANTS

Vegetatively propagated plants are usually considered to be the true copy of the mother plants. However, certain variations may occur in them in due course of time. These variations may take place through mutations or chimeras.

Mutations

The term mutation was first coined by De Varies in 1905, who designated mutations as the sudden heritable changes occurring in the cells of an individual, which may bring permanent genetic changes in a part of a clone. A plant that arises as a result of mutation is known as *mutant*. Mutations may occur due to duplication or inversions of some parts of chromosomes or due to addition or deletion of the chromosomes. In some cases, changes may be brought about by changes in the DNA of plastids (plastone) and mitochondria (chondrianome), which may result in the development of albino (no chlorophyll) or variegated plants.

Mutations may occur spontaneously or can be induced artificially through the use of mutagens. Its rate primarily depends upon the location of the cell in the meristem tissue. Cells in the apical meristem are more stable and are likely to mutate less. The changes usually go unnoticed until and unless they occupy a large part or an entire branch of a plant such as a *bud sport*. The changes in colour (variegation) can, however, are detected easily. Usually, the results of mutations are defective and are more often inferior as in citrus. However, some mutants have been found useful for selecting new clones of superior varieties in horticultural crops. For instance, pink fleshed grapefruit, a superior type, is a bud sport of white fleshed grapefruit. 'Washington Navel' orange is a bud mutant of 'Laranga Selecta' Brazilian orange. 'Dashehari-51' is a prolific and regular bearing mutant of 'Dashehari' mango.

Chimeras

Sometimes, mutations result in the development of plants with two or more genetically different tissues growing separately but adjacent to each other, usually called as *chimeras*. Most chimeras usually originate from mutations either from spontaneous or induced in one of the cells of the shoot apex. However, chimeras may also result from hybridization. These tend to be unstable but their stability depends on their structure and genotype of the plant. The chimera may also result in the production of variegated foliage in the plants. Citrus, *Vitis,* hydrangea, dahlia, coleus, pelargonium, sansevieria, etc., are some examples of chimeras. In all these plants, plastids in some portions of leaves is unable to produce chlorophyll but is able to do so in other parts, resulting in the formation of variegated leaves. Further, mixture of sour and sweet type apple, fuzzy and smooth peaches and thornless blackberries are some other examples of chimeras. Similarly, occurrence of red colour in outer skin of apple and potato with mutated inner layers are also the examples of chimeras. Depending on the structure, the chimeras are of four different types (Figure 25.2).

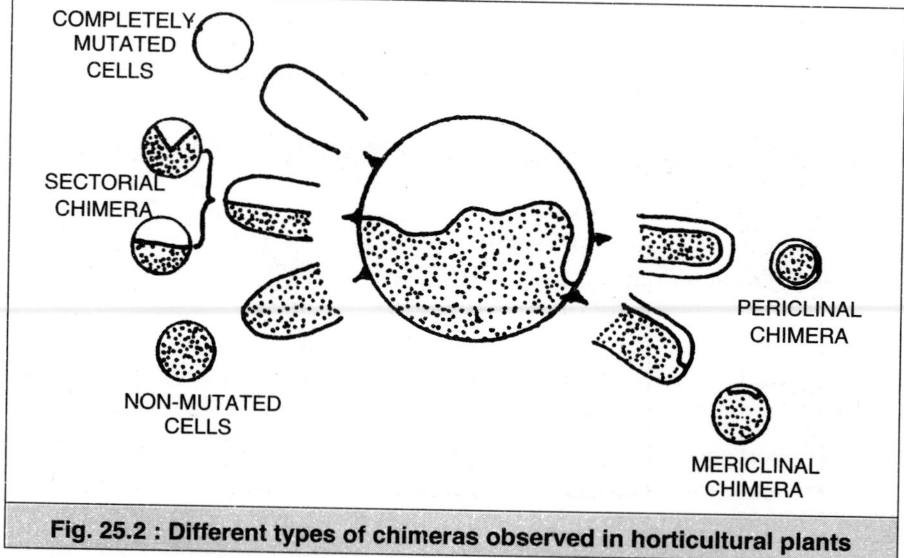

Fig. 25.2 : Different types of chimeras observed in horticultural plants

Periclinal chimera : In this chimera, a tissue with one genotype occurs comparatively as a thin layer around genetically different core. This is most stable form of chimera if propagated through stem or grafting. A typical example is the genus *Rubus,* which has many thornless blackberry types. The epidermal layers in these types lack the gene for thorniness. If such types are propagated through stem cutting or tip layering, the new plant will be thornless. However, the roots of such thornless plants arise endogenouly from the non-mutated tissues and such roots are used for subsequent propagation, the plants will be thorny. Similarly, if seeds of thornless plants are used for propagation, the seedling plants will be thorny because the gametes are produced from the cells of the tissues, which originated from non-mutated layer. However, rarely some thornless forms may also be produced. interestingly, propagation of periclinal chimera through leaves may result in non-mutated forms. For example, if variegated forms of sansevieria are propagated through leaves, it results in the production of green sansevieria.

Mericlinal chimera : In mericlinal chimera, the mutated tissue occupies only a part of one layer and does not extend completely around the shoot as in periclinal chimera. It occurs more frequently because it is the resultant of mutation of a single or a few cells. It is not stable but with continued vegetative propagation, it may either convert to periclinal chimera or to non-mutated tissue or a shoot. For example, a lateral bud arising from a mericlinal chimera would become a periclinal chimera (shoot) and from a normal part to develop in to a normal shoot.

Sectorial chimera : In this type, a shoot may be composed of two genetically different tissues, occupying two distinct sectors in a shoot. Thus, the leaves or buds arising from such shoots will also be of two different identities. These types of chimeras are, however, common in roots than in shoots. It is also an unstable form of chimera but with continuous vegetative propagation, the developed plant may either be normal, mutated or periclinal chimeras. Sectorial chimeras may sometimes results in the development of variegated plants. The variegation in plants is caused by mutations on plasma genes, which affect the chlorophyll development, as a result of which, some parts of the leaves produce albino patterns. However, variegation may sometimes be caused by the attack of viruses.

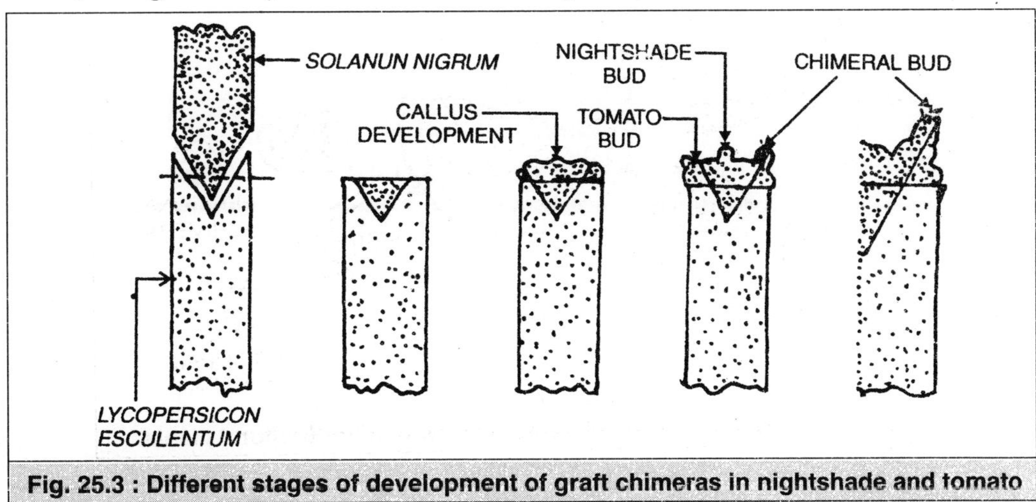

Fig. 25.3 : Different stages of development of graft chimeras in nightshade and tomato

Graft chimeras : Most chimeras occur naturally, however, these can be induced artificially too. It is particularly true with grafting. As and when a scion grafted on the stock is severed back to the region of callus formation (graft union), the resultant plant may be entirely different to stock and scion. Various artificially induced chimeras have been reported in horticultural crops. A naturally bizzaria orange, having a fruit half-orange and half citron was discovered as early as 1644 in Florence, Italy. It originated from a bud arising from callus at the junction of sour orange grafted on citron. Hawmedlars are the resultant graft chimeras of medlar grafted on hawthorn. Similarly, Winkler, a world famous scientist, has produced numerous graft chimeras of tomato many years ago by grafting tomato on *Solanum nigrum* and *vice versa* (Figure 25.3).

Transposons

Genetic variation in some plants may take place due to transposable elements also. This variation typically involves pattern of coloured structures and dots of different sizes and arrangements. For example, variegated seeds of Indian corn, veriegation in poinsettia, and sansvieria. The work of Dr. B. McClintock, a Nobel prize winner shows that these patterns are produced by unique mobile genetic factors, which have the capacity to move to different portions of the chromosomes and then grow on and off. Such genes have been called as 'jumping genes'.

Somaclonal Variations

This type of genetic variation has been observed in plants under *in vitro* conditions. In fact, this type of variation occurs in the cells in the deeper areas of growing point below the meristem, which often become polyploid. This variation has been referred to as a boon to the breeders but a curse to the propagators.

• • •

CHAPTER 26

Propagation by Cuttings : Principles

Raising of plants through cuttings is a simple, quick, and cheap method of plant propagation. Many kinds of plants can be propagated through cuttings. Cuttings are so easy to take that any visit to a friend's garden can yield a few new specimens with no detrimental effect on either the parent plants or your friendship. A cutting is any part of a plant, which when put in suitable propagation medium, becomes an individual plant in its own roots similar to the mother plant. Most plants are raised by stem cuttings, but in some cases plants can also be propagated from their leaves, buds or root cuttings. The plant part to be used as cutting chiefly depends upon the growing habit of the plant. However, any plant part taken for cutting, it must reproduce into full plant. As the rooted cutting will be identical to its parent, great care should be taken while selecting the parent plant from which cuttings are to be taken. A poor growing, spindly plants are unlikely to produce a prized specimen from a cutting. Similarly, diseased or unhealthy plants also do not produce good cuttings. Therefore, healthy and strong growing plants should be selected if these are to be utilized for the preparation of cuttings.

PROPAGATION BY CUTTINGS : ADVANTAGES

Propagation through cuttings has numerous advantages as enumerated below :

- It is a simple, cheap and convenient method of propagation of plants.
- It involves minimum expenses.
- Much technical skill is not required.
- As the plant grows on its own system, the complex problems of stock-scion relations, which exist in budded/grafted plants, can be avoided.
- Because rootstocks are not involved, so there is no problem of graft incompatibility, which otherwise is encountered in many stock-scion relations.

- Rapid multiplication is possible within a short time.
- Plant produced through cuttings are uniform with not much genetic changes.
- Much space is not required.
- Transportation of cuttings is easier.

PROPAGATION BY CUTTINGS : DISADVANTAGES

This method of propagation has certain disadvantages, like :

1. Cuttings of some plants do not root easily.
2. The desirable benefits of rootstock on a scion variety/cultivar can not be exploited.

ROOTING OF CUTTINGS

In propagation through cuttings, it is highly essential that proper root and shoot system should develop. In fact, regeneration of a plant part (cutting) into an entire plant structure basically depends on two fundamental properties of a plant cell. One is totipotency, which means the capacity of an individual living plant cell to reconstitute all plant functions and the second is dedifferentiation, *i.e.*, the capacity of mature cells to return to a meristematic condition and develop into a new growing point. Thus, for transformation of a plant part into a full-fledged plant, a plant propagator has to do some manipulations for inducing root and shoot system in the cuttings. In general, there exists greater variability amongst plant species in the ability to root in their cuttings. In some species, cuttings root very easily as in grape, in some species, rooting takes place only after some treatments, while others, do not root at all. Therefore, it becomes mandatory to study the anatomical and physiological basis involved in the process of rooting of cuttings.

A. Anatomical Basis of Rooting of Cuttings

The roots may be developed on stem, root or leaf cuttings but they all have internal origin in their parent structures. Thus, internal origin of roots is called as endogenous and is reported in most of the plant species. However, in some species (*e.g. Tamarix*), roots may first develop exogenously on the stem and then these are connected to the internal tissues of the stem. First study on anatomical basis of rooting was done by a French Dendrologist, Duhamel du Monceau in 1758 and later various research workers emphasized that in the process of rooting of stem cuttings, generally four anatomical changes are observed. These anatomical changes are (*i*) dedifferentiation of specific mature cells, (*ii*) formation of root initials from the dedifferentiated specific mature cells, (*iii*) development of root initials into organized root primordia and formation of vascular connection between root primordia and conducting tissues of the cuttings, emerging through the cortex and epidermis, and (*iv*) emergence of roots outside the cuttings.

However, herbaceous cuttings have different anatomical changes during the rooting process, because of entirely different structure of the stem. The herbaceous stem generally has four major areas *viz.*, a large pith in the centre, a ring of vascular bundles outside the pith, a cortex outside the vascular bundles and a thin epidermis. In such plants, adventitious roots generally appear just outside and between the vascular bundles but the tissues involved at the site of origin of roots vary widely, depending upon the kind of plant. For example, adventitious roots in tomato and pumpkin arise in phloem parenchymatous region, epidermis in *Crassula,* from pericycle in coleus, and in castor these arise from vascular bundles.

In woody plants, one or more layers of xylem and phloem are present and adventitious roots are formed in the stem cuttings from the living parenchyma cells. Generally, the origin and development of adventitious roots takes place next to and just outside the central core of vascular tissues. After emergence, the roots develop root cap and other tissues of the root. Adventitious root and shoot usually arise within the stem (endogenously) near vascular tissue, outside the cambium.

Callus formation and rooting : In the early part of nineteenth century, research work was initiated on how the actual rooting process is initiated in the cuttings. Later, it was observed that when cuttings are placed in a suitable medium, a mass of undifferentiated parenchymatous cells, called as *callus,* is usually developed at the base of the cutting and only then the root initiation process takes place. Then, it led to belief that callus formation is necessary for rooting of cuttings but now it has been established that callusing and root initiation are two independent phenomenon and can occur simultaneously. Though, the excess callus formation may hinder root initiation in some species, it may, however, be a precursor for root initiation in others as in *Sedum, Hedera helix* etc.

Presence of preformed root initials : Root initials are sometimes developed in the intact stem of certain woody plants, even before cuttings are made from them. These root initials remain dormant in the stem. These dormant root initials are called as *preformed* or *latent root initials*. When the cuttings made from such stems are placed in favourable environmental conditions, these root initials become active and roots are developed rapidly and easily from them. Occurrence of the root initials is quite common in willow, hydrangea, poplar, jasmine, currant and citron. In some clonal apple and cherry rootstocks and old trees of apple and quince, the preformed root initials show a swelling (outgrowth) on the stem and are more often called as *burr knots*. Cuttings taken from such plants (having burr knots) usually root better and early, compared to those having no burr knots.

Stem structure and rooting : It has been found that certain type of stem structure or tissue relationship plays a vital role in adventitious root formation in the cuttings. The development of continuous sclerenchyma ring between the phloem and cortex of the stem is generally considered as an anatomical barrier to the rooting of cuttings as in olive and in some leaf cuttings. It has been observed that stems of certain kinds of plants have lignified tissue, which may also act as a mechanical barrier to rooting. Similarly, presence of sclerenchyma fibres in the cortex of stem may cause difficulty in the rooting of cuttings in some species as in English Ivy and *Hedera helix*. Further, in some species, some structures within the stem favour initiation of root primordia than others. For example, citron (*C. medica*) produces root profusely from preformed root initials within a short time compared to sour orange (*C. aurantium*), which forms only a few roots that too after several weeks.

B. Physiological Basis of Rooting of Cuttings

Many physiological factors have been reported to be associated with rooting cuttings. These are :

(a) Hormonal factors

Various classes of growth regulators, such as auxins, cytokinins, gibberellins, abscisic acid and ethylene influence rooting of cuttings. However, of these, auxins are known to have greatest effect on root formation in the cuttings. In addition, various other natural occurring promoters and inhibitors may also take part in the root initiation process as discussed briefly hereunder :

Auxins : During 1934, indole-3-acetic acid (IAA) was identified as the first naturally occurring compound having auxin activity, which takes part in inducing rooting in the cuttings. Subsequently, synthetic auxin was tested for its ability to induce rooting in the stem segments and various research workers demonstrated the practical use of synthetic auxin in 1935. Later, indole butyric acid (IBA) and napthaleneacetic acid (NAA) were found to be more effective in inducing rooting in cuttings. For initiation of rooting in the cuttings, a continuous supply of auxins is basically required for the first 3-4 days. However, discontinuation in auxin supply during root elongation process does not affect the rooting process at all. Usually, the cuttings do not respond to exogenous application of auxins once the rooting process has initiated.

Cytokinins : Cytokinins are responsible for growth and differentiation of the cells. Various natural and synthetic chemicals like kinetin, zeatin and 6-benzyl adenine (BA) have cytokinin activity. In general, cuttings having high endogenous level of cytokinins have more difficulty to root than those having low cytokinin activity. Similarly, if cytokinins are applied artificially, root initiation process is inhibited. However, cytokinins at a very low concentration promote root initiation process, while high concentration inhibits it. Further, cytokinins strongly promote bud initiation in root cuttings of some species as in *Bryophyllum* and *Convolvulus*.

Gibberellins : Gibberellins are a group of naturally occurring compounds first isolated in Japan during 1939. They are primarily known for their elongation effects in the intact stems. The higher concentrations of gibberellins (10^{-3} M) usually inhibit root initiation process in cuttings but lower concentrations (10^{-11} to 10^{-7} M) promote it. The inhibitory effect of higher doses of GA_3 is mainly due to prevention of early cell division, involvement in transformation of mature stem tissues to the meristematic conditions. Thus, many gibberellin antagonist chemicals like alar, arest or abscisic acid usually promote rooting in cuttings.

Abscisic acid : The role abscisic acid in rooting of cuttings is not very clear and reports are highly contradictory. However, its affects on rooting depend upon concentration and nutritional status of the stock plant from which the cuttings have been prepared.

Ethylene : Ethylene, a gaseous hormone produced by the plants, has a promoting effects on rooting of cuttings at a low concentration. Ethylene application at 10 ppm causes root formation in stem and root cuttings but at a higher concentration (100 ppm and above) decreases or inhibits these processes. Root promotion activity of ethylene is suppose to be due to its synergistic effects on auxin synthesis, which ultimately leads to root initiation process.

(b) Role of Vitamins

Endogenous optimum level or exogenous application of vitamins promotes root initiation process in the cuttings. The vitamins like thiamine (Vit. B_1), pyridoxine (Vit. B_6), naicin, biotin and vitamins K or H all are known to stimulate the rooting process in different plants. Vitamins, if applied with some auxins like indole acetic acid (IAA), naphthalene acetic acid (NAA) or indole butyric acid (IBA) have synergistic effect on root initiation process in cuttings of different plant species.

(c) Presence of buds and leaves

It has been proved experimentally that presence of leaves on the cutting exerts a strong stimulating influence on the root initiation process in cuttings. The leaves produce carbohydrates, which are undoubtedly translocated downward to the base of stem, where it promotes root formation. Further, the leaves and buds are powerful auxin producers and their transport to the basal parts may initiate rooting faster as in poplar, grape and currant. Further, in many cases, if the buds are removed from the cutting, the root formation may be hindered adversely.

(d) Rooting co-factors

Many rooting co-factors have been isolated from cuttings of different plants. These co-factors are natural occurring substances that appear to act synergistically with auxins primarily indoleacetic acid (IAA), for root formation in cuttings. Bouillenne and Went scientists were the first to name the root-forming factor of leaf, cotyledon and buds as rhizocaline in 1933. Later, a new hypothesis was proposed by Bouillenne and Bouillenne-Walrand in 1955, highlighting that rhizocaline is a complex of three components *viz.* (*i*) a specific rooting factor, which is translocated from buds, leaves, characterized chemically as ortho-dihydroxy phenol, (*ii*) a non-specific factor (auxin) in low concentration, which is also translocated from leaves and the buds, and (*iii*) a specific enzyme, probably polyphenol oxidase, present in pericycle, phloem or cambium. According to this hypothetically scheme, ortho-hydroxyphenol reacts with auxin, giving rise to rhizocaline. The formation of rhizocaline leads to the reactions involved in the root initiation process (Figure 26.1). Another hypothesis proposed by Hess (1968) indicated that a number of co-factors interact with IAA, resulting in the formation of IAA co-factor complex, which subsequently initiates rooting in the cuttings. Usually, the easily rooted plants have a large number of co-factors and in higher concentration than in difficult-to-root plants. One of these co-factors (No. 4) consists of highly active substance, characterized as oxygenated terpenoids, and another factor, (No. 3) was identified as chlorogenic acid. Later, it was postulated that phenolic compounds, more especially catechol, reacts synergistically with indole acetic acid in root production. This oxidation of ortho-dihydroxy phenol is one of the first steps leading to root initiation process. Similarly, abscisic acid was also found to be associated as one of the co-factors in root initiation process. The action of phenolic compounds in root initiation is mainly in protecting the natural occurring auxin, indole acetic acid from destruction by indole acid oxidase. Rooting co-factors have been reported to occur in many plant species such as in *Hedera helix,* Crab C, MM_{106} and M_{26} apple rootstocks. These were also found in easy-to-root 'Old Hume' pear cultivar but not in difficult-to-root 'Bartlett'.

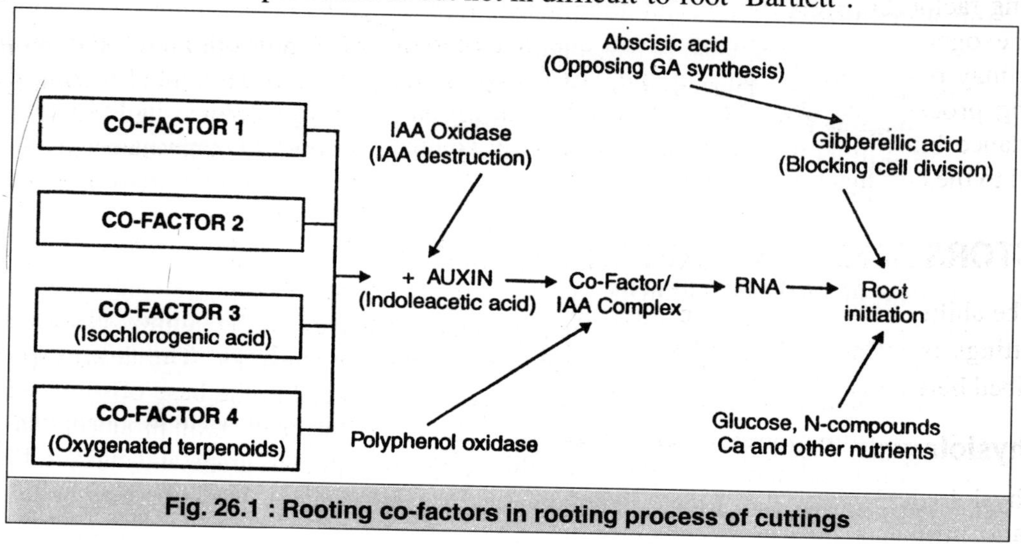

Fig. 26.1 : Rooting co-factors in rooting process of cuttings

(e) Nutritional factors

The nutrient status of plant from which cuttings are taken, also plays a vital role in root initiation process of the cuttings. However, among different nutrients, nitrogen and carbohydrate reserve are important. Low nitrogen and high carbohydrate reserve usually favour root initiation process compared to high nitrogen and low carbohydrate reserve. High N content usually favours luxuriant growth and hinder root initiation process but extreme N deficiency also hinders it. Thus, high C : N ratio should be maintained in the stock plants before taking cuttings from them.

(f) Endogenous rooting inhibitors

In additions to root promoting substances, certain endogenous root inhibiting substances are also present in some plant species, which inhibit the process of root initiation in the cuttings. Therefore, cuttings of certain plants do not root easily mainly because of presence of endogenous inhibitors. For example, the cuttings of *Vitis berlandieri* do not root easily as compared to the cuttings of *V. vinifera* because of presence of higher amount of inhibitors in them. The cuttings of Langra mango also do not root easily because of presence of high concentrations of abscisic acid. Similarly, hardwood cuttings of Bartlett pear do not root easily as compared to Old Hume cultivar. Placing the cuttings in water before planting helps in leaching of inhibitors, which enhances the root initiation process and subsequent development of the roots in them.

CLASSIFICATION OF PLANTS ON THE BASIS OF EASE-TO-ROOTING

The plants can be divided into four classes in relation to adventitious root initiation : *(a)* plants in which relative proportion of all endogenous substances (growth promoters and inhibitors, nutrients etc.) required for rooting is optimal and root initiation occurs readily when cuttings made from them are placed under favourable environmental conditions, (*b*) plants in which natural occurring co-factors are present in adequate amount but auxin is a limiting factor, however rooting initiation takes place easily when cuttings are treated with auxin exogenously, (*c*) plants in which one or more co-factors of rooting are sub-optimal, auxin may or may not be present and exogenous application of auxin is unable to initiate rooting process, and (*d*) plants in which endogenous rooting inhibitors are present in abundance but root initiation takes after giving some treatments (*e.g.*, placing cutting in water) to the cuttings for leaching of inhibitors.

FACTORS AFFECTING ROOTING OF CUTTINGS

The ability of the rooting of cuttings in different plant species varies widely. The rooting of cuttings is influenced both by internal and external factors. These factors are briefly described hereunder :

A. Physiological Condition of the Mother Plant

Physiological condition of the mother plant from which cuttings are made, can exert a dramatic influence on rooting of cuttings. For example, cuttings taken from the plants deficient in water often show reduced rooting. Cuttings root better if taken in the early morning hours, when plants are in turgid condition than those taken from water-deficient plants. Similarly, the nutrient status of the stock plant also exerts a strong influence on development of roots and shoots from the cuttings. It has been reported in apple and raspberry that their cuttings root and sprout better if taken in fall, the time at which carbohydrate content is the highest and survival was very poor when cuttings are taken in summer, the time when carbohydrate storage is less. Plants supplied with excessive N have luxuriant growth and if cuttings are taken from plants supplied with excessive N, they produce poor roots. Thus, low N and high carbohydrate balance in stock plant is necessary if cuttings are to be prepared from them. For blocking the carbohydrate reserve in the stock plant and to have better rooting from the cuttings made from them, the technique of girdling has been perfected. Girdling blocks the downward movement of carbohydrates, hormones and other root promoting factors and thereby induce rooting process in cuttings readily. Girdling has been found to be very useful in rooting of cutting in citrus, hibiscus, water oak, apple etc. Internal factors of the stock plant, such as, auxin level, rooting co-factors and carbohydrate etc., also influence the root initiation process of the cuttings. Thus, the girdling of shoots prior to their use as cuttings, is useful for stimulation of rooting.

B. Juvenility Factor

Age of the stock plant from which the cuttings are prepared, influences the rooting of cuttings. In most plant species, the cuttings taken from juvenile (young) growth phase often root better than those taken from adult phase, indicating that the ability of cuttings to form adventitious roots decreases with the increase in the age of the plant. In fruit plants like, mango, apple, apricot, olive, citrus, pear, peach etc., the cuttings often root better if these are taken when plants are in juvenile period (pre-bearing period). Cuttings taken from old and non-bearing plants rarely root better. Scientifically, it is proven that plants produce more rooting inhibitors with the growing age and thus inhibit root initiation process in the cuttings taken from them. Moreover, the phenolic level – a synergists for root initiation, also decreases with the increase in the age of plants, which may possibly reduce the rooting ability of the cuttings taken from the mature plants. The juvenility in mature plants can be induced artificially by heading back, spraying GA_3, forcing of sphaeroblasts, by stooling, grafting adult forms on to juvenile form and/or through dis-budding. Some of these practices are very common and followed commercially by the nurserymen throughout the world.

C. Type of the Wood

The propagators have many choices in terms of selecting type of bud for making cuttings, ranging from very succulent terminal shoots (current season's growth) to mature hardwood cuttings. However, this factor is less important in the easy-to-root plants but in difficult-to-root plant species, a proper choice of wood is an important factor because the wrong choice may be quite harmful as it may result in complete failure of rooting of cuttings. In general, the capacity of the cuttings to root, depends upon the type of wood taken by the propagator. Usually, the cuttings taken from the lateral shoots root better than those taken from terminal shoots. It is particularly true with plum, spruce and pines etc. It may be due to the reason that the lateral shoots have more stored food (carbohydrate), which facilitates better rooting in the cuttings. In hardwood cuttings, more roots are develop in the basal portion of the shoot than the cuttings taken from mid or top of the shoot. Cuttings of basal parts root better because of accumulation of carbohydrates and root promoting substances are more in the basal parts. Similarly, formation of some root initials in the basal portions probably under the influence of root promoting substances from bud and leaves may be responsible for it. Further, cuttings taken from a vegetative shoots root better than those taken from flowering shoot, perhaps due to presence of higher level of rooting co-factors (hormones) in the vegetative shoots than in the flowering shoots. Similarly, cuttings with a heel (a small piece of old wood) or mallet (a short segment), usually initiate rooting faster as in quince. Better rooting of cuttings with mallet or heel is probably due to presence of preformed root initials.

D. Presence of Leaves and Buds

There is a considerable evidence to show that the presence of leaves and buds is essential for the rooting of cuttings in many plant species. In most species, the rooting process is inhibited if leaves and buds are removed from the cuttings. The promoting effect of leaves and buds in rooting initiation is due to the fact that these are the primary source of carbohydrates and auxin synthesis and other root promoting co-factors in the plants. These root promoting co-factors are transported by the leaves and buds to the basal portion of the cuttings for root initiation. The effect of buds on rooting may vary with the time of the year, when cuttings are prepared. It also depends upon the species and variety of the species. However, in some cases, the leaves are removed to some extent to reduce the loss of water due to transpiration. Usually, growing buds promote rooting and dormant buds inhibit it. Removal of buds or girdling of the stem below the bud promotes the rooting process during dormant period. Presence of buds in easy-to-root species stimulate rooting but in difficult-to-root species, rooting is inhibited. The promoting effect of buds on rooting during dormant period may be due to the higher auxin and low inhibitor supply to the basal portion of the cuttings.

E. Presence of Viruses

The process of root initiation in cuttings taken from plants infected with viruses is inhibited or reduced as compared to those taken from virus-free stock plants. It is particularly true with apple and Chinese gooseberry, where virus-free clones have been reported to root better than the infected stocks. Viruses not only reduce rooting percentage but also the root number.

F. Position of the Basal cut in the Cutting

Sometimes, the position of the basal cut with reference to the node, has been found to affect root initiation process in the cuttings. In some plants, rooting is better when the cut is made just above or below the node, in others rooting is better if cut is made at the node and still in others, position of the cut has no effect on root initiation process.

G. Time of the Year

In some instances, season of the year, when cuttings are produced, has significant influence on rooting of the cuttings. In deciduous plants, the hardwood cuttings could be made in dormant season and semihard wood or softwood cuttings in the growing season. The evergreen plants usually have one or more flushes of growth in a year and thus cuttings should be prepared at various times in relation to growth flush but more especially in spring or late fall, depending upon the species. However, the response of the rooting of cuttings in different plant species may vary widely. For example, the cuttings of privets root easily when taken at any time of the year, whereas leafy cuttings of olive root better under mist if produced during late-spring and summer but rooting is completely inhibited if taken in mid-winter. Cuttings of pilkhan (*Ficus infectoria*) root best if prepared during spring and summer but rooting is completely suppressed in fall and winter. Softwood cuttings of deciduous plants taken during summer or spring usually root readily than hardwood cuttings procured in winter. For softwood cuttings of deciduous plants, the best rooting is obtained if the cuttings are prepared in the spring, when leaves are fully expanded and shoots have attained some degree of maturity. For best results in broad-leaved evergreen plants, cuttings should be prepared after a flush of growth has been completed and wood is partially matured, particularly during spring to late-fall. Similarly, in narrow-leaved evergreen plants, the results are better if cuttings are prepared during late-fall to late-winter.

H. Treatment of Cuttings

Various treatments have been standardized to treat cuttings before planting for root initiation. These are :

***(a)* Use of growth regulators :** Treatment of cuttings for ease in rooting is an old horticultural practice. In the early historic days, people used to treat cuttings with manganese, boron and phosphorus but better results were used to obtain with potassium permagnate. However, the discovery of auxins in 1934 for their stimulative effects on root initiation was a major milestone in the history of plant propagation. Among growth regulators, indole butyric acid (IBA) has been found to be the most ideal for PGR promoting rooting in cuttings in most of the plant species. The concentration varies from plant-to-plant and type of the cuttings used. Naphthalene acetic acid (NAA) and 2, 4-D(2, 4-dichlorophenoxy acetic acid) are the other root promoting auxins. Their concentrations and degree of success also depend upon species and type of the cuttings used. However, 2, 4-D is potent weed killer and it may inhibit shoot development in certain species. Thus, it should not be used frequently for obtaining beneficial effects on root initiation unless very necessary. These compounds are usually insoluble in water and can be dissolved in a few drops of alcohol or ammonium hydroxide, before adding water. For general use in rooting of stem cuttings in most plant species, IBA or sometimes NAA is recommended. Plant growth regulators may loose their strength as these may be destroyed by bacteria. For example, *Acetobacter* bacteria destroys IAA but not IBA, NAA or 2, 4-D. Further, IAA is highly sensitive to light and is destroyed by strong sunlight but NAA and 2, 4-D are stable. Thus, to have desirable effects, the solutions of growth hormones should always be prepared fresh and under the advice of a technical expert. In propagation with leaf or root cuttings, production of buds and shoots is necessary. For this, treatment of cuttings with cytokinin, such as, kinetin or 6-benzyl adenine (BA) is useful.

(*b*) Use mineral nutrients : Different mineral nutrients may also have significant influence on rooting process of cuttings of some plant species. Treatment of cuttings with nitrogenous compounds (organic and inorganic) usually promotes rooting in cuttings in several plants. Among different organic forms of nitrogen, aspergine and adenine are most effective in doing so. Boron also plays important role in rooting process of certain plant species. It promotes root growth rather than root initiation. The combination of nutrients (N, B) with auxins (IBA) is the most effective treatment for root initiation and development process in many plant species. Concentrations, of course, varies from plant-to-plant and type of the cutting to be used.

(*c*) Use of fungicides : In some instances, the rooting of cuttings initiates at a better rate but their survival is low because these are attacked by pathogens. Thus, treatment of the cuttings before planting, with fungicides like captan and benomyl, gives better results. The fungicides may be used as a powder dip, following IBA treatment or the fungicides may be mixed with IBA powder before treatment. Out of these two fungicides, captan in more suitable for treatment of cuttings because it does not decompose easily and has long residual effect.

(*d*) Wounding of stem : Basal wounding of the cuttings, in some instances, is helpful for inducing rooting in cuttings. Wounding is helpful in cuttings having older wood at the base. Wounding promotes rooting in cuttings in several ways, *i.e.*, (*i*) it tears sclernchyma rings of tough cells in the cortex, exterior to the point of origin, (*ii*) it helps in better absorption of moisture and growth substances from the rooting medium, (*iii*) due to higher excretion of hormones in the wounded area, respiration rate is increased, and (*iv*) division of cells in the wounded area and adjacent cells is faster, and (*v*) wounded cells release ethylene, which act as root promoting hormone. All these conditions induced by wounding help in better root initiation process in the stem.

***(e)* Etiolation treatment :** Etiolation is the oldest horticultural practice in which light is excluded from the plant or its part for sometime during the growing season. Sachs, a renowned scientist was the first to demonstrate the effect of light exclusion on adventitious root initiation during 1864, which was later, confirmed by many other research workers. In the actual practice of etiolation, the basal portion of the newly emerged shoots is kept under complete darkness by covering them with a black polyethylene cover and the terminal portion is allowed to grow in normal light conditions. After sufficient time, which varies from species-to-species, the shoot is cut down, its basal portion is treated with auxins (IBA) and cuttings so prepared are planted in the nursery.

The trench and mound layering and stooling methods of propagation in which basal portion of the shoots are kept in darkness by soil cover, are based on the principle of etiolation. Experiments have shown that etiolation is useful in inducing better rooting in cuttings of difficult-to-plant species like mango, jackfruit and avocado.

***(f)* Girdling :** In girdling, a ring of bark, preferably 2.5 to 3.0 cm is removed from the base of the shoot, which is to be used for the preparation of cuttings. It can also be done by making a notch or tying a wire around the shoot. This process helps to initiate the rooting process much early in some plant species. Girdling basically blocks the downward movement of carbohydrates, hormones and other root promoting factors. When to girdle and from where to girdle, differs widely among different plant species. However, cuttings of mango, litchi, guava and citrus root better if the shoots are girdled about 10-15 days before preparing the cuttings from them.

I. Environmental Conditions

Environmental conditions like availability of water, temperature and rooting medium, may also affect the root initiation process in the cuttings. Some of the important environmental factors affecting rooting process are briefly discussed below :

***(a)* Water conditions** : Presence of leaves on the cuttings is desirable for rooting in them as these supply carbohydrates and growth hormones for better rooting of cuttings. But the loss of water through leaves may be equally harmful because for better rooting, particularly in leafy cuttings, turgidity of leaves should be maintained. Similarly, there should be sufficient moisture in the medium, and if rooting medium is deficient in water and the leaves continue to transpire, it may result in loss of turgidity and the water potential may come down. All these conditions inhibit rooting process adversely and further development of the roots. Therefore, it advisable to remove some leaves in the leafy cuttings before planting them in the rooting medium to maintain proper turgidity in the cuttings and to keep the cuttings alive until root formation. Similarly, the loss of water from the leaves should be reduced by placing the cuttings in covered propagation frames having automatic humidifiers and ventilation system, laying a thin layer of polyethylene over the bed of leaf cuttings by using mist propagation units.

***(b)* Temperature** : A day temperature of 21-24°C and night temperature of 13-15°C are considered quite satisfactory for cuttings in most plant species. Very high or very low temperature may inhibit the root initiation process.

(*c*) **Light :** For proper growth and development, light is the major source of energy, as it is mainly required for carbohydrate synthesis during photosynthesis. In rooting process of cuttings, the formation of carbohydrates is most important. Effects of light on rooting of cuttings may be due to its intensity (irradiance), day-length (photoperiod) or light quality. Usually, the rooting is better in cuttings taken from stock plants grown at a low light intensity than those obtained from plants grown at a higher light intensity. It has been demonstrated experimentally that the etiolated shoots of mango, apple, hibiscus, jackfruit, and avocado root better than the non-etiolated shoots. Under etiolated conditions, the level of certain growth inhibitors is reduced, which then facilitates the rooting process. The photoperiod under which the stock plant is grown, may also exert a dramatic influence on the rooting of cuttings taken from them. The rooting of cuttings is better under photoperiods promoting carbohydrate synthesis. Some plants manufacture better carbohydrates under short-day conditions and others at long-day or neutral-day conditions, and accordingly, the process of rooting in the cuttings is influenced widely. The orange red light of the spectrum seems to favour rooting of cuttings than the blue region. Further, red light (680 nm) has more inhibitory influence on rooting than the blue or far red light spectrum.

(*d*) **Rooting medium :** An ideal rooting medium must provide sufficient porosity to allow free passage of air, should have better water holding capacity, yet well drained and must be free from harmful pathogens. It has to perform three important functions *viz.*

1. It has to hold the cuttings properly.
2. It has to provide adequate moisture to the cuttings.
3. It has to permit free air passage to the base of the cuttings.

Thus, type of the medium may also significantly influence rooting of cuttings. Some cuttings when rooted in sand, produce long, unbranched and brittle roots but produce well branched, slender and flexible roots in perlite or peat mixtures. In addition to the texture and structure. the pH of the rooting medium above 7 may be harmful for root initiation process in cuttings of the most of plant species. Generally, a pH near 7 is considered ideal for rooting process in cuttings. Further, the level of exchangeable calcium in the rooting medium, especially peat moss, should not be too high as it may also effect rooting process adversely. Similarly, the availability of proper concentration of oxygen (1-10 ppm) in the medium is essential for root initiation process.

• • •

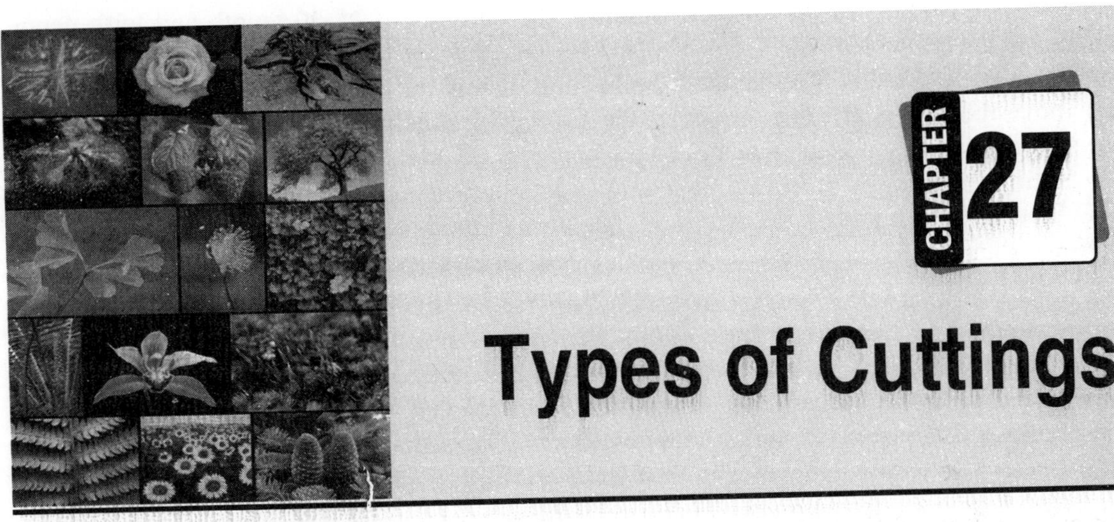

Types of Cuttings

Cuttings are prepared from vegetative portion of plants such as stem, leaf, root and are classified into various groups according to the plant part used. Different types of cuttings used by plant propagators and nurserymen are as described below :

A. STEM CUTTINGS

Next to seed, propagation through stem cuttings is the most convenient and popular method of plant propagation. Most cuttings fall into this category. A stem cutting is any cutting taken from the main shoot of a plant or any side shoot growing from the same plant or stem. The facility with which different species can be propagated in this way is largely determined by experience of a propagator. There are, however, a few general considerations, which help in selection of suitable cutting. First of all, it is essential for the cuttings to have a sufficient reserve food to keep tissues alive until roots and shoots are produced. The shoots with high carbohydrate content usually root better. To maintain high carbohydrate content in a shoot, ringing or notching stems down to the wood are useful practices. As a general rule, cuttings from young plants root better, but if older shoots of the plants are cut back hard, very often they can be induced to produce suitable shoots for rooting. Broadly, there are four types of stem cuttings, namely hardwood, softwood, semi-hardwood and herbaceous cuttings, which are used for multiplication of plants as discussed hereunder :

Hardwood Cuttings

Hardwood cuttings are made from the mature and lignified stems of shrubs and trees. It is the simplest method of plant propagation. Such cuttings are easy to secure and can be easily handled, stored and transplanted. Hardwood cuttings are prepared during dormant season, usually from one-year-old immature shoots of previous season's growth. Only healthy shoots are selected and the weak, fast growing shoots with long internodes should be avoided. The length of the cuttings varies from 10 to 45 cm in length and 0.5 to 2.5 cm in diameter, depending on the species. Usually, the cuttings of 25-30 length and with pencil thickness are preferred. Each cutting should have atleast two buds. While preparing the cutting, a straight cut is given at the base of shoot below the node while a slanting cut, 1 to 2 cm above the bud, is given at the top of the cutting. This helps in maintaining the polarity of the shoot and if rain occurs, water does not accumulate on the tip of the cutting, which saves the cutting from fungal infection. It is often advantageous to take hardwood cuttings with a heel, that is, with a piece of older wood, attached to the base of a cutting. Mallet hardwood cuttings should also be preferred. Similarly, stem cuttings will more often root better if their bases are etiolated, notched, girdled or ringed before being removed from the parent plant.

A number of deciduous fruit plants like are grape, hazelnut, chestnut, fig, quince, pomegranate, mulberry, plum, olive, tamarix, gooseberry, and apple etc., are commercially propagated by hardwood cuttings. Some broad-leaved evergreen plants like olive, deciduous ornamentals like privet, wisteria, honeysuckle, rose rootstock, *Rosa multiflora* and certain tree species like willow, poplar are also propagated through hardwood cuttings.

Semi-hardwood Cuttings

The semi-hardwood cuttings are prepared from partially matured, slightly woody shoots. These are succulent and tender in nature and are usually prepared from growing wood of current season's growth. Usually those shoots, which snap clean when broken, are considered ideal for preparation of semi-hardwood cuttings. These types of cuttings are mostly used in evergreen fruit plants like mango, guava, lemon, jackfruit and some ornamental shrubs (camellia, rhododendron etc.). The length of the cuttings varies from 7 to 20 cm. The cuttings are prepared by trimming the cuttings with a straight cut below a node and removing a few lower leaves. However, it is better to retain two to four leaves on the cuttings. Treating the cuttings with IBA (5,000 ppm) before planting gives better results. Similarly, while planting, one quarter of their length should be inserted in the soil. The best time for taking such cuttings is summer, when new shoots have emerged and their wood is partially matured.

Softwood Cuttings

Softwood cuttings is the name given to any cuttings prepared from soft, succulent and non-lignified shoots, which have not become hard or woody. Usually, the cutting size is 5.0 to 7.0 cm, but it varies from species-to-species (Figure 27.1). Usually, some leaves should be retained with this type of cuttings. Before planting, treatment with auxins (IBA) is beneficial. The best time for preparing softwood cuttings is late-summer. The plants like juniper, azalea and lilac are easily propagated by softwood cuttings.

Herbaceous Cuttings

The terminal leafy portion of the stem of a plant is used for preparing herbaceous stem cuttings. These cuttings are soft and succulent with length varying from 7-15 cm. These cuttings are rooted under the same conditions as the softwood cuttings, but require high relative humidity than softwood cuttings. Bottom heat is also useful for initiation of rooting process in such cuttings. Under proper conditions, the rooting is very high. However, these are liable to wilt soon if proper humidity is not maintained. Therefore, much attention is required for working with herbaceous cuttings. Herbaceous cuttings of some plants exude a sticky sap (as in geranium, pineapple, cactus etc.), that interferes with root initiation process. In such cases, basal ends of cuttings should be allowed to dry for few hours before planting. Generally, fruit plants are not propagated by herbaceous cutting. However, some flowering plants like chrysanthemum, dahlia, coleus, carnation etc., are propagated through herbaceous cuttings.

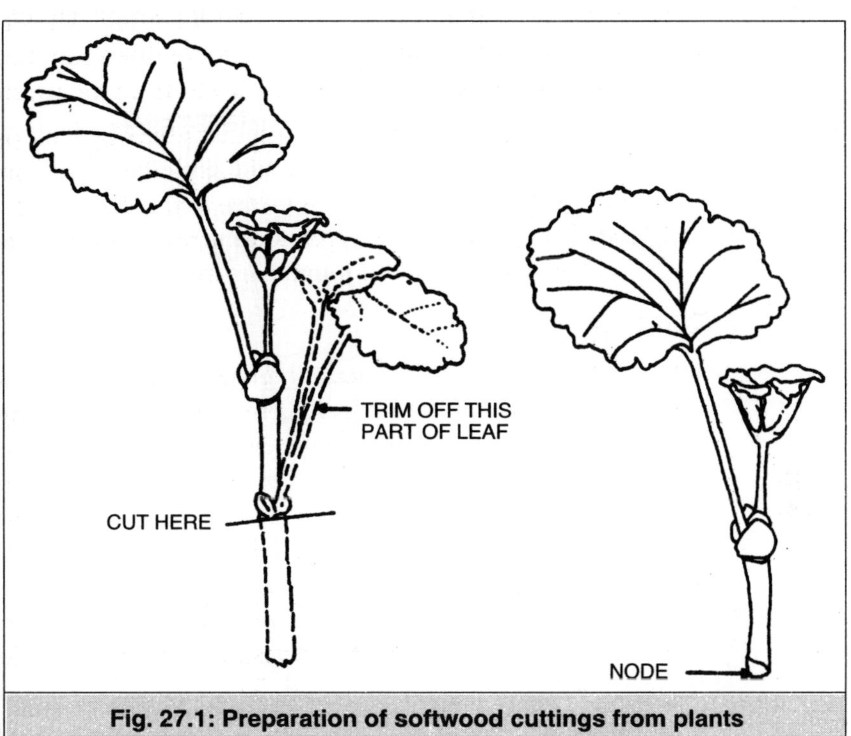

Fig. 27.1: Preparation of softwood cuttings from plants

B. ROOT CUTTINGS

Propagation by means of root cuttings is also a simple and cheap method of vegetative propagation in species, which are difficult-to-propagate with other methods (Figure 27.2). In general, the plants, which produce suckers freely, are easily propagated by root cuttings. For preparation of root cuttings, roots which are of 1 cm thickness and 10-15 cm long are cut into pieces (Figure 27.3). The best time for taking root cuttings is late-winter or early-spring, when the roots are well supplied with stored food material but before the new growth starts. However, in temperate fruits, root cuttings are prepared in the month of December and are kept in warm place in moss grass or wet sand for

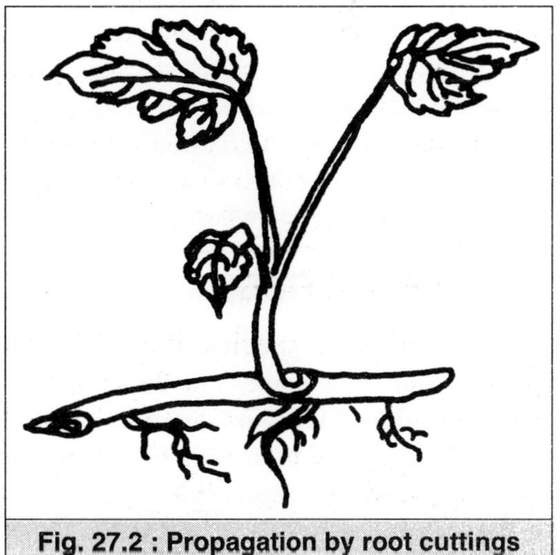

Fig. 27.2 : Propagation by root cuttings

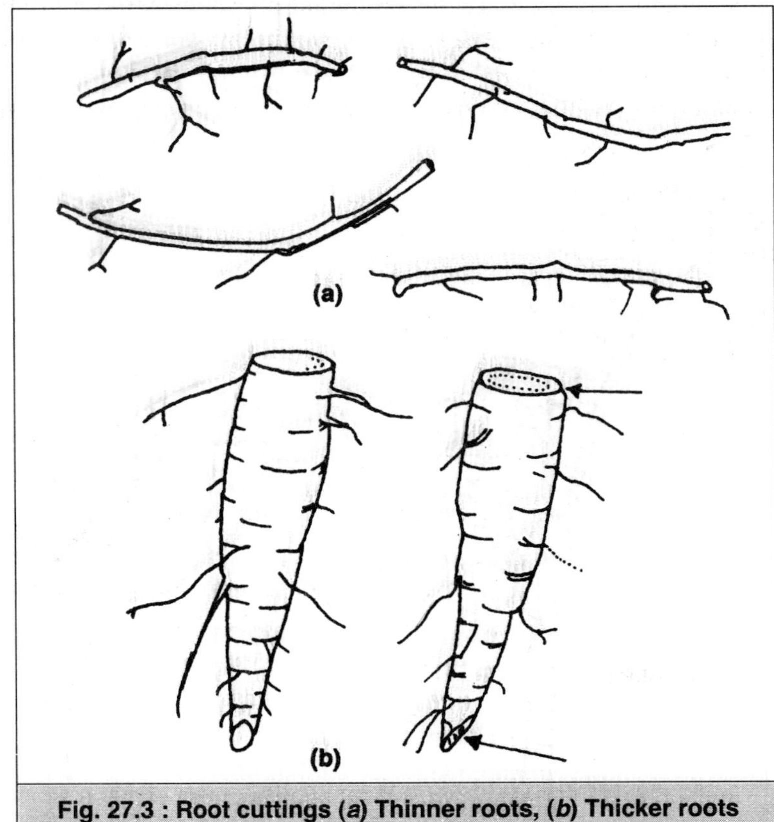

Fig. 27.3 : Root cuttings (*a*) Thinner roots, (*b*) Thicker roots

callusing and then transplanted during February-March in the open beds. While using root cuttings as the propagating material, it is important to maintain the correct polarity. Thus, to avoid their planting upside down, the proximal end of the cutting should have straight cut and distal end a slanting cut. The proximal end should always be kept up, while planting, insert the cuttings vertically so that the top is above the soil level. However, in some cases, horizontal planting gives satisfactory results as in sweet potato. blackberry and raspberry are commercially propagated by this method. However, kiwifruit, breadfruit, fig, rose, mulberry, apple, pear, peach, cherry and persimmon are also propagated by root cuttings. In some variegated foliage plants (*e.g.*, pelargonium, bougainvillea), the new plants more often loose their variegated forms when propagated through root cuttings.

C. LEAF CUTTINGS

In some plant species, the thick and fleshy leaves have the capacity to produce new plants from their leaves. In these species, leaf blade or leaf sections are used for further multiplication. The roots develop on the leaf blade or leaf section. Some greenhouse plants can be propagated by means of various types of leaf cuttings. In leaf cuttings, the formation of root and shoot is essential to get a complete plant. In many cases, only roots develop but not the shoot system (*e.g.*, camellia, rhododendron etc.), as result of which, the cuttings die. Leaf cuttings should be made from fully expanded recently mature leaves and old leaves may be avoided. Leafy cuttings require environmental conditions similar to herbaceous cuttings but for high success with leaf cuttings, relative humidity should be a little high. In some plant species, particularly variegated sansevieria, the new plants don't come true-to-type if propagated by leaf cuttings.

The different types of leaf cuttings used for propagation are leaf petiole cuttings, leaf sections and leaf slash etc. In leaf petiole cutting, complete leaf with its stalk is used for propagation. This is an easy and simple method of plant propagation but plants produced by this method are very few. Saintpaulia and kalanchoe are propagated through leaf petiole cuttings. In leaf sections, a section of the leaf about 1-10 cm long is cut at right angles to the veins and planted in the rooting medium by maintaining the polarity. Sansevieria is mainly propagated through leaf sections. Leaf slashing is commonly used in plants having capacity to regenerate new plants from the leaf veins. In this method, the cut leaf is placed flat on the medium and the leaf is pinned down with a thin wire. The new root and shoot system develops on each cut vein. It is commonly followed in *Begonia rex*.

D. LEAF BUD CUTTINGS

Propagation through leaf bud cuttings is partially useful in species where leaves develop root system but die because of non-development of shoot system. Leaf bud cuttings are particularly useful when planting material is scarce because the each node in leaf can be used as cutting. Leaf bud cutting should preferably be prepared during growing season because if buds enter in dormancy it may be difficult to force them to active stage. A leaf bud cutting consists of a leaf blade, petiole and shoot piece of stem with attached axillary bud of actively growing leaves. In leaf bud cuttings, 1-15 cm stem portion is used when propagating material is small. These cuttings should be made in growing season. It is an useful method of propagation in blackberry, raspberry, lemon, camellia, etc.

E. SPECIAL KINDS OF CUTTING

Some different kinds of cuttings are also used by the propagators and nurserymen, which are described here under :

Basal Cuttings

A basal cutting is a young shoot cut or gently pulled or removed from the base of a parent plant. Generally, such cuttings need little preparation, only the removal of a fewer lower leaves before planting is required. Such cuttings should be taken when they are young. A cutting length of 4-8 cm is the most ideal. These cuttings require some form of protection from drying out while rooting. Many perennials and alpine trees are propagated from basal cuttings.

Heel Cuttings

A heel cutting is usually a side shoot of a hardwood or semi-hardwood cutting pulled from the parent plant with a heel of bark from the main stem. *Jasminum* roots better from such cuttings. These cuttings should be prepared and planted as an ordinary hardwood or semi-hardwood cuttings without doing anything to the heel at the base of cuttings other than trimming the loose ends with a sharp knife before planting.

Bud Cuttings

Taking a bud cutting is similar to preparing a bud or budding, the main difference being that with the cutting, the leaf is retained attached to the slice of the stem. A semi-hardwood cutting should be choosen to supply the bud from the current season's growth in the late-summer. A slice of stem around 2.5 cm in length is cut from the parent plant, each cutting containing a leaf axil with a bud. The cutting is planted in such a way that the bud and leaf are just above the surface of soil. Rose and camellias are among the plants propagated through bud cuttings.

Eye Cuttings

This is a very successful method of propagation of vines. An eye cutting is a hardwood stem cutting taken from the parent plant during the dormant period of autumn or winter. The best cutting is one, which is 4-5 cm long having an eye or a bud. The pruned wood of vines is cut into many sections, each containing a bud. With the sections so cut, a strip of bark is removed from the opposite side of the stem to the bud and the cutting is pegged down in the soil. Lightly pegged down the cutting, sprinkle water and press it so that only the bud protrudes above the compost/sand. A bottom heat of 24°C is required for rooting, making a propagating frame. With this temperature, the compost may dry out rapidly. So, check it periodically to make sure that it retains moisture and if dry, water may be given frequently.

Internodal and Nodal Cuttings

The node is that part of plant where the leaf stalk or petiole joins with the stem. Most cuttings root better if they cut across in a straight line immediately below a node, and are therefore termed as *nodal cuttings*. Nodal cuttings are specially advantageous to those plants having hollow stems, as the node generally leads to a branch of solid tissue throughout such stems. Some plants although appear to even root better if the stem is cut midway between a pair of nodes. These cuttings are therefore termed as *internodal cuttings*. Clematis is one plant propagated through internodal cuttings.

Irishman's Cuttings

An Irishman's cutting is a basal cutting, which is pulled from the crown of a parent plant having a few roots already attached to it. It is perhaps a mild form of division. This type of cutting is quite often used to propagate early flowering chrysanthemums and some herbaceous perennials.

Piping Cuttings

A piping cutting is simply a stem cutting taken from the tip of carnation. It is very easy to take the tip of any non-flowering shoot for cutting. For taking piping cutting, hold the selected stem with one hand and pull the tip carefully with the other. This piping is just dipped into a rooting compound and dibbled into sandy compost for rooting. A cold frame is suitable for proper rooting of these cuttings. Different types of cuttings and their relative position on a plant given in Figure 27.4.

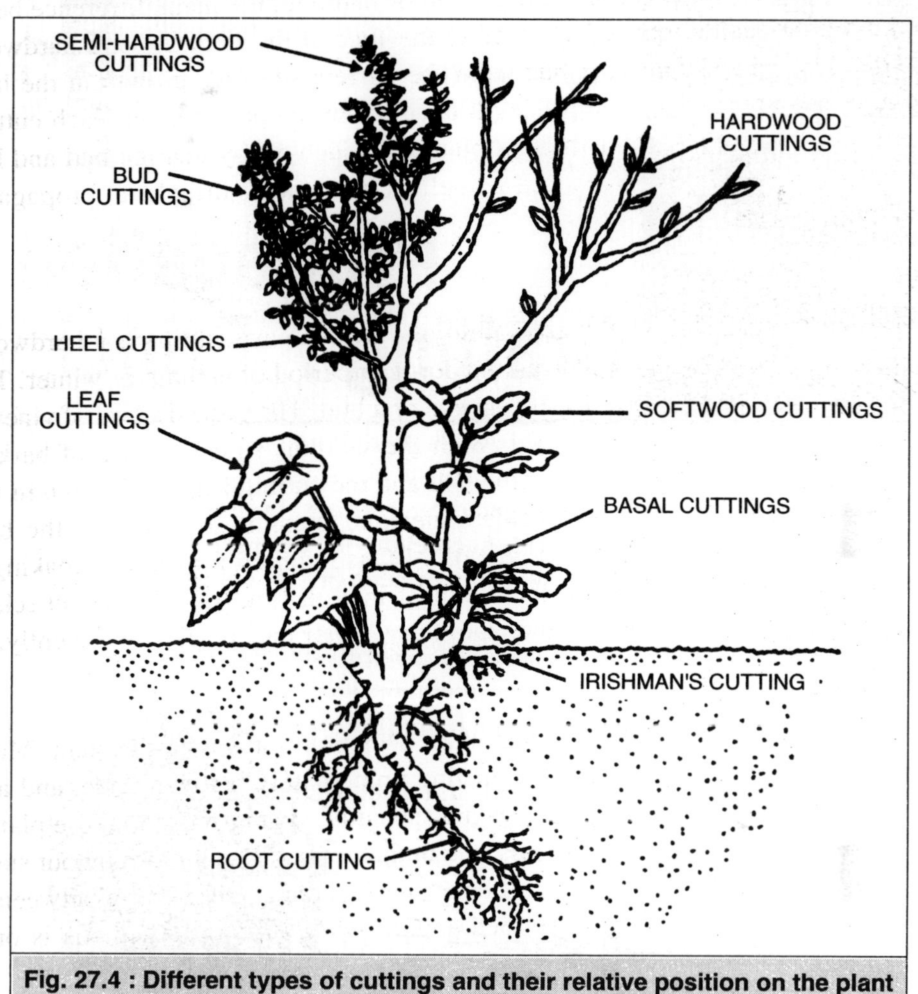

Fig. 27.4 : Different types of cuttings and their relative position on the plant

• • •

Propagation by Grafting : Principles

Many horticultural plants are propagated by grafting. In grafting, the desired cultivar can be raised on other plants (rootstocks) for achieving the desirable benefits. Propagation by grafting also involves some principles, which have been briefly discussed hereunder :

GRAFT UNION

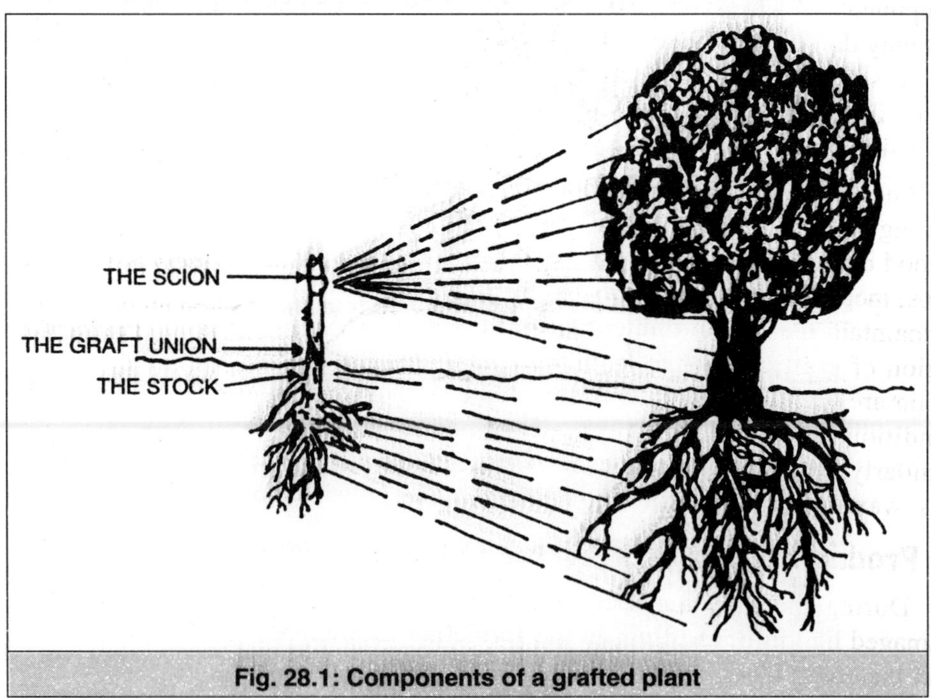

Fig. 28.1: Components of a grafted plant

The graft union is the basis of a successful graftage. In grafting, the stock and the scion are placed in a close contact with each other and held together firmly, until they unite to form a composite plant (Figure 28.1).

The formation of the graft union is basically a process of healing of wounds. In this process, the parts of the graft that are originally prepared and placed in close contact don't themselves move about or grow together but the union is accomplished entirely by the cells that grow and develop after the actual grafting operation has been done. Further, in a graft union, there is no intermingling of the cell contents of both stock and scion and cells maintain their own identity.

FORMATION OF A GRAFT UNION

The principal steps involved in the healing process and formation of the graft union are discussed hereunder :

A. Establishment of a Direct Contact between the Cambial Region of the Stock and Scion

Establishment of a direct contact between the cambial layers of both stock and scion is the first and foremost step in the healing process of the graftage *i.e.,* the cambial layers of stock and scion must be matched closely. It is necessary that the cambial layers of stock and scion be close enough so that the parenchymatous cells of both stock and scion produced in this region can become interlocked. Two badly matched or mismatched cambial layers may prevent the formation of a successful graft union. Thus, it is essential that the two components of a graftage (stock and scion) must be tied firmly with a wrapping material so that they don't dislodge. For this, both stock and scion should be of same thickness, and the cut on stock and scion should be smooth and uniform to avoid any free passage between them. In this way, both stock and scion will match each other firmly and lead to the formation of a successful union.

Environmental factors, especially temperature and relative humidity, also favour the healing process. Usually, a temperature range of 15-30°C is considered ideal for establishing a good cambial contact and for rapid growth of the cambial cells Under greenhouse conditions, there is a rigid control over temperature and relative humidity, however, it is difficult to maintain these environmental conditions in the field. Further, it is essential to keep the region of graft union free from the pathogenic organisms as the thin-walled cells of graft union are highly susceptible to the attack of pathogens under high rainfall and temperature conditions, which are highly detrimental to the successful healing of the graft union. Similarly, it is important that cut surfaces of both stock and scion are not dehydrated. For this, waxing of graft union is essential to prevent desiccation of cells and disease infection.

B. Production and Interlocking of Parenchymatous Cells

During the grafting operation, the living cells of both the stock and the scion are damaged by knife, resulting in the formation of a necrotic plate, which separates the stock and the scion from each other. However, under the dead/damaged cells, some living cells also exist, which initiate the formation of callus from the parenchymatous cells of phloem and xylem tissues. These cells mainly comprise of spongy tissue, which penetrates in the thin necrotic layer and soon fill up the gap between the stock and the scion and become intimately interlocked. This interlocking of the cells provides a mechanical support to the graftage and also allows the passage of nutrients and water between the stock and the scion. Initially, there is a development of a brown line between the callus of stock and scion, but it disappears gradually. Similarly, the old vascular tissues of both the components are sealed off with a deposit of gum, which accelerates the healing process.

C. Production of New Cambial Cells

In the callus bridge of the stock and the scion, there are certain parenchymatous cells, which differentiate into new cambial cells after 2-3 weeks of grafting. The development of these cells result in the formation of continuous cambial connections between stock and scion. This is a good indication of formation of a successful graftage.

D. Formation of New Vascular Tissues

After the formation of continuous connection between the stock and the scion through the callus bridge, new cambial activity results in the formation of new xylem towards inside and phloem outside the original vascular cambium of the stock and the scion. The new xylem tissues generally originate from the activities of the scion tissues rather than from the stock. Thus, the production of new xylem and phloem tissues permits vascular connection between the stock and the scion, which continue throughout the life of a grafted plant.

FACTORS AFFECTING THE HEALING OF THE GRAFT UNION

As stated earlier, he formation of the graft union is basically a process of healing of the wounds. This healing process is influenced by many factors, thereby affecting the success or failure of the graft union. Different factors, which influence the formation of a successful graft union are described briefly hereunder :

Compatibility between the Stock and the Scion

The compatibility between the stock and the scion is the first and foremost pre-requisite for the formation of a successful graft union. More often, compatible stock and scion form perfect union, whereas incompatibility between them results in complete failure of the graft union. It has been demonstrated in many fruit plants that the ability of two kinds of plants to form a successful graft union is related largely to their natural relationship. In general, distantly related plants show incompatibility or very slow healing of graft union and *vice versa*. The fruit plants like apple, pear, mango, sapota etc., can easily be grafted, whereas walnut, hickories, oaks are difficult-to-propagate by grafting. However, sometimes, incompatible partners also make satisfactory union but it eventually fails in the later stages.

Plant Type

The success or failure of grafting also depends upon the kind of plant. Some plants form successful union easily, whereas others don't, although they are compatible. For example, hickories and oaks usually don't form successful graft union initially but once the union is established, they grow very well. Further, top working of apple and pear with any simple technique do well but difficulties appear while top working the stone fruits like peach, plum and apricots. Strangely, top working of peaches with compatible species of plum or almonds is more successful than top working with peaches itself. Similarly, it is difficult to propagate Muscadine grape (*Vitis rotundifolia*) and mango (*Mangifera indica*) with simple budding and grafting techniques, for which approach grafting is often used to have better success. In some plants, as in walnut, desiccation and death of tissues near the graft union takes place readily, resulting in the graft failure. Similarly, some plants produce wound gum, which plugs the xylem elements after grafting operation and prevents the excessive desiccation and death of the tissues. However, exudation of latex or resin may be auto-inhibitory in other (*e.g.*, walnut), which prevents the healing process. Thus, variation among different plant species in their ability to form successful graft union is related to their ability to produce callus, which is a pre-requisite for the formation of a successful graft.

Environmental Conditions

Optimum temperature, moisture and O_2 supply is necessary during and after the grafting operation for the formation and healing of the graft union. These environmental factors are necessary for initiation and production of callus tissue. Optimum conditions usually vary from plant-to-plant. For example in apple, little or no callus is formed below 0°C and above 40°C, the optimum being 24-27°C. For walnut, it is 25-30°C. Similarly, for better healing, high humidity is essential to avoid drying and desiccation of thin walled and tender parenchymatous cells of callus and better growth of the callus, which is important for better healing. Highly turgid cells favour proliferation of callus mass than those in wilt conditions. Thus, it is advisable to maintain 100 per cent relative humidity for abundant callus formation and healing of the graft union. In addition, optimum temperature and supply of O_2 is equally important. Optimum O_2 supply helps in rapid cell division, which is necessary for callus enlargement and healing process. Waxing and white washing of the graft union also help in healing process by protecting the graft union from the infection of pathogens and extremes of heat. Thus, optimum temperature, relative humidity and O_2 during and after the grafting should be maintained for better and early healing of the graft or bud union.

Growth of Stock and Scion

The growth condition of stock and scion during budding or grafting also plays a vital role in the healing process. If plants are in active growth, they would heal soon otherwise they may take some time for healing. For example, the success in bark grafting chiefly depends on the slipping ability of the bark *i.e.*, if the cambium cells are actively dividing, the bark slips easily and if grafting or budding is done at that time, the success would definitely be more. Similarly, the callus proliferation occurs more readily in spring. In contrast, there are certain fruit plants, such as, walnut, maple and grape, which bleed very badly when scion sticks are prepared from them during their active growth periods. It hinders the healing process of their graftage adversely. Although, it can be avoided to some extent by making slanting knife cuts below the graft around the tree. Further, in veneer grafting of mango, the scion-sticks are prepared by forcing (defoliation) them on the mother plant for a week or so to encourage bud break. Keeping the active growth of plants in mind, spring and monsoon seasons are considered best for grafting in India. Thus, the growth activity of both stock and scion has strong influence on the success or failure of a graft or bud union.

Grafting Method

Budding or grafting technique also influences the healing of the graft or bud union. The grafting methods vary in different plant species. A highly successful method of grafting for one plant species may be a completely failure for the other species. For example, stone grafting is a highly successful propagation method for mango in *Konkan* region but not in north India. Inarching is the best propagation method of mango in south India but veneer grafting or softwood grafting is the most successful in north India. Thus, a particular propagation technique may be successful for a particular crop of a particular locality because successful union takes place with that method in that particular locality. Further, errors in grafting or budding technique, such as, poor or delayed waxing, loose tape, large or uneven cuts or desiccation of the scion etc. can, also help the failure or the success of a graft union.

Incidence of Insect-pests and Diseases

Infection of the scion sticks by viruses, insect-pests (*e.g.*, borer) and diseases may result in poor healing of the graft or bud union. Thus, it is advisable not to take scionwood from the infected stock. Similarly, if any of them appears during or after the grafting operation, these should be controlled adequately.

Growth Substances

Exogenous application of growth substances, particularly auxins and kinetin, although don't promote healing process in the graft or bud union but these compounds are considered beneficial in callus formation in micropropagation.

Polarity

Maintenance of proper polarity is important for proper healing and formation of a successful graft union. In grafting, the proximal end of the scion should be inserted into the distal end of the stock (Figure 28.2). However, in root grafting, the proximal end of the scion

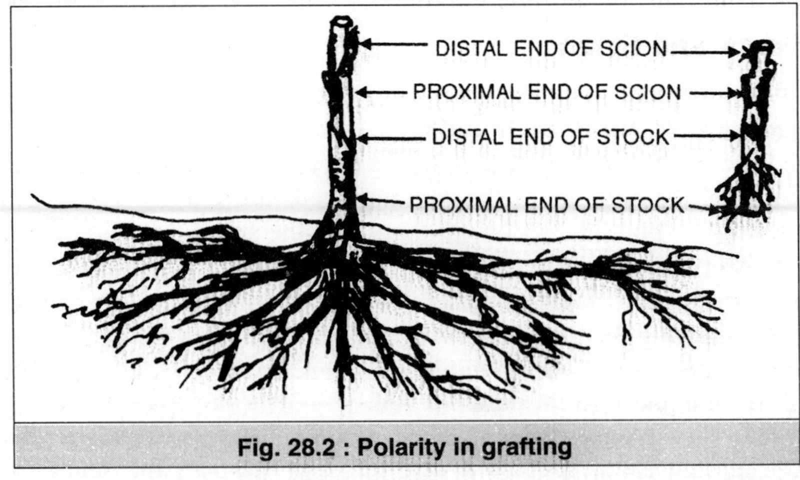

Fig. 28.2 : Polarity in grafting

should be inserted into the proximal end of the root pieces. If the order is reversed *i.e.*, if the scion is inserted upside down, the union may stay alive for some time, but will not grow afterwards. Similarly, in nurse-root grafting, the rootstock is knowingly grafted on to scion with reverse polarity.

LIMITS OF GRAFTING

As a general rule, grafting is more successful in plant species, which are more closely related to each other. Moreover, there are always better chances of formation of a successful union if both the partners (stock and scion) have the better capacity of producing parenchymatous cells near the cambial layers. Thus, grafting is generally confined to dicotyledogenous plants and to some extent in gymnosperms because they have a vascular cambium layer between the xylem and the phloem. Monocotyledenous plants don't have the cambium tissues and thus, grafting is not successful in them, although, some success has been achieved in some grasses and in tropical orchid, *Vanilla*. Thus, before starting any grafting operation, it may be made clear in mind that the plants to be used in grafting are able to form a successful union or not. However, it is very difficult to predict whether the proposed pair of the stock and the scion is going to form a successful graft union, but it is thought that more closely the plants are related botanically, better are the chances for the graft union to be successful. Although, there are many exceptions to this rule.

Grafting within a Clone

A scion can be grafted on the plant from which it has been taken. Similarly, a scion from a plant of a clone can be grafted on any other plant of the same clone. For example, a scion prepared from a 'Dashehari' mango tree can be grafted successfully on the same 'Dashehari tree or any other 'Dashehari' tree located anywhere.

Grafting between the Clones within a Species

The clones within a species can be successfully inter-grafted in most of fruits and nut crops. However, due to incompatibility problems, clones of same species in some conifers can't be grafted.

Grafting between Species of a Genus

Inter-grafting between species of the same genus is successful in some but unsuccessful in the others. For example, grafting between most species of citrus is highly successful and used commercially. The almond, apricot, Japanese and everbearing plum are grafted commercially on peach, a different species. However, almond and apricot both in the same genus, can't be inter-grafted successfully. Further, the Beauty cultivar of Japanese plum (*Prunus salicina*) makes a successful union when grafted on almond but 'Santa Rosa' does not. Similarly, the Japanese plum cultivars can successfully be grafted on European plum cultivars but the reciprocal grafts are not successful. Thus, the compatibility between the species in the same genus depends on the particular cultivar combination of stock and scion.

Grafting between Genera within a Family

Grafting success in plant species of a family with different genera is almost negligible, although some exceptions are there. For example, Trifoliate orange (*Poncirus trifoliata*) is used commercially as a dwarfing rootstock for orange (*Citrus sinensis*) and other citrus species. Similarly, the quince (*Cydonia oblonga*), a deciduous plant is used as a dwarfing rootstock for pear (*Pyrus communis*) and loquat (*Eriobotrya japonica*), a deciduous and evergreen plant, respectively.

Grafting between Families

It is usually considered impossible to make successful grafting in plants of different families, although some instances have been reported in short-lived herbaceous plants. For example, successful grafts have been obtained with white clove (Leguminosae) as scion and sunflower (Compositae) as stock. However, there is no such report of grafting in woody perennials belonging to different families.

• • •

CHAPTER 29

Graft Incompatibility

The inability or failure of two different plants grafted together to produce a successful graft union is called as *graft incompatibility*. In contrast, if the graft union is successful, it is called as *graft compatibility*. The ability of two plants to unite successfully into a graft is due to their natural relationship. For example, stock and scion of closely related species unite readily and grow as a composite plant and that of unrelated plants do not. Even sometimes stock and scion of unrelated species unite initially and develops symptoms of incompatibility later and eventually die. Thus, plants differ widely in their ability to produce a successful graft union. For example, some pear cultivars are successfully grafted on quince rootstock, whereas, the others may die soon. However, the reverse combination *i.e.,* the quince on pear rootstock is always a failure. Further, plum grows well on peach rootstock but peach graftage on plum is always a failure. The incompatibility symptoms may appear soon after grafting or may be delayed for several years.

SYMPTOMS OF INCOMPATIBILITY

The following symptoms are known to be associated with graft incompatibility :

- Complete failure to form a successful graft or bud union by some species, varieties or clones.
- Very low percentage of graft success.
- Union takes place and initial growth also occurs but the tree dies pre-maturily either in the nursery or in the field.
- Degeneration of the tissues at the graft union, decline in vegetative growth and premature defoliation.
- Marked differences in the growth or vigour of stock and scion.
- Appearance of deficiency symptoms or nutritional disorders.

- The plants may have stunted growth.
- There may be yellowing of foliage in later part of the season.
- Appearance of outgrowth at, above and below the graft union (Figure 29.1).
- Excessive swelling at the graft joint, and
- Graft components 'break off' cleanly at the graft union.

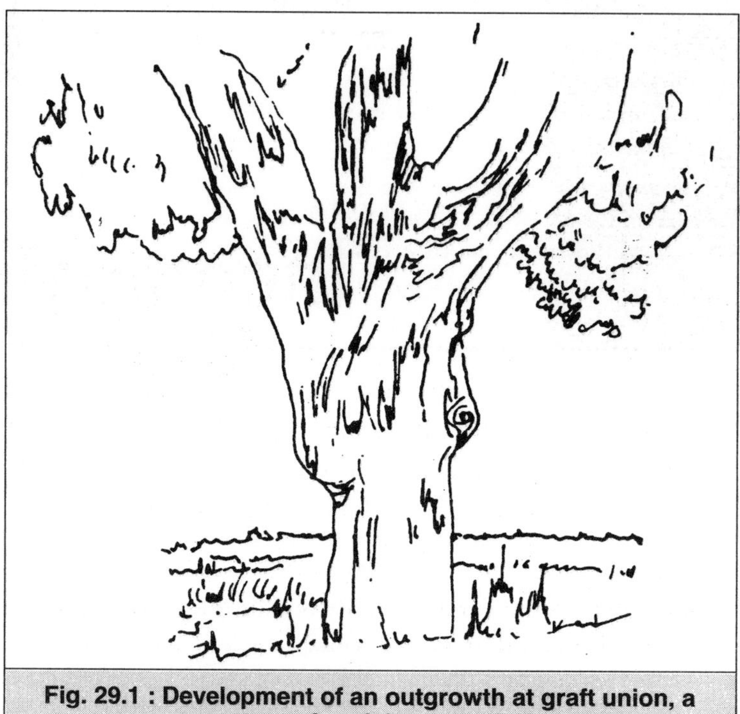

Fig. 29.1 : Development of an outgrowth at graft union, a symptom of graft incompatibility

However, one or combinations of these symptoms does not mean that the combination is incompatible because the above symptoms may also result from nutritional deficiency, insect-pest and disease incidence or poor budding or grafting technique. Similarly, swelling or outgrowth at, below or above the graft union is not sometimes considered to be associated with incompatibility because some compatible forms also develop such symptoms. For example, 'Kinnow' mandarin budded on 'Troyer' citrange usually develops an outgrowth above the bud union. Similarly, 'Allahabad Safeda' scion of guava also develops such swellings above the graft union on dwarfing 'Pusa Srijan' rootstock. Thus, the symptoms of outgrowth at or above the graft union are not a reliable indication of incompatibility reaction. However, incompatibility is clearly indicated if the plant 'breaks off' clearly at the joint after a few months or years of satisfactory growth and the break is smooth and clean.

TYPES OF GRAFT INCOMPATIBILITY

Graft incompatibility is of two types: (1) *Localized incompatibility,* and (2) *Translocated incompatibility.*

Localized Incompatibility

This type of incompatibility includes combinations in which the incompatibility reactions apparently depend upon the actual contact between stock and scion. In this incompatible combination, the union structure is oftenly weak resulting in the discontinuity of cambium and vascular tissues of stock and scion. It may result in poor translocation of the metabolites across the graft union. The external symptoms appear slowly depending upon the degree of anatomical differences at the graft union. Usually, masses of undifferentiated parenchymatous tissues are commonly found at the graft union in this type of incompatibility or even inclusion of bark tissue may develop, which may disrupt the normal vascular connection between stock and scion. Localized incompatibility reactions are quite common in apple when grafted on pear or when plum is grafted on cherry. This type of incompatibility can be overcome by an insertion of a mutually compatible interstock between stock and scion. An example of this category of incompatible reaction is that of 'Bartlett' pear grafted on 'Quince' rootstock. With the insertion of a compatible interstock 'Old Home' pear, the combination becomes compatible and satisfactory tree growth takes place.

Translocated Incompatibility

Sometimes, the incompatibility is not overcome by an insertion of a mutually compatible interstock, because apparently some labile influence can move across stock and scion. In this type, usually phloem degeneration takes place, resulting in the formation of a brown line and a necrotic area in the bark. It imposes restriction on the movement of carbohydrates through the graft union, with the accumulation above and reduction below. Most important example of this category of incompatibility is that of 'Hale's Early' peach grafted on 'Myrobalan B'plum rootstock. This forms a week union in which distortion of tissues takes place with the accumulation of starch at the base of peach. If a mutually compatible interstock, 'Brompton' plum is inserted between these two, the incompatibility still exists, with the accumulation of starch at the base of 'Brompton' interstock. But, if the same combinations are tried at cotyledonary stage, these are highly compatible, indicating that some factors responsible for incompatibility reactions are not present at a juvenile or a nursery stage of the plants. Similarly, the combination of 'Non Pareil' almond on 'Mariana 2624' plum shows complete phloem breakdown, although the xylem tissues are normal. However, the Taxas' another almond is highly compatible on 'Mariana 2624'. However, if a piece of 'Taxas', almond is inserted as an interstock between 'Non Pareil' and 'Mariana 2624', bark disintegration occurs, resulting in incompatible graft union. The compatibility reactions change with the time also. For example, the pear cultivar 'Bristol Cross' when grafted on 'Quince' in 1932 made good growth but required interstock 'Beurre Hardy' 30 years later to form an acceptable union. Such changes in plants may result from mutations or from latent viruses.

CAUSES OF GRAFT INCOMPATIBILITY

In general, incompatibility is the result of genetic difference between stock and scion. However, the mechanism by which a particular case is expressed is not fully known till now. Many proposals or attempts have been made to explain the causes of incompatibility but the evidence supporting them are inadequate and conflicting. However, of the possible causes, structural, physiological, biochemical, incidence of diseases, insects or combination of all these factors may be responsible for incompatible reactions. Such possible causes or reasons have been discussed briefly hereunder :

(a) Structural or Anatomical Reasons

In histological studies, it has been found that stock and scion may not differ structurally but abnormalities may be found due to the presence or development of parenchymatous cells at the graft union, preventing the formation of vascular continuity between stock and scion. Sometimes, a bark layer develops at joint, which may result in the graft failure. Distortion of vascular tissues between stock and scion also takes place due to the development of some whorls or loops, which restrict the movement of essential nutrients and water across the graft union, resulting in poor growth or failure of the graft union. Similarly, from microspectrographic examination of the cell walls of incompatible graft union, it has been found that the cell wall in the adjoining cells are lacking in lignin, whereas the cells in the compatible forms were lignified. Thus, any reaction inhibiting the formation of lignin and development of middle lamella between stock and scion results in a weak union. The incompatibility reactions in plum, pear and peach are primarily due to structural abnormalities. Usually, the development of undifferentiated tissues at the graft union causes mechanical weakness, resulting in phloem disintegration and uneven or poor growth of the graftage.

(b) Physiological and Biochemical Reasons

Sometimes, there is insufficient supply of necessary components either by stock or scion, which may cause incompatibility in stock and scion. Usually, in an incompatible unions, the water supply is impeded. It may be due to failure of phloem or xylem at the union, resulting in root starvation and subsequent wilting and death of the scion. In compatible unions, translocation of sugars from scion-to-stock is normal. In many cases, there is an accumulation of assimilates in the scion of incompatible combination. For example, when certain pear cultivars are grafted on to Quince, a cyanogenic compound (prunasin) found in Quince (not in pear) is translocated to pear phloem. The pear tissues breakdown prunasin with hydrocyanic acid as one of the products. This acid accumulates near the graft union, resulting in tissue breakdown and lack of cambial activity and pronounced anatomical disturbances in the phloem and xylem of the union. As a result, conduction of water and other materials is seriously hampered. Thus, presence of a toxic chemical may inhibit the growth of other or may kill it altogether. It is a world known fact that dwarfing effects of rootstock on scion cultivar are mainly due to the restricted supply of water and nutrients to scion cultivar. It is a common observation that there is a development of an outgrowth at or above the graft union, if budded or grafted on-to a dwarfing rootstock. This out-growth is considered to restrict the supply of nutrients and water to the plant top, resulting in dwarf stature of scion cultivar.

(c) Nutritional Deficiency

Deficiency of certain nutrients may also result in an incompatible union. It has been found that when 'Jonathan' apple is grafted on to 'EM-IX' rootstock, the scion develops molybdenum (Mo) deficiency. It may be due to the inability of 'EM-IX' rootstock to absorb 'Mo' in sufficient quantity to supply to scion cultivar. However, 'Johathan' apple on other rootstocks does not show such deficiency symptoms. Similarly, deficiency of P, K and Mg in peach has been reported if grafted on to 'Myrobalan B' plum rootstock, resulting in incompatibility symptoms.

(*d*) Presence of Viruses

Presence of latent viruses and mycoplasma like pathogens may result in failure of the graft union. Pear decline disease on 'Bartlett' trees grafted on *Pyrus pyrifolia* rootstock is quite common due to the presence of viruses at the graft union. However, it never appears if *Pyrus communis* is used as a rootstock. Further, the incompatibility reactions in citrus are primarily due to infection by viral diseases like tristeza, psorosis and xyloporosis etc. The failure of rough lemon rootstock for many sweet orange varieties in Punjab (India) has been due to the occurrence of bud union crease, a viral disease. Similarly, black-line of walnuts, a delayed incompatibility reaction, is believed to be due to a virus and not due to a rootstock failure.

• • •

Methods of Grafting

Grafting is an art and science in which two living parts of different plants are joined together so that they unite together and subsequently grow into a composite plant. Usually, graft has two parts, the scion and stock. Scion refers to that part of a graft combination that becomes the top of the plant, while stock (rootstock) is the lower portion of the graft combination, which becomes the root system of the composite plant. Grafting has certain advantages and disadvantages over other methods of vegetative propagation. These are as under :

ADVANTAGES OF GRAFTING

Grafting has following advantages over the other methods of propagation :

- It is a better method of propagation for plants, which are not easily propagated through cuttings and layering.
- Grafted plants are precocious in bearing *i.e.,* they come in bearing much early than seedling plants.
- The desirable benefits of a rootstock are best utilized through grafting.
- Old or declining orchards of a inferior variety can be rejuvenated with superior cultivars by top working, using different grafting techniques.
- A number of varieties can be grafted on one plant by grafting, which creates novelity.
- The damaged plant parts can be repaired easily by grafting.
- Grafting is helpful in supplementing pollination of some dioecious plants and overcoming self-incompatibility in temperate fruit plants like apple, pear etc.
- Viral diseases are studied through indexing of plants by grafting on a rootstock.
- Special forms of plant growth such as 'tree' roses or 'weeping' cherries etc., can only be obtained by grafting.

DISADVANTAGES OF GRAFTING

Grafting has the following disadvantages as well :

- All the plant combinations are not compatible and thus it is not possible to use grafting for their propagation.
- Grafting techniques are successful only in dicots and gymnosperms.
- Monocots do not respond to grafting, primarily because of absence of cambium layer.

The success of grafting primarily depends on the successful union of stock and scion. However, there is a great variability in different plant species in their ability for forming a successful graft union. Some plants can be grafted easily, while the others form such union with a great difficulty. There are five general requirements for any successful grafting operation. These are :

- The stock and scion must be compatible. Usually, closely related plants form graft union easily than distantly related plants.
- The cambial region (exposed portion) of the scion must be placed in intimate contact with stock. The cut surfaces of both stock and scion must be matched properly and held together tightly with the help of a suitable wrapping or tying material.
- The grafting operation should be done at the right time of the season when both stock and scion are at proper stage of their maturity and growth.
- After grafting operation, the cut surfaces should be covered with a tape or wax to avoid desiccation.
- The grafts should be protected from the wagaries of weather and attack of insect-pests and diseases. Similarly, 'off shoots' appearing below the graft union should be removed or 'rubbed off' as they appear.

METHODS OF GRAFTING

A. Attached Methods of Grafting

In approach or attached method of grafting, the scion is attached with the stock, when both the components are on their own root system. After the union takes place, the top of the stock above the graft union and the base of scion plant below the graft union are removed. The stock and scion are 'looped off' gradually rather than at once. In this method, seedling plant growing in a pot or in a container is placed under/near to the established plant, which would be the scion part of the grafted plant. Such methods are particularly useful in plants, which are difficult to be propagated by other methods. The various forms of approach grafting are described below :

Inarching : It is a synonym of approach grafting in the sense that both stock and scion plants are on their own roots at the time of grafting. However, it differs from approach grafting in that the top of the rootstock plant does not extend above the point of graft union as in approach grafting. It is generally used for repairing or replacing damaged root system and hence also called as *repair grafting*. Selection of parent tree for taking the scion is an important factor for its success. The scion plant should be healthy, vigorous and high yielding. Old plants selected as scion usually thrive for a shorter period. For the actual inarching process, the stock is brought close to the scion (Figure 30.1) and then a thin slice of bark (about 6 to 8 cm in length and) about 1/3 inch in thickness) at about 20 cm above the ground level is removed from the stock with a sharp knife. A similar cut is made in the scion. Thus, the cambium layers of both stock and scion are exposed. These cuts are brought together and tied firmly with the help of an alkathene strip. After the successful union, stock above and scion below the graft union are 'looped off' gradually. In low rainfall areas, it should be done with the onset of rains, while in regions of heavy rainfall, it should be done

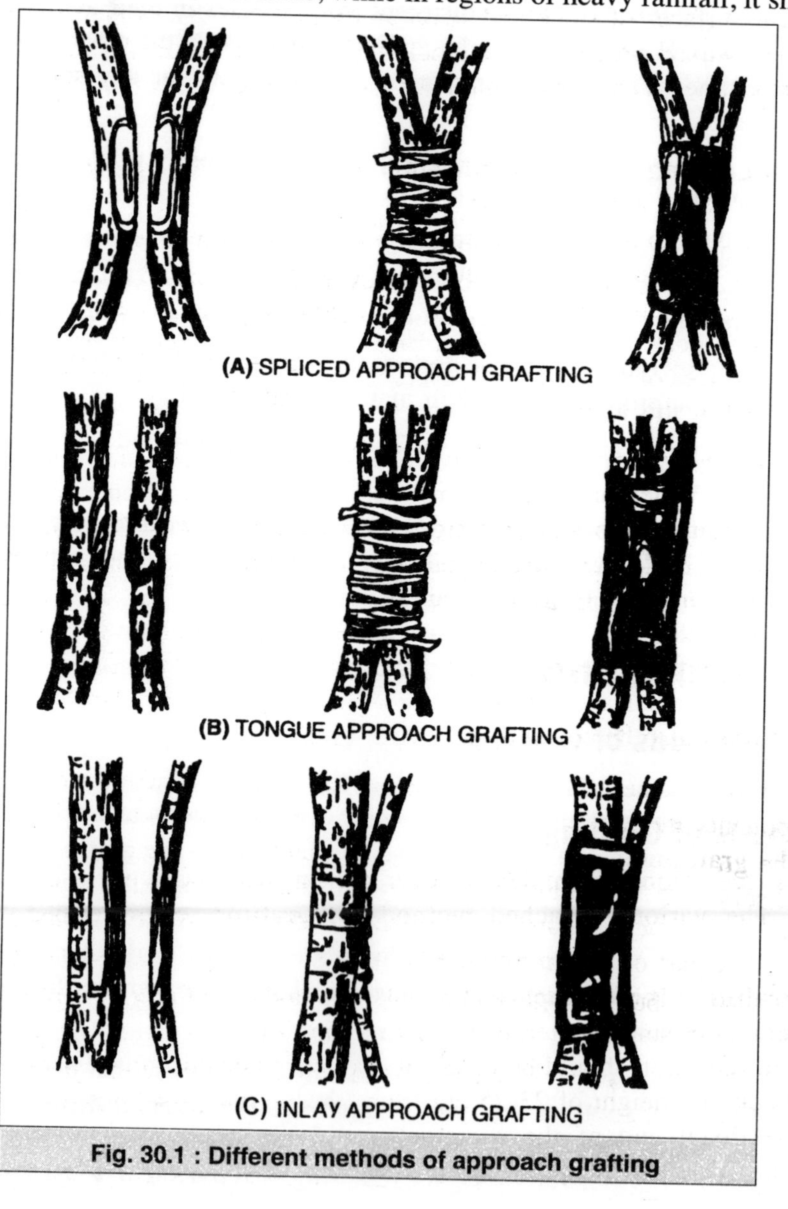

Fig. 30.1 : Different methods of approach grafting

soon after the rainy season is over, provided temperature of the locality do not fall below 15°C. Heavy rainfall, hot and dry period adversely affects its success. Inarching is also employed to rejuvenate the old, weakly growing, diseased or damaged tree to give a new lease of life, which is known, as *adjuvant grafting*. For it, a good number of seedlings are planted around the trunk of damaged tree. When the seedlings get established, these are inarched to the old plant as discussed above.

Splice-approach grafting : In spice approach grafting, both stock and scion should be of same thickness. In this method, a splice of 4-5 cm bark and wood is removed both from stock and scion (Figure 30.1). The cut on both the partners should be of same size and should be smooth too. The cut surfaces of stock and scion are bound together with some suitable tying material and then waxed properly. The stock above the union and scion below the union are cut after some time when proper union has taken place and sprouting of scion has started.

Tongue-approach grafting : It is a modification of simple inarching and is generally practiced with thicker rootstock and scion. It is almost similar to splice-approach grafting in which a second cut is made into the wood of stem about half-way down to the first cut in a sloping direction pointing tongue upwards (Figure 30.1). A similar cut is made in the scion shoot. Both cuts are made in such a manner that one fits tightly into the other. Tongue grafting is little more complicated than inarching. However, tongue grafting provides better success because three surfaces of the cambium layers are involved in the union. Apple, pear and walnut are propagated by this method.

Bridge grafting : Bridge grafting is basically not a method of propagation but is a form of repair grafting in plants, which have been damaged either by frost, rodents or insects. It may, however, be kept in mind that bridge grafting is only helpful if the trunk is damaged but the root system of the plant is all right. In this method, the damaged portion of the stock is first cleaned and the irregular edges of the girdled area are cut evenly. The scion of desired variety is then inserted in such a way that it is attached at both upper and lower ends into the living bark (Figure 30.2). Similarly, it is important that scions be inserted right side up. After the scions are inserted, the cut surfaces must be thoroughly covered with grafting wax. The exposed injured wood must also be covered otherwise it may serve as an entry channel for decay organisms.

B. Detached Methods of Grafting

In these methods, the scion is completely severed from the parent plant and grafted onto the rootstock. The various detached methods of grafting are described briefly hereunder :

Splice or whip grafting : It is a simple and popular method of grafting. It is particularly useful for grafting relatively small material. In this method of propagation, it is essential that both the stock and the scion should be of equal diameter. For this, about one-year-old rootstock is headed back at a height of 23-25 cm from the soil and a diagonal cut (3-4 cm long) is made at the distal end of the rootstock with the help of a sharp knife. A similar slanting cut is made on the proximal end of the scion (Figure 30.3). The cuts

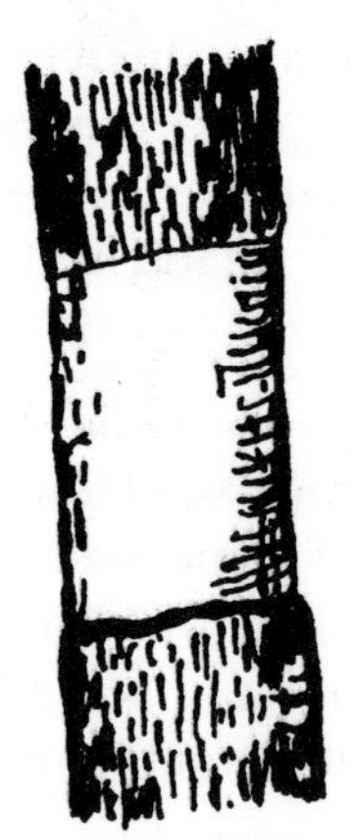

(A) PREPARATION OF THE BARK OF THE TREE

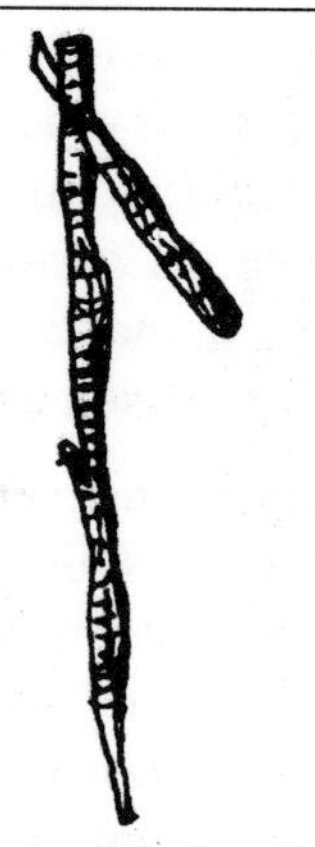

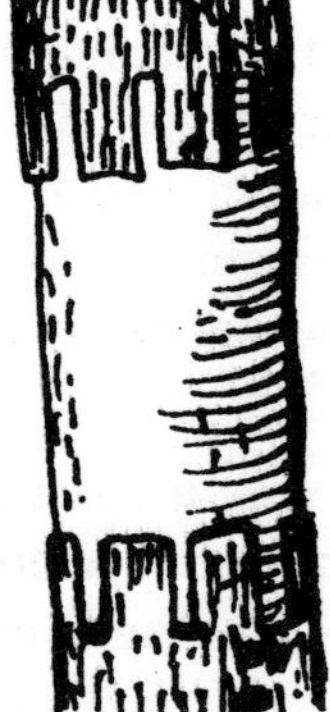

(B) MAKING CUTS IN THE BARK

(D) INSERTION OF SCION STICKS

(E) SEALING OF UNION WITH GRATING WAX

Fig. 30.2 : Different steps involved in bridge grafting

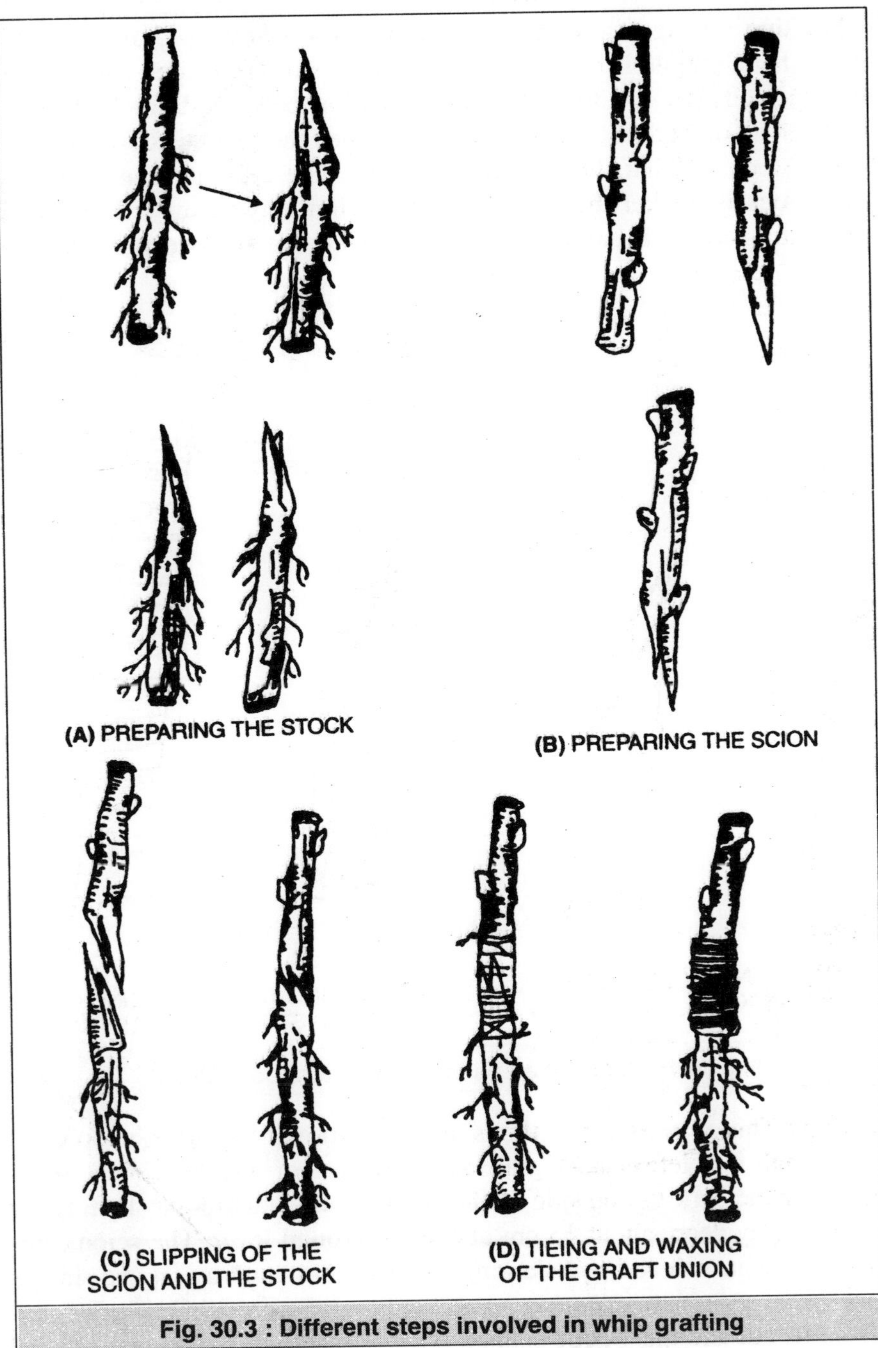

Fig. 30.3 : Different steps involved in whip grafting

should be smooth and even otherwise there won't be proper cambial contact and ultimate the union. The cut surfaces of both the rootstock and scion are bound together and tied firmly with an alkathene strip. After the union has taken place, the rootstock above the union is loped off gradually. Similarly, the sprouts arising below the graft union must be removed regularly.

Tongue grafting : Tongue grafting is a modified form of whip grafting. It differs from whip grafting in that a tongue is made in the cut surfaces of both stock and scion to provide better filling and rigidity (Figure 30.4). For this, a shallow cut in addition to the one given for whip grafting is given on both the stock and scion. When the tongue is prepared, the scion is slipped into the stock, so that they interlock easily. The graft is then tied and wrapped securely. If the stock is thicker than the scion, the scion may be moved to one side of the stock. It gives better success because of better cambial contact between stock and scion due to the formation of tongue.

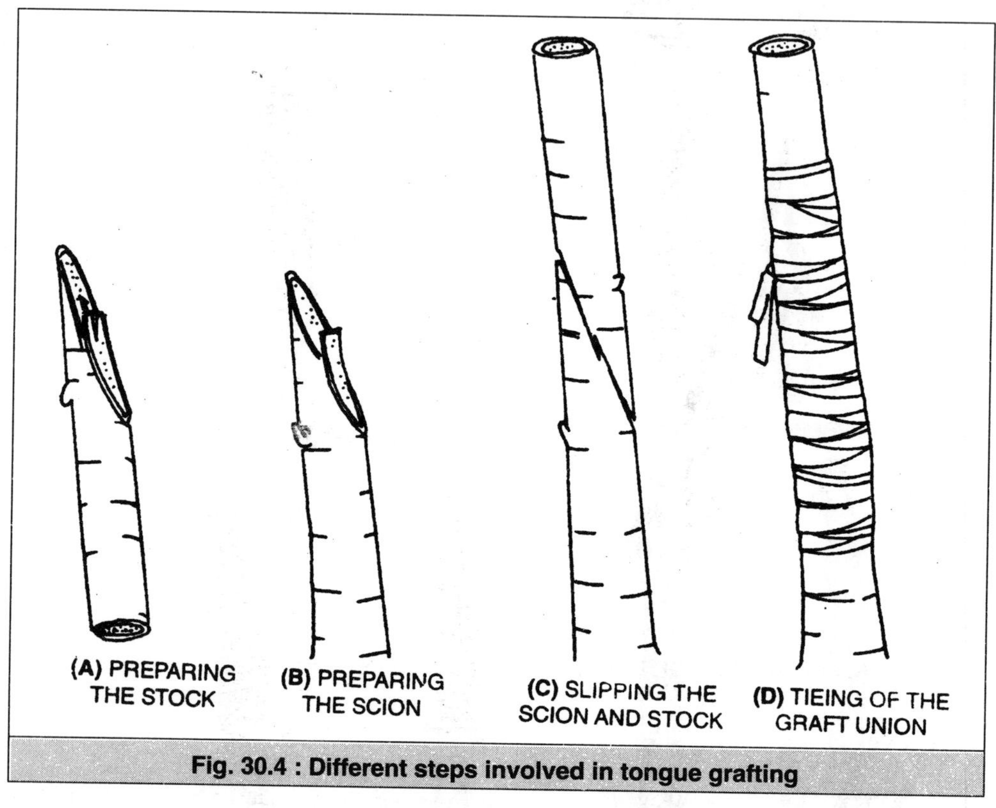

Fig. 30.4 : Different steps involved in tongue grafting

Side grafting : This is an useful methods of propagation if the stock is too large for whip grafting yet too small for cleft or bark grafting. It consists of insertion of the scion base for about 2.5 cm under the bark in one side of the stock. In this method, a slanting cut (2-4 cm long) is given half way deep, about 15 cm above the ground level. The scions are collected from current season's growth and two slanting cuts (one 2-4 cm long on one side and other 1 cm on the other side) are given, exposing cambium layer (Figure 30.5). The stock is made to bend away from its cut and the wedge shaped scion is inserted into the cut of stock and the shorter cut is covered by the bark of stock. The stock and scion are wrapped and tied with a polyethylene strip tightly. The union takes place in 6-8 weeks, depending on the environmental conditions. After the union has taken place, the stock is cut step-by-step. It is commonly followed in mango. Side grafting differs with veneer grafting in the manner that in case of veneer grafting, only one side of the scion is sliced away in a sloping manner, whereas in side grafting, the scion is sliced on both sides of the lower portion in the form of a wedge. It is a popular method of grafting and has many modifications. However, two types of side grafting *i.e.* side grafting and side tongue grafting are commonly used by the nurserymen and propagators throughout the world.

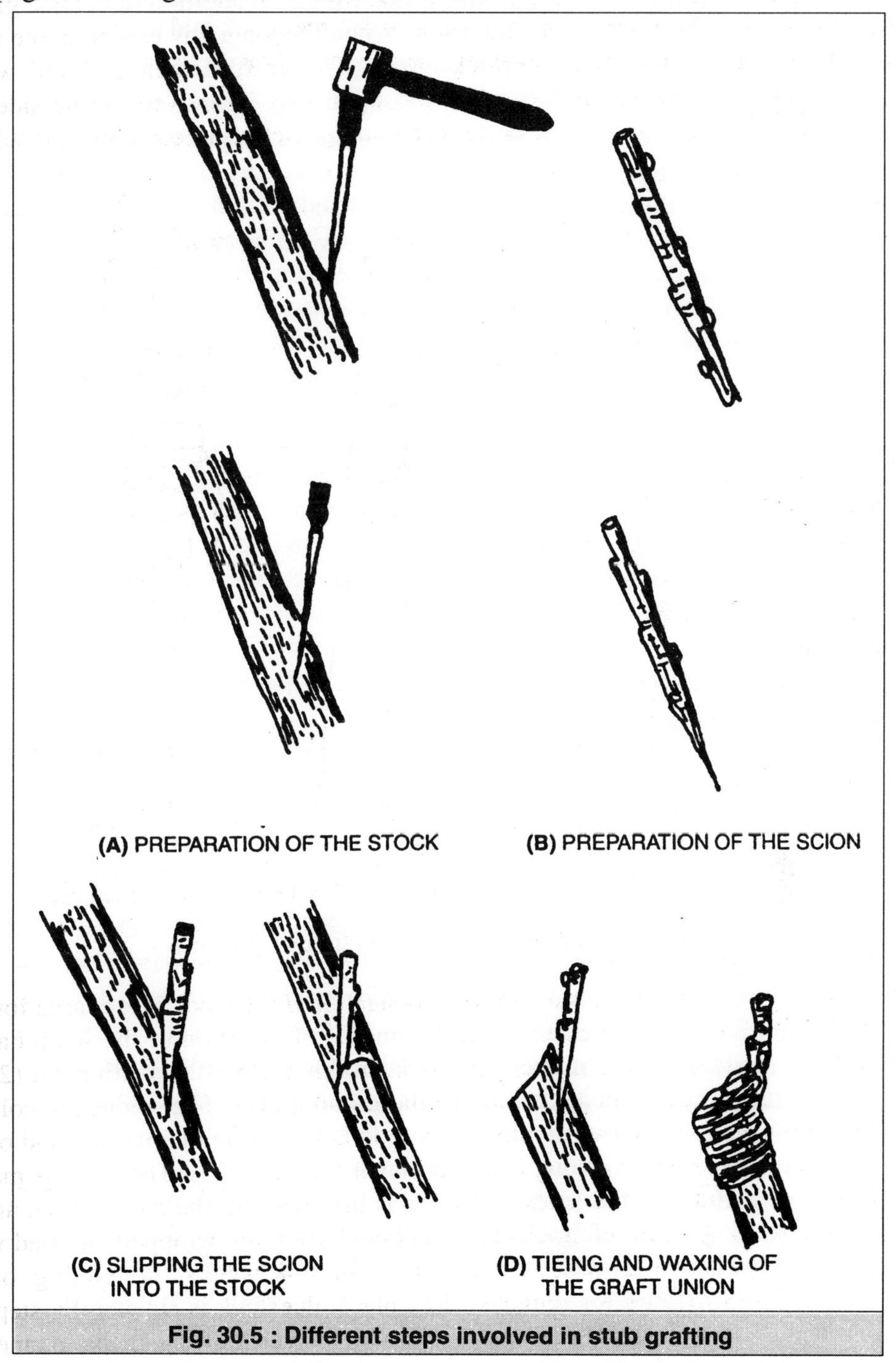

Fig. 30.5 : Different steps involved in stub grafting

Notch grafting : It is also called as *inlaying*. It is particularly useful in humid tropical areas, when girth of the rootstock exceeds 8 cm. In this method, the rootstock is decapitated 45 cm above the ground level and V-shaped notches (four or five) are made at equal distance from the periphery of the cut surface, depending upon the thickness of rootstock. The proximal end of the scion is fitted into the notch of the rootstock and scions are sliced onto the notches made in them. Both stock and scion are tied firmly with a polyethylene strip.

Veneer grafting : It resembles side grafting, which is being used for the propagation of many fruit plants by the propagators. It is a simple method of propagation and can be used in one-year-old rootstock seedlings having a diameter of 1.0 to 1.5 cm. For veneer grafting, 3-6 months old scion shoots with lush green leaves are selected. Usually, the terminal and next to terminal shoots are most ideal. The selected shoots are defoliated 5 to 10 days prior the grafting (depending upon the season) leaving the petiole attached. Usually, scions are ready for grafting after a week of defoliation. However, defoliation of the shoots is not required under humid condition. The rootstock is prepared by making a slanting cut (5 cm long) and an oblique cut is then made at the base of the first cut so that a piece of wood alongwith bark is removed. Generally, about 1/4th splice of bark and wood is removed from the stock and in scion 3/4th portion of bark with wood is removed. The base of the scion wood is then fitted into the rootstock in such a manner that the cut surfaces including the cambium layers of scion and rootstock face each other. The rootstock and scion are tied together with a polyethylene strip. When scion growth begins, the shoot of rootstock is removed above the graft union in one or two stokes. The technique is easy to operate, more economical and gives high degree of success. The best time for veneer grafting in north India is March-April and July-August.

It is highly successful for establishing *in situ* orchards and top working of old unproductive orchards. Besides, scion can be easily stored for 8-10 days at ambient temperature (25 to 35°C) in moist sphagnum moss grass covered with polyethylene. The added advantage of this technique is that the scion of desired variety can be procured from distant places and grafted in a desired manner. Mango is commercially propagated by veneer grafting in north India.

Cleft or wedge grafting : It is an old method of grafting and is particularly suitable in rootstocks having diameter greater than the scion. Rootstock with 5-7 cm or more girth is selected for this purpose. The rootstock is cleft grafted after decapitating the stock 45 cm above the ground level. The beheaded rootstock is split to about 5 cm deep through the center of stem with a sharp knife or a chisel (Figure 30.6). When the knife is removed, a hard wooden wedge is inserted to keep it open for the subsequent insertion of the scion. The scion of 15 to 20 cm size is taken from a terminal shoot, which is more than three months old and then it is wedged securely (6-7 cm). The cleft of the scion is slipped into the split of the stock. In thicker stocks, more than one scion can be inserted. The graft should be thoroughly waxed to prevent wilting of the scion. The scion starts growing after 2 to 3 months of grafting. The right time for cleft grafting is the later part of the dormant season or just before the start of the active growth. Walnut, hazelnut, pecan nut and grape are propagated by this method. It is also used for rejuvenating the old orchards of temperate fruits by top working.

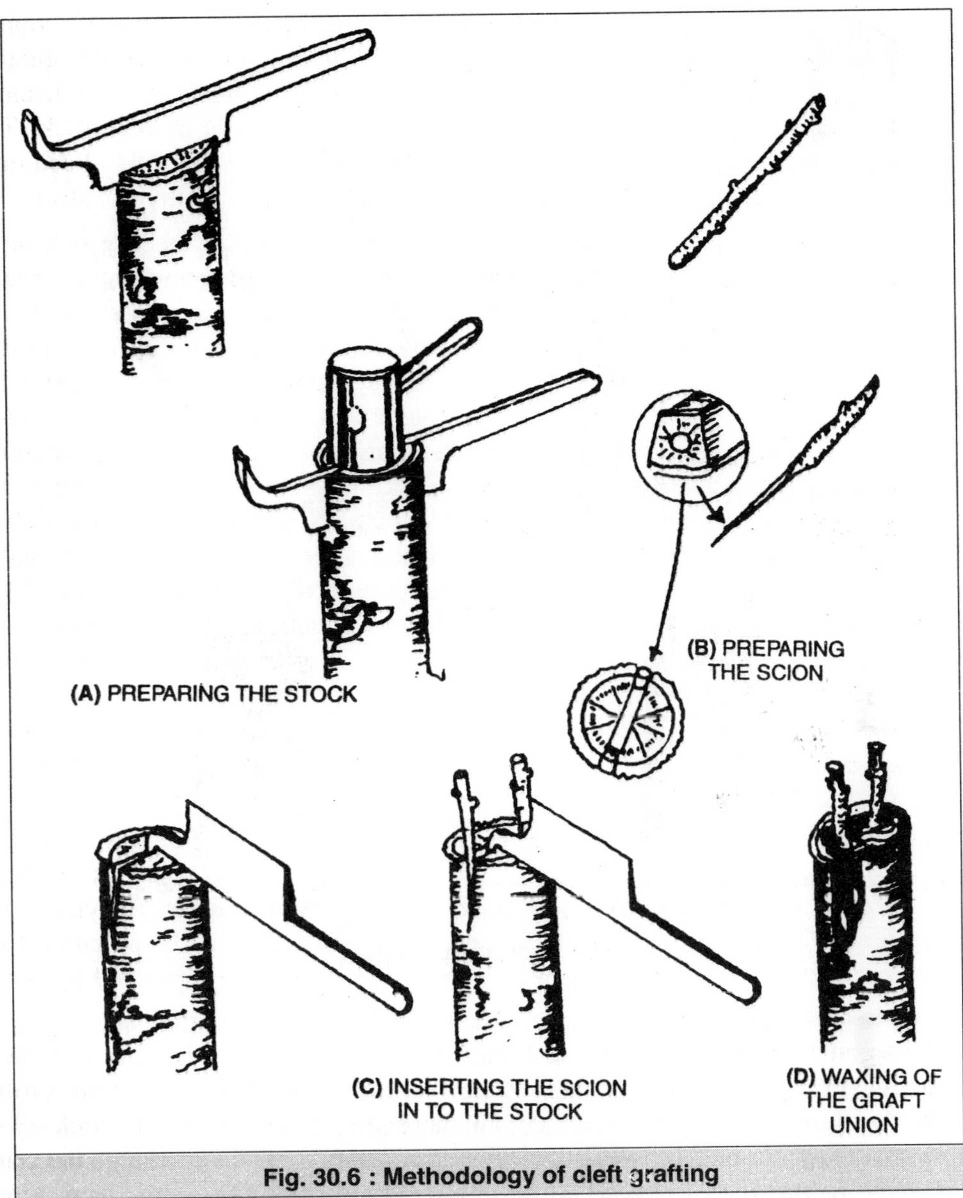

Fig. 30.6 : Methodology of cleft grafting

Bark grafting : It is a simple and rapid method of propagation and gives high success. Moreover, it does not require special instruments also. Some nurserymen prefer it, because it can be done without splitting the stub, which prevents the entry of decay organisms. Similarly, it can even be done in branches, which are 25 to 30 cm long (Figure 30.7). It should be done in spring when the bark of the stock slips easily. It is important of note that the scion used in bark grafting should be dormant. In actual practice, the stock is first sawed off at a point, where bark is smooth. If the stock is thick, many scions can be inserted. For each scion, bark is split downward, about 5 cm from the top end of stub. Scions of 10-12 cm long, containing 2-3 buds are collected from dormant wood and are prepared by giving a

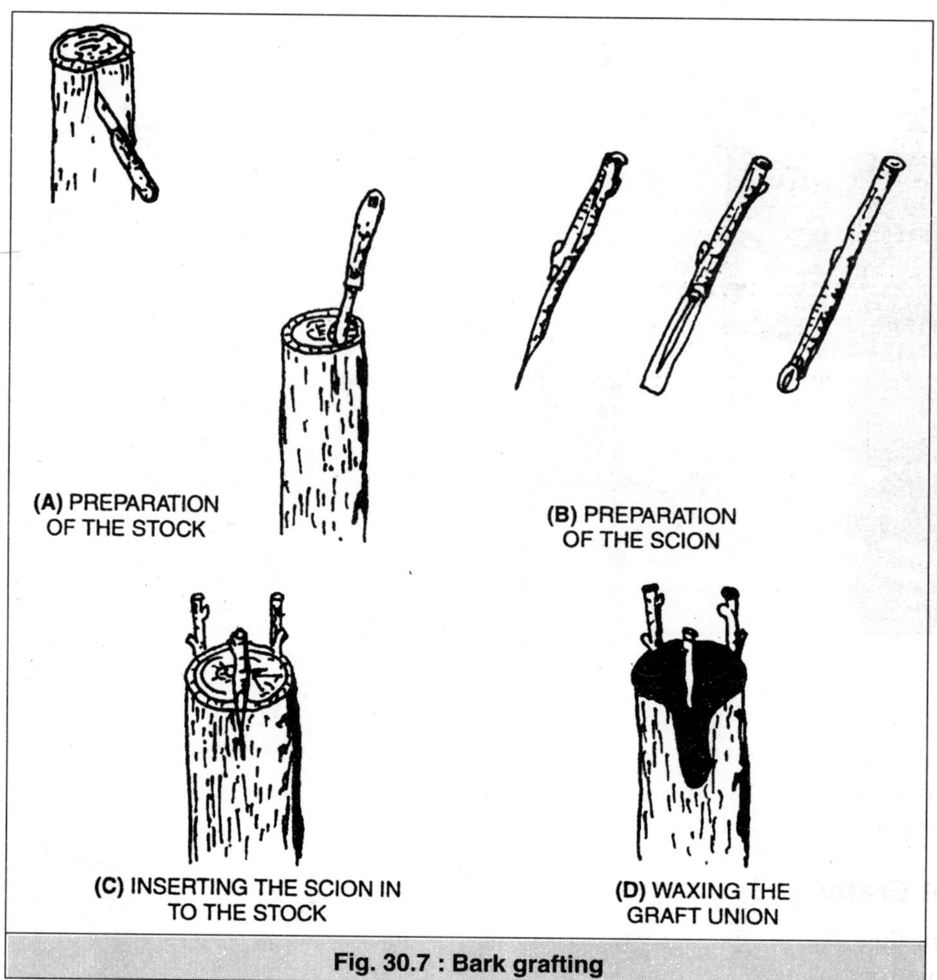

Fig. 30.7 : Bark grafting

slanting cut (5 cm) downward along one side of the base. On the other side, a small cut is given. The scion is then inserted in the centre of split between the bark and wood of the stock. The longer cut of the scion is placed against the wood. The scion is kept firmly by using adhesive tape or string. Afterwards, all the exposed portions of the stub and scion should be covered with melted wax. If grafting is successful, the scion would start sprouting after 6-8 weeks of grafting.

Saddle grafting : In saddle grafting, the rootstock is beheaded 25 cm above the ground level. Two upward sloping cuts are made on the side of the rootstock to form a 5-7 cm long wedge on its cut end. On the stem of the scion shoot, a tongue is made. The wedge of the rootstock is then fitted into the groove of the tongue and tied firmly with a polyethylene strip.

• • •

CHAPTER 31

Special Grafting Techniques

Several fruits are propagated commercially by different methods of grafting. In most cases, grafting is done at various places on the top portion of the plant. However, there are instances, wherein, the scion may be placed on a root or crown also. Some examples special grafting are briefly described hereunder :

Root Grafting

In this method, the roots are used as a stock for the grafting. The entire root system (whole root graft) or the pieces of the root (piece-root graft) may be used as a rootstock. In root grafting, whip method of grafting is used. It is usually performed indoors during late-winter or early-spring. The scion wood is collected well in advance like the rootstocks (roots), which are dug in late fall and kept under cool and moist conditions (35-40°F) until use. The term *bench grafting* is also used for root grafting because it is practically done by propagators on benches. Apple, pear, plum, filbert, wisteria etc., are propagated by root grafting.

In actual practice, the roots are cut into pieces of 7-15 cm size. The scions of same length containing 2-4 lateral buds are used. The whip grafting is done in late-winter or early-spring and the grafts are bundled in groups of 50-100 and stored for callusing in damp sand or peat moss at a temperature of 7°C for about 2 months. The callus formation and healing process can be accelerated or retarded by temperature regulation. Ideal temperature for the healing process and callusing ranges between 7-21°C. Usually, the higher temperature accelerates and the lower temperature retards the healing process. Similarly, adequate aeration and moisture supply is equally important for healing process. After proper healing, the grafts are stored in a cool temperature (2-4°C) for breaking the rest period of the buds and to keep them dormant until planting. Afterwards, the grafts may be planted in the nursery during spring or early-summer. These grafts should be planted in such a way that the graft union remains well above the ground level. If the joint remains below the soil, it is likely that roots may develop in the scion, which is not desirable because one may not exploit the favourable influences (such as dwarfing or disease resistance etc.) of the rootstock on the scion cultivar.

Nurse-root Grafting

Sometimes, it is desirable to have a stem cutting of a difficult-to-root species on its own roots. This can be done by root grafting, using the plant to be grown on its own roots as scion and a root of a compatible species as the rootstock. The procedure is the same as root grafting except that the scion is a little longer than usual and the grafts are planted deep in the field with major portion of the scion below the soil. In this method, scion rooting is promoted by using some auxins like IBA on the cuts made above the graft union before planting. After 2-3 seasons, the scion develops proper roots. The temporary nurse rootstock is then 'cut off' and top is reduced in proportion to the root system. The rooted scion is replanted in the soil on its own roots. It is also a type of mound layering as the new scion shoot grows, soil is mounded up around to a height of 13-15 cm and terminal leaves are never covered. It is comparatively a cumbersome and laborious method of grafting because grafts are to be planted for 2-3 seasons for rooting and then these are to be dug up after the rootstock is cut off. Now, the following methods have been standardized to eliminate this cumbersome and laborious method.

(*a*) By grafting rootstock on to scion in an inverted position. In this way, the union heals and the inverted stock sustains scion until roots are formed and stock will die as it won't receive food and water from the scion, thus leaving scion growing on its own roots.

(*b*) Using incompatible rootstock, the scion will grow properly and the incompatible rootstock won't.

(*c*) By wounding the scion above graft union by suitable wrapping material to girdle and cut off the rootstock.

Crown Grafting

Crown grafting is done at the stem-root transition region, which is called as the crown of the stock. It is specially performed in deciduous fruit plants in late-winter to late-summer. The scion of desired cultivar should be prepared from well-matured wood of previous season's growth. Different methods may be employed in crown grafting but whip, side or cleft grafting gives good success, though it also depends upon the species and size of the rootstock. The grafting may be done just above, below or at the soil level. After grafting, the union must be wrapped tightly to hold both the components in position and avoid desiccation so that the healing of union takes place at an early date.

Double Working

Double working is a specialized technique of grafting in which the composite plant has three different components, the rootstock, interstock and the scion *i.e.*, the desired cultivar or variety. Thus, double worked plants have two graft joints, one between the rootstock and interstock and the other between interstock and the scion (Figure 31.1). Double working is used with one of the following objectives :

- To overcome the incompatibility between the desired cultivar and the stock.
- To provide a cold or disease resistant trunk to the desired cultivar.
- To make use of a interstock for using its desirable effects like dwarfing etc., and
- To obtain desirable crotch system characteristics of certain cultivar on desirable rootstocks.

Double working has been exploited commercially in apple and pear. Propagation of 'Bartlett' pear on 'Quince' rootstock by using mutually compatible rootstocks like 'Old Home' or 'Buerre Hardy' pear is one of the most important examples of double working.

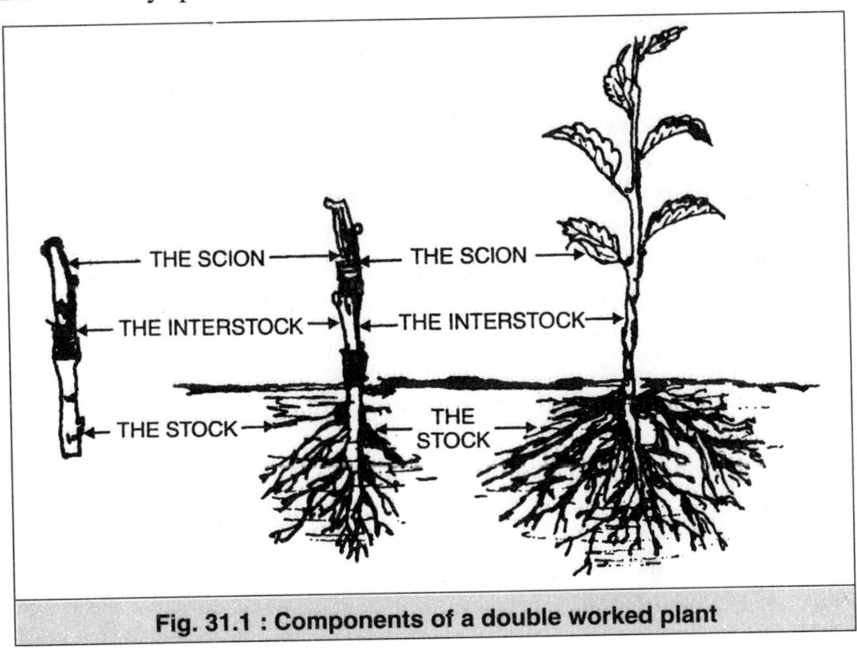

Fig. 31.1 : Components of a double worked plant

Herbaceous Grafting

Herbaceous grafting is done in some cucurbitaceous vegetables in seedling stage. Splice method of grafting is used with a diagonal cut made through the seedling stock, just above the cotyledons. In this method, a small piece of thin polyethylene tubing of proper size is slipped over the cut end of the stock. The scion is prepared from the desired cultivar with cut and size similar to the one given on the stock. The scion is then slipped down in to the polyethylene tubing so that the two cut surfaces make intimate contact. The polyethylene tubing is removed and cut away with a razor blade after healing (12-15 days) of the graft union. Herbaceous grafting is done in most advance countries in herbaceous plants mainly to study the virus transmission, stock-scion physiology, stock-scion compatibility and commercial greenhouse production of cucurbits.

Green-wood Grafting

A method similar to herbaceous grafting, commonly called as 'green wood' grafting is used in grape to establish the desired cultivar on *Phylloxera* (root louse) or nematode resistant rootstock. In this method, the scion taken from new growth is placed on a new green shoot developed from the rootstock plant. A simple, 2.5-4.0 cm long splice cuts are given both on stock and scion of equal size. In this case, the scion has only one bud. The cuts of stock and scion are matched together closely and graft union is wrapped carefully with some smooth and soft wrapping material like plastic rubber. Both stock and scion are very fragile and hence these must be tied to a stake. The healing may take place within two weeks or so after grafting.

Nurse-seed Grafting

Nurse-seed grafting is used in woody species in which the seed germination is hypogeal *i.e.*, the cotyledons remain below the ground in the seed coat and shoot tip appearing above the ground. In this method the petioles of the cotyledons are cut off transversally just cut at the seed, leaving the cotyledons inside (Figure 31.2). A knife point is then inserted between the cut petioles for making an opening for inserting the scion. The scion is prepared from previous season's growth by making a wedge at the base. The scion is inserted between the cotyledons so that the exposed cuts of the scion and stock (seeds) make direct cambial contact. The grafts are wrapped carefully with a suitable material like rubber strips or wax strings. These seed grafts are then placed in a suitable rooting medium with graft union about 4 cm below the surface. The healing takes place within 2-3 weeks and roots arise from the petioles of cut cotyledons. Nurse-seed grafting is done in early-spring and is successful in chestnut, avocado and camellia. A little modification of nurse-seed grafting has been very

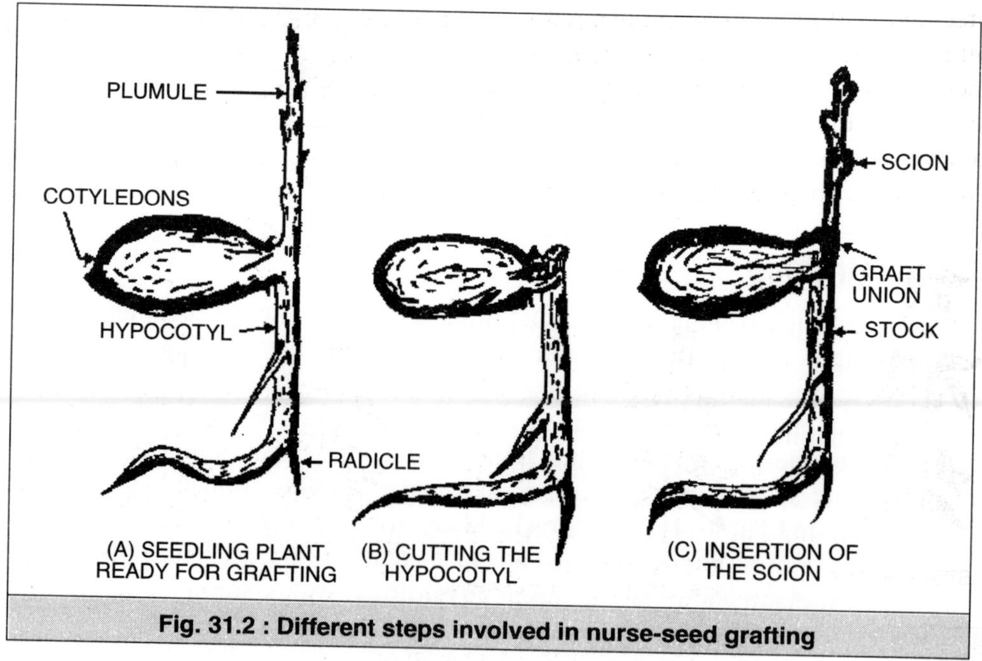

Fig. 31.2 : Different steps involved in nurse-seed grafting

successful in propagating chestnuts, pecans and walnuts. In this method, the grafting is done on the seedling hypocotyl at the point of attachment of the cotyledons. The hypocotyl is cut longitudinally and split is made for insertion of the scion. The scion is prepared from the previous season's growth and is cut to a thin wedge. The scion is inserted into the split made in the hypocotyl, after which, the union is wrapped firmly with a suitable wrapping material like rubber strips.

Cutting-grafts

In this technique of grafting, a leafy scion is grafted on to a leafy, unrooted stem piece, which would be a future rootstock. After grafting, the components are placed in a rooting medium under intermittent mist for healing of graft and rooting of the stock. For better rooting, the leaves must be retained on the stock. Splice grafting procedure is usually followed in cutting grafts. The slope of the cut is at 30° angle and the cut made is usually 1.5-2.0 cm long. The union is tied carefully with a rubber band. The base of the stock is kept in the rooting medium (IBA) for sometime and grafts are then placed in mist for healing process and rooting of the stock. After proper rooting and healing, the grafts are removed from the mist for hardening for 2-3 weeks. The grafts are usually ready for planting in pots or containers after two weeks of hardening process.

Initially, this technique was devised for studying the stock scion physiology but now it is commercially used for propagation of citrus and clonal dwarfing rootstocks. Similarly, it is of commercial significance for difficult–to-root species like conifers and rhododendrons.

Softwood Grafting

This technique of grafting is commercially used for raising mango, sapota, tamarind and cashewnut in western India. In this technique, grafting is done with mature, precured scion on the emerging soft, coppery-red shoot of the rootstock, which is 60-70 days old. The basic technique involves the beheading of rootstock with a sharp knife. Then, a slit is made on the beheaded stock to insert the precured scion (Figure 31.3). The lower portion of scion is prepared into wedge shape so that both the faces of scion fit well with the stock. Both stock

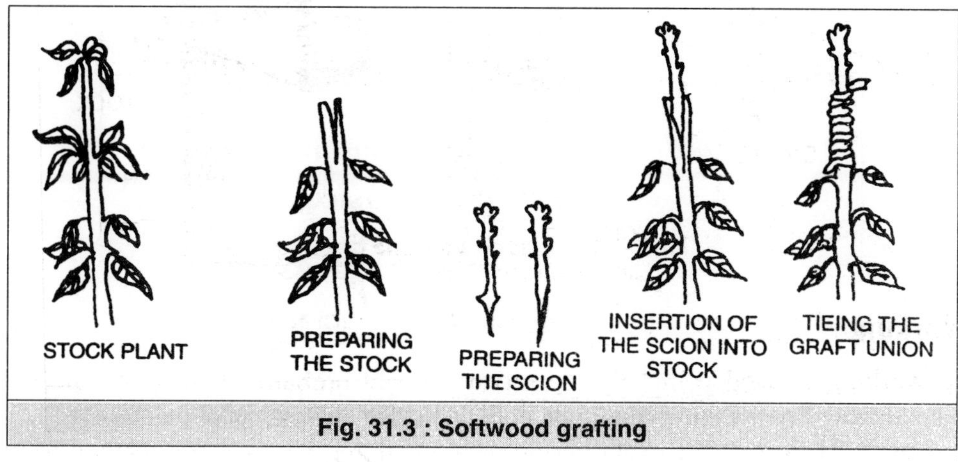

Fig. 31.3 : Softwood grafting

and scion are tied with help of a polyethylene strip. The scions are then covered with a polythene bag (100 gauge) and tied with a thread to keep the scion fresh till the complete union is formed. After sprouting, this bag is removed. To have better success, the leaves on the stock must be retained. This technique is effective in dry, hot weather or in areas of low precipitation. It is also used for *in situ* grafting especially in the drylands and areas experiencing hot weather with low precipitation where mortality of nursery raised grafts is very high.

Epicotyl or Stone Grafting

In this method, germinating seeds of less than two weeks old are wedge or splice grafted with mature scion. Moderate temperature and high relative humidity are major factors related to success of epicotyl grafting. For epicotyl grafting, germinating seeds (less than 2 weeks old) are used as a rootstock. A 2-3 cm long slanting cut is made in the epicotyl with a matching cut on the proximal portion of the scion and then these are tied firmly with a alkathene strip (Figure 31.4).

In wedge grafting, the selected 2 weeks old seedling are headed back by retaining 6-8 cm long stem with the stone. The longitudinal transverse cut running 4 6 cm centrally down on the beheaded rootstock is made with the help of a grafting knife. A wedge shaped cut, slanting from both sides in made on the lower side of scion stick. The scion stick is then inserted into the saddle-like cut made on the rootstock and pressed properly so that cambium tissues of the rootstock and scion overlap each other. The union is then tied with the help of 150 guage-polyethylene strip. This technique has been commercialized for rapid multiplication of mango in Maharashtra in general and Konkan region in particular.

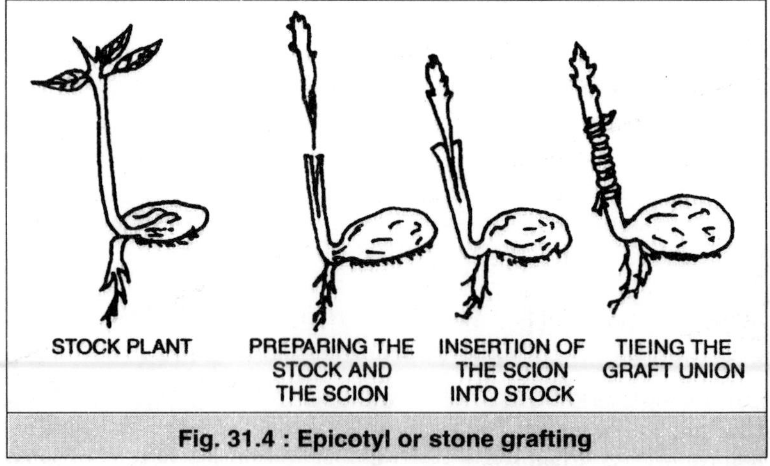

Fig. 31.4 : Epicotyl or stone grafting

Top Working

Top working is used primarily to convert an old orchard of inferior variety with a desirable cultivar. Top working should be done in spring season, shortly before the beginning of new growth. However, the exact period depends upon the crop and method to be used.

Top working is an extremely severe pruning operation and one must decide before pruning, the number of scaffold branches to be retained, for maintaining a considerable balance between root and shoot system of plant. The branches to be grafted should be well distributed around the tree. In general, top working is most successful if grafting is done on scaffold branches, which are more than 10 cm in diameter. It is always better to retain some nurse branches while doing pruning in plants as these will be required for food preparation by the plants. In general, plants are severely pruned and new growth is allowed to come. Only few selected branches are retained and others are removed. Any grafting method can be used for top working. However, cleft, side and veneer grafting are most commonly used. Mango and cashewnut are commercially top worked by side veneer grafting in India.

Renewal Grafting

It is often used on fruit trees when they have passed their normal productive life. New varieties are grafted onto the mature trees to produce a large fruiting tree in a very short time. It is almost similar to top working (Figure 31.5).

Micrografting

Grafting of tiny plant parts under aseptic and controlled environmental conditions is called as *micrografting*. In this method, small grafts are grown in closed containers until they grow large enough to be transferred to the open conditions. Micrografting has been used mostly in citrus, apple and plum. The primary use of this technique is to produce virus free plants of woody plant species, which oftenly do not regenerate into root and shoot system from shoot tips in culture medium. This technique was first developed by Spanish scientists, Navarro and his co-workers in late seventies for producing virus-tested plants of citrus. Now, this method is widely used for producing plants in adult phase, which is helpful in avoiding the long juvenile phase that result when nucellar seedlings are used to produce virus-free plant of established cultivars.

In shoot tip grafting, embryos are excised from the rootstock seeds. These embryos are sterilized with some sterilent and then cultured in standard inorganic salt medium supplemented with one per cent agar. The cultured embryos may germinate within two weeks if kept in dark. The seedlings are then removed and decapitated to 1 to 1.5 cm length. The cotyledons and the lateral buds are excised with a sharp blade. A 0.14 to 0.18 mm tip having three leaf primordia is used as a scion. After this, an inverted 'T' cut is given in the seeding rootstock by cutting 1 mm down the stem followed by a horizontal cut at the bottom. The excised shoot tip is then placed inside the flap, next to the cambium. Grafted plants are placed in a liquid medium. A filter paper bridge with a centre hole is made to support the stem. The cultures are kept in the light for 3-5 weeks for healing. When two expanded leaves appear on the scion, the grafted plant is pre-conditioned and hardened in the incubation chamber. Afterwards, it is transplanted in the field.

The non-aseptic micrografting is also a promising procedure in some plant species for the production of virus-free plants. In this method, propagation of small mini nursery plants is done by grafting a tiny seedling with scion of the same size. The minute grafts are then allowed to grow enough to have a viable plant. This technique is useful in tropical areas, where nursery plants are sent to distant places either by ship or air because more number of plants can be accommodated in a smaller space.

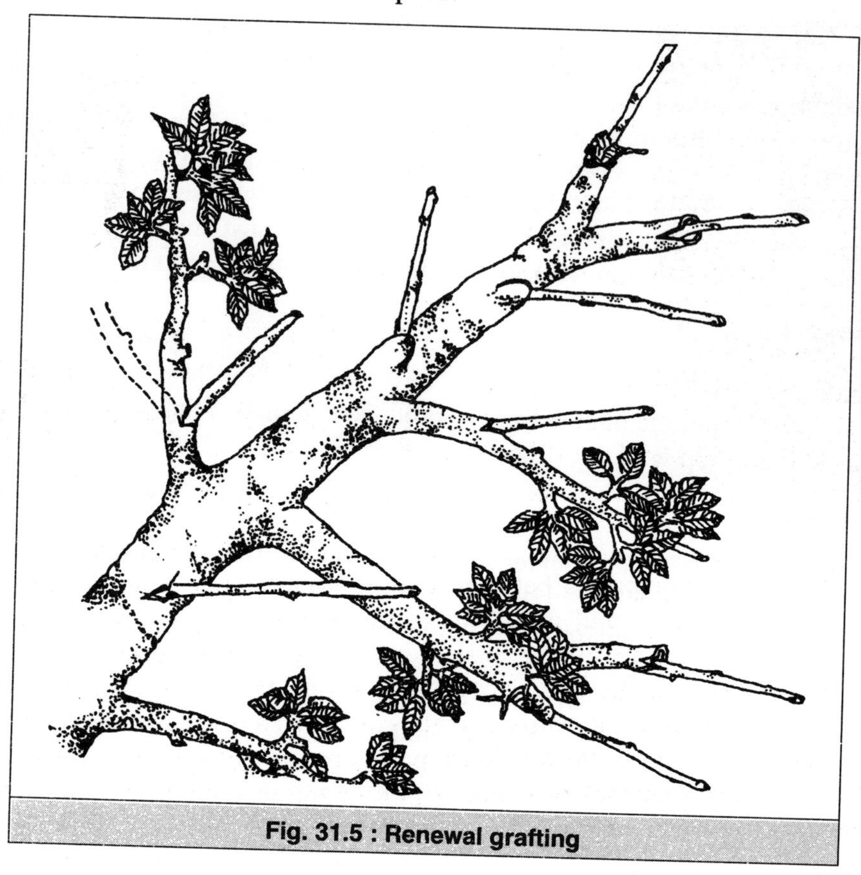

Fig. 31.5 : Renewal grafting

PROBLEMS ASSOCIATED WITH MICROGRAFTING

Under certain conditions, shoot tip grafting techniques may be a complete failure because many problems like desiccation of rootstock, drying of meristems, swelling of rootstock, browning of culture medium, size of the shoot tip, pre-treatments of the scion, sucrose concentration in the medium and hardening process etc., are associated with it. Selecting rootstocks of ideal age, keeping the dissected rootstocks in covered petri dishes, adding activated charcoal in the medium, selecting shoot tips of 0.1-0.3 mm size, adding sucrose @ 75 g/L of medium and proper pre-hardening of the micrografts help to get better success with micrografting.

• • •

CHAPTER 32

Stock-Scion Relationships

The relationship and interactions between two or more components of a budded or grafted plant are quite variable and interesting. Some of these interactions have major horticultural value, whereas the others, may not have any significance at all. These interactions are known to alter the tree-size, growth, productivity, fruit quality and other attributes of different horticultural corps and thus it has commercial significance. The different stock-scion interactions or the relationship have been divided into the three different categories : (*i*) influence of rootstock on scion, (*ii*) influence of scion on rootstock, and (*iii*) influence of interstock on stock, scion or both as described hereunder :

A. EFFECTS OF ROOTSTOCK ON THE SCION

The effect of rootstock on the scion cultivar is one of the most significant influences of stock-scion relations. The rootstock influences the scion cultivar in many ways. The effects are variable, depending on type of stock and scion. Some of the influences of rootstock on scion are as under :

(*a*) Influence on Size and Growth Habit

The effect of rootstock on the size and growth of scion cultivar is the most pronounced influence of rootstock. The rootstocks may affect the growth, and growth habits of the scion cultivar invariably. There are many examples in the literature to quote these effects of rootstocks on scion cultivar but the most important example is the rootstocks of apple. In fact, with the proper selection of rootstocks in apple, the complete range of tree size *i.e.* from very dwarf to very large has been obtained in the given scion variety, grafted or budded on these rootstocks. The Malling series of rootstocks (16) were selected by R.G. Hutton at East Malling Research Station, Kent (England) from old European clones like French Paradise and Doucin and were assigned the Roman numeral, prefixed with the word Malling (M).

Later, MM series of rootstocks (MM_{101}, MM_{102}, MM_{106}, MM_{109}, MM_{111}) were developed by crossing Malling series rootstocks with 'Northern Spy' to get nematode resistant rootstocks. Out of these, MM_{111} and MM_{106} produce trees of medium vigour under high-density planting systems. At present, many more rootstocks have been developed, which influence the tree canopy invariably. 'Mazzard' (*Prunus avium*) rootstock usually produces vigorous trees of any scion variety of sweet cherry. whereas 'Mahaleb' (*Prunus mahaleb*) produces small trees of the scion cultivar grafted or budded on it. Similarly. Stockton Morellgo also induces dwarfness in cherries. 'Quince C' is a known rootstock for inducing dwarfness in pears. In citrus, 'Dragon Fly' in being used universally under high-density plantings in citrus because it has very high potential to induce dwarfing effect in scion, which is the first and foremost requirement in high-density plantings. In India, 'Troyer' citrange rootstock has been recommended as a dwarfing rootstock for Kinnow mandarin, whereas 'Karna Khatta' and 'Sohsarkar' are semi-vigorous and vigorous rootstocks, respectively. Similarly, 'Creeping' and 'Vellai Collumban' rootstocks have been reported to induce dwarfness in vigorous mango cultivars like 'Banganpalli' or 'Alphonso'. Similarly, for guava, 'Pusa Srijan' has been recommended as a dwarfing rootstock for 'Allahabad Safeda' cultivar.

(*b*) Influence on Flowering and Fruiting

The effects of dwarfing apple rootstocks for inducing precocity in bearing are world known. The dwarfing apple rootstocks like M_7, M_9 and M_{27} have pronounced effects on precocity (earliness) in bearing in all the cultivars grafted or budded on them. These rootstocks also enhance the fruit set in the scion cultivar. Pears usually come into bearing quite early if grafted on 'Quince C' rootstock. Similarly, different rootstocks have been reported to stagger the fruit maturity in Kinnow mandarin. With the use of three different rootstocks, the availability of Kinnow in the market can be enhanced for a longer period. 'Troyer' citrange hastens the maturity in Kinnow, whereas 'Sohsarkar' delays it. Thus, Kinnow fruits on 'Troyer' citrange are ready for harvesting by the end of November (quite early), whereas these are ready for harvesting by mid-January on 'Sohsarkar', thereby making its availability for $2\frac{1}{2}$ to 3 months.

(*c*) Influence on Fruit Size and Quality

Rootstocks also influence the fruit characteristics of the scion cultivar invariably. For example, pear fruits have pronounced tart and astringent flavour, if grafted on Quince rootstock, which is otherwise not present if no rootstock is used. The dwarfing apple rootstock although do not influence fruit size but colour development in the scion cultivar is much better as compared to the standard or seedling rootstock. In citrus, the striking effects of the rootstock appear in fruit characteristics of the scion variety. Sour orange rootstock produces thin skinned, juicy and excellent quality fruits of scion varieties like sweet orange, tangerine and grapefruit and rough lemon rootstocks usually produces thick, coarse skinned fruits of inferior quality of the scion cultivar. Similarly, the bitterness of the grapefruits disappears if budded on 'Trifoliate' orange rootstock but it never disappears on rough lemon rootstock. 'Troyer' citrange rootstock usually reduces fruit size in Kinnow but 'Karna Khatta' and 'Sohsarkar' improves it. Further, the fruits of Kinnow on 'Troyer' citrange are thin skinned with attractive bright orange colour having good TSS, compared to those produced by 'Karna Khatta' and 'Sohsarkar' rootstocks. Rootstocks also affect the storage quality of the fruits. For example, the fruits produced on the dwarfing apple rootstocks have poor storage life as compared to those produced on standard rootstocks.

(*d*) Influence on Insect-pests and Disease Resistance

Dwarfing apple rootstocks such as M_7, M_9, M_{27} etc. are susceptible to wooly apple aphid and crown gall but Malling-Merton (MM) series of rootstocks (MM_{104}, MM_{106}, MM_{109}, MM_{111} etc.) impart resistance to the scion cultivar against woolly aphids. French pear rootstock imparts resistance to the scion cultivar against fire blight disease. Similarly, 'Old Home' pear rootstock is resistant to pear decline and Mahaleb rootstock of cherry imparts resistance against buck skin virus. Interestingly, a considerable variability exists in citrus rootstocks in their capacity to induce resistance against many diseases. For example, sour orange is highly susceptible to tristeza virus, whereas rough lemon and Cleopatra mandarin imparts resistance against it. *Citrus macrophylla* is a very good rootstock for lemons because it is known to impart resistance against gummosis, the most serious disease of lemons. *Poncirus trifoliata* rootstock imparts resistance against triserza and root rot disease.

(*e*) Influence on Winter Hardiness

Some rootstocks impart resistance to scion cultivar for cold tolerance. Apple rootstocks like EM-IX, EM-XVII and Alnarp 2 have been found to be resistant to winter injury. EM IX is resistant to early-winter frost, although it is susceptible to late-winter frost. Among citrus rootstocks, Trifoliate orange is considered as the hardiest rootstock to low temperature injury followed by sour orange and sweet orange.

(*f*) Miscellaneous Influences

Different rootstocks have been recommended for different fruit corps and even for different cultivars of the same species. One rootstock recommended for the one cultivar in a particular locality may not be suitable for same cultivar in the different locality. Thus, selection of a proper rootstock as per scientific recommendation for different regions having different soil and climatic conditions is very important, because the success or failure of a particular fruit crop in that particular locality would be decided by it. For example, 'Jatti Khatti' citrus rootstock performs well in semi-arid zones of Punjab and Rajasthan (India) whereas 'Karna Khatta' is suitable in Indo-Gangetic plains of Uttar Pradesh (India) but not the other localities. 'Myrobalan' plum is highly suitable rootstock for regions of excessive soil moisture, whereas almond is most susceptible to wet conditions. Sour orange, Rangpur lime and Cleopatra mandarin rootstocks of citrus have been reported to be resistant for higher salt concentrations. In general, 'Rangpur' lime has shown wider soil adaptability as compared to other citrus rootstocks. 'Kurukkan' has been recommended as a salt tolerant rootstock for mango soils having higher salt concentrations.

B. EFFECTS OF SCION ON ROOTSTOCK

It is not only the stock, which influences the scion but the top (scion) can also influence the stock in many ways, although, examples are very less. Influence on the vigour is the major effect of the scion on the stock. If a vigorous cultivar is grafted on a weak rootstock, the growth of the stock is also stimulated. Further, if a dwarf or a weak scion is grafted on the vigorous rootstock, the growth of the stock is also reduced. Similarly, the size, nature and form of the root system of the stock are also influenced by the scion cultivar. If the rough lemon seedlings are budded with 'Malta' and 'Blood Red' sweet oranges, a more fibrous root system is developed by the stock as compared to the non-budded plants. Similarly, 'Red Astrachan' apple scion produces a fibrous root system with a few tap roots, while 'Oldenburg' scion produces two or three prolonged deep tap roots but not fibrous root system. A strong scion seems also to cause the roots to be more resistant to wet, poorly aerated soil than the same roots would be if the scion is of a weaker growing variety.

C. EFFECTS OF INTERSTOCK

It use of certain dwarfing rootstocks, as interstock between vigorous top for producing a dwarfing and early bearing tress is in practice. For example, 'Paradise' apple is widely used as interstock between scion cultivar and 'Crab' apple rootstock, for inducing precocity in bearing. Similarly, use of Malling IX dwarfing rootstock as interstock is a commercially viable technology. In mango, 'Anupam' has been recommended as interstock for 'Amrapali' mango grafted on 'Mallika' rootstock to induce dwarfness, precocity in bearing and improvement in fruit quality. However, sometimes interstock may not result in favourable or desirable influences.

• • •

CHAPTER 33

Budding : Theory and Principles

In grafting, the scion is a detached piece of shoot or stem with several buds. In budding, scion consists of only one bud and small portion of the bark. Thus, budding is a form of grafting and often termed as *'bud grafting'*. Like grafting, the scion is inserted into the stem tissue of the rootstock so that the two components unite together, grow and eventually develop into a composite plant.

ADVANTAGES OF BUDDING

Budding has several advantages over other methods of propagation, as described briefly hereunder :

- It is a quick and efficient method of propagation as compared to grafting.
- It is the best propagation method if propagating material is scarce.
- Budding is useful in plants, which release excessive wound gum (*e.g.,* stone fruits) from injury carried to xylem or wood portion of stem at the time of grafting.
- Budding results into stronger union compared to grafting thus injury to the budded plants due to storms and strong winds is less.
- It is comparatively a simple method of plant propagation than grafting and can be done by amateurs.

In general, the success of a particular budding method depends upon the bark's slipping ability of stock and scion. It denotes that period of the year when plant is in its active phase of growth, cambium cells are actively dividing and the bark can be easily separated from the wood. However, the factors, which reduce the growth and results in tissue tightening like moisture stress, infection by insect-pests and diseases, defoliation or even low temperature etc., can interfere with the budding process. Among different methods of budding, chip budding can only be done when bark is not slipping. Like grafting, rootstock plays prominent role in budding also. At the same time, selection of bud-wood is equally important. Always use vegetative buds rather than the flowering buds for budding. The vegetative buds are usually small and pointed, while the flower buds are large and plumpy.

TIMES OF BUDDING

The time at which budding is to be done usually depends upon the availability of well developed bud-wood of a desired cultivar, when the plant is in its active growth, cambium cells are actively dividing and bark separates easily from the wood. In most plants, these conditions exist at three different times of the year as described below :

1. Fall Budding

In fall budding, the budding is actually done in late-summer rather than in fall *i.e.*, from mid-June to mid-September. In fall budding, the bud sticks, consisting of current season's shoots are obtained at or near the time of budding. The bud sticks should be vigorous and contain healthy vegetative buds. Short, diseased shoots should be avoided as these may contain floral buds. Every effort should be made to select the bud-wood from healthy plants as the infected bud-sticks may infect all the budded plants in nursery with one or the other disease. It may be noted that vegetative buds are small and pointed, whereas flower buds are round and plumpy. Some species have mixed buds and can be used in budding. Budding in fall is the best time for various kinds of budding to be done. Most of the tropical, sub-tropical and temperate fruit plants like *aonla*, mango, guava, *ber*, *bael,* jackfruit, stone fruits, walnut, chestnut and hazelnuts etc., are propagated during this period.

2. Spring Budding

It is similar to fall budding except that budding is done in spring when the bark separates easily from the wood. Similarly, budding is done when rootstock is also in its active stage growth. This period is very limited and budding should be completed soon before the rootstock has made much growth. Budwood is collected from healthy plants as in fall budding. In spring budding, the buds must be dormant and the stock must be in active growth. Thus, it must be assured that bud sticks are collected well in advance of actual budding operation and stored at 32-40°F temperature to hold them dormant by wrapping them in moist newspaper or sphagnum moss grass.

In spring budding, the actual operation should be done as soon as the bark slips easily on the rootstock. Due to the combination of dormant bud and active growing rootstock, the bud sprouts and develops easily. Like fall budding, the rootstock top is cut after 2-3 weeks after budding when healing has taken place. The stock should be cut by giving a slanting cut. After it, the latent buds may sprout and hence these should be removed. Spring budding is commonly followed in citrus.

3. June Budding

June budding is done in the early part of the growing season *i.e.*, in June and the inserted bud is forced into growth immediately in the same season. June budding is usually done to get one-year-old budded tree in one growing season. The budwood used in June budding consists of current season's growth *i.e.*, of new shoots, which have developed in the spring. At this time, the buds don't enter rest period and continue to grow in summer, producing the top of the budded plant.

The actual operation of budding and post-budding operation in June budding are different to that adopted in fall or spring budding. In June budding, the rootstocks are smaller and have less stored food. Further, the budding is done at a relatively higher height of the root-stock stem. Usually, 3-4 leaves are also retained below the bud. In most cases, the buds on the basal or central portion should be used. After the budding, no special care of the stock is done up to spring except to cut the stock above the union. It is done only after the bud has started sprouting. The healing of the bud union usually takes place within 2-3 weeks under normal environmental conditions.

If leaf stalks or petiole drops off cleanly, it is a good indication that the bud has united. If it has adhered strongly and starts shriveling and darkens, it may be assumed that budding operation has failed. The healing process is faster due to high active growth of the stock at a relatively high temperature and favourable growing conditions. The rootstock is immediately cut after healing process, leaving atleast one leaf above the bud and several below. It can be done by giving a slanting cut to the stock. It will force the buds to grow and will check terminal growth of the stock. The shoots arising from the stock should be headed back to suppress their growth. Following heading back of the stock, the latent buds of the stock may also sprout and it is always necessary to rub off these sprouts as and when they appear. June budding is common in stone fruits like peach, plum, apricot, cherry, nectarine and subtropical fruits like *ber* and *aonla* etc.

FORMATION OF BUD UNION

Like graft union, a series of changes take place in the formation of successful bud union also. In general, four stages *viz*. per-callus, callus, formation of cambial bridge and healing process etc. take place for the formation of successful bud union, which are described hereunder :

Pre-callus Stage

This is called as the first stage of bud union formation. During this stage, there is no sign of callus formation and scion (bud) falls if the wrapping material is removed from the bud union. This stage lasts for 5 to 8 days after budding operation and is different for different species.

Callus Formation

After 5th to 8th day of budding, the callus formation may take place in the bud union. The process of callus formation is fast in the firmly attached stock and scions, and if the stock and scion are not in a close proximity, the callus formation and proliferation process is inhibited. The initiation of callus formation takes place either from stock, scion or both. Usually, living cells of recently formed xylem and phloem play active role in the callus formation but sometimes cells in the cortex or pith may also produce callus.

Formation of Cambial Bridge

The parenchymatous cells of the callus form a cambial bridge within 12-15 day after budding. Similarly, after the formation of a cambial bridge, de-differentiation of secondary xylem and phloem takes place after 36-48 days of budding. After that, there is a partial movement of water and nutrients from soil to the stock and from stock to the scion starts, which establishes a better contact between the scion and the stock. Thus, the bud union in complete within 6-7 weeks if conditions are favourable. However, if there is no better cambial contact with the stock and scion, the union may be weak and it may break after a few days of budding or may be broken by strong winds.

Healing of the Bud Union

After de-differentiation of secondary xylem and phloem of both stock and scion, there is formation of many new xylem and phloem tissues after 6-8 months of budding. the callus unites completely with the scion pith and the union heals completely. Thus, like grafting, the success in the budding is also related directly with the proportion of live cells, environmental conditions, alignment of the cambial layer, uninjured cells near the proximity of both stock and scion. In general, better contact and alignment of the cambial cells of stock and scion helps in development and proliferation of callus. It prevents excessive accumulation of wound periderm and infiltration of resinous material, which hinders the healing process adversely.

• • •

Chapter 34 Methods of Budding

Nurserymen employ various methods of budding but according to convenience in performing the operation and percentage of success, the following methods are commercially advocated in propagating various horticultural crops :

Shield or "T" Budding

As the name indicates 'shield' is the shape of the bud and 'T' is the shape of cut given on the rootstock for the budding operation. 'T' budding is the most common propagation method of budding used by the nurserymen throughout the world. For shield budding, one-year-old rootstocks seedling of 25-35 cm height and 2-2.5 cm thickness is selected. The bark of seedling should slip easily. The selected and well-prepared bud-wood of desired cultivars is inserted on the stock at a height of 15-20 cm above the ground level and tied tightly with a thin polyethylene strip.

For actual operation of budding, a 'T' shaped cut is made on the selected portion of the stock with the help of a sharp budding knife (Figure 34.1). Propagators must be careful to give incisions through the bark and not through the wood. The two flaps of the bark are then loosened slightly with the help of a budding knife. From the bud-wood, which is selected from a healthy shoot of current season's growth, the buds of the middle portion are selected. These are removed from the bud wood by cutting shallowly about 5-6 mm below and 2-3 cm above the bud. This shield piece containing a bud is inserted carefully in the 'T' shaped incision made on the rootstock. This bud is pressed firmly and tied with a polyethylene strip, tape, paraffin tape or raffia fiber or polyvinyl chloride (PVC) sheet to hold the two components firmly until healing is completed (Figure 34.2). To achieve desired results, the shield should be covered by two flaps of bark but bud itself should be exposed.

After the bud has sprouted, the stock is cut to about 10-15 cm above the bud. When the healing process is over and bud has attained the length of 15-20 cm, the remaining portion of stock is cut right above the joint. Plants with thin bark, with sufficient flow of the sap like apple, pear, plum, apricot, cherry, rose and citrus are propagated by this method.

Inverted 'T' Budding

As the name indicates, the cut to be given on rootstock is reverse to that of 'T' *i.e.* the inverted 'T' cut is given on the rootstock. This kind of budding is widely used in areas of high rainfall because in such localities, the water running down the stem of rootstock may enter the 'T' cut, which may cause decaying of both shield and stem bark, resulting in the poor survival. Similarly, some plants (*e.g.*, chestnut) bleed badly during budding and thus inverted 'T' budding is used for their propagation, which allows better drainage and healing of the bud union in such plants. Technically inverted 'T' budding is the same as shield budding except that the 'T' cut on the stock is inverted. In this method, the shield bud is prepared in a fashion similar 'T' budding. The shield containing the bud is inserted into the stock by pushing upward beneath the flaps of the inverted 'T' cut. However, shield containing a bud in not inserted in an inverted position on the stock because the budwood then have reverse polarity, which may result in poor development of the shoot.

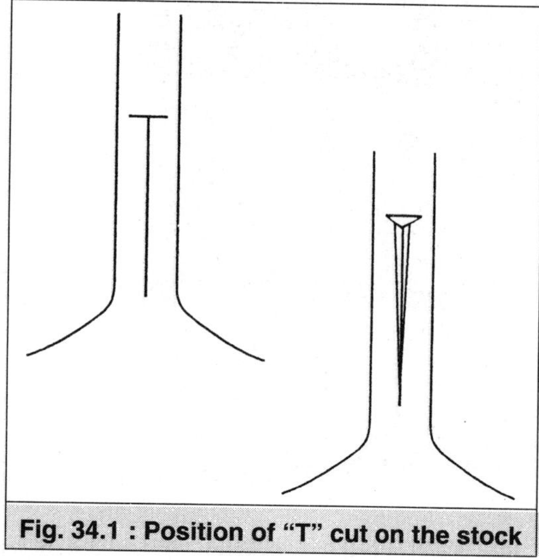

Fig. 34.1 : Position of "T" cut on the stock

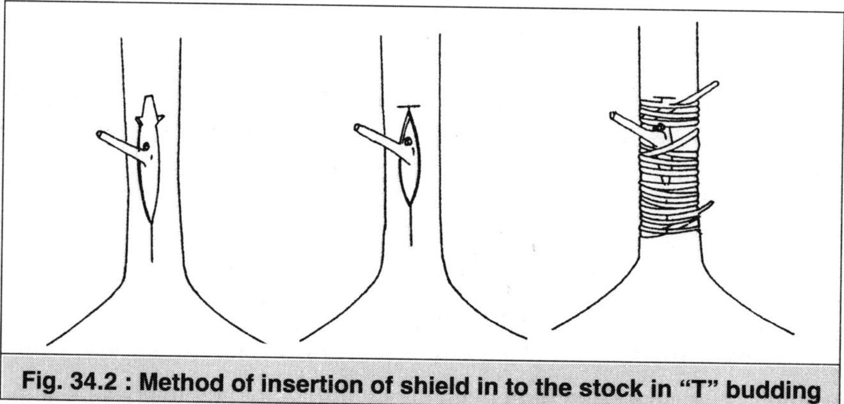

Fig. 34.2 : Method of insertion of shield in to the stock in "T" budding

Patch Budding

An important feature of patch budding is that a rectangular patch of bark is removed completely from the stock and replaced with a similar patch of bar (Figure 34.3), containing a bud of the desired variety. This kind of budding is slow and difficult to perform. However, it is successfully used in plants having thick bark such as walnuts and pecans, where 'T' budding gives poor results. The essential requirement of patch budding is that the bark of both the stock and scion should slip easily. Thus, it is usually done in late-summer or early-fall and rarely in summer. However, the right time may be decided by the propagator depending on the species and cultivar to be propagated.

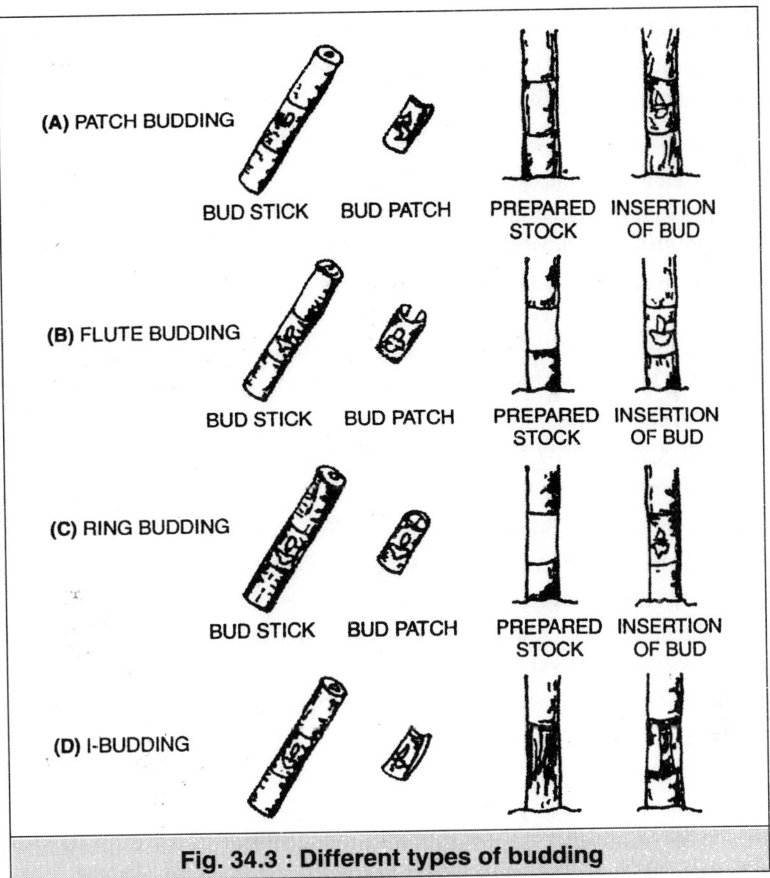

Fig. 34.3 : Different types of budding

In patch budding, the stock and scion should preferably of soma thickness (20-25 mm). For actual operation of budding, first a rectangular piece of bark (25 mm long and 10-15 cm wide) is removed from the stock and a similar patch, containing a bud is removed from the scion of the desired cultivar by making two horizontal cuts above and below the bud and then two vertical cuts connecting the horizontal cuts given. After removing the patch, it is immediately inserted on to the stock. The patch containing bud should fit tightly at the top and bottom. It is them wrapped either with an adhesive tape or a polythene strip, keeping the bud uncovered. The wrapping material should hold the bark tightly and cover all the cut surfaces to prevent free entry of air or water. After the bud starts sprouting, the stock above the bud union may be cut off step-by-step. In addition to walnut and pecannut, mango, rubber plant, *anola,* jackfruit and *jamun* are also propagated by this method.

Ring or Angular Budding

In ring budding, a complete ring of bark is removed from the stock and it is completely girdled. A similar ring of bark containing a bud is removed from the bud-stick and inserted on to the rootstock (Figure 34.3). In ring budding, both stock and scion should be of same thickness. However, it is applicable in stock having thickness below 2 cm. It has been utilized in *ber,* peach and mulberry because the newly emerged shoots from the heavily pruned plants are capable of giving such buds for budding, which can be easily separated. In this method, since the stock is completely girdled and if the bud fails to heal in, the stock above the ring may eventually die.

Chip Budding

Chip budding is successful method of budding when the bark of the stock does not slip easily and plants are not in active growth. The poor slipping of bark may be due to poor sap flow or adverse growing conditions, such as, lack of water, defoliation or low temperature, etc. These conditions may lead to tightening of the bark, which seriously interfere with the budding operation. Under such circumstances, 'chip budding' is done in the early-spring, summer or autumn. It is not as simple as is 'T' budding but is similar to side veneer grafting, wherein the scion is reduced to a small piece of wood containing only a single bud. Grape, apple and pear are propagated by this method. Besides, grape in dormant stage is also budded on nematode or phylloxera tolerant rootstock by this method.

In actual operation of chip budding, a chip of bark and wood is removed from the smooth surface between the nodes of the stock. A chip of similar shape and size is then removed from the bud-wood of the desired cultivar. For which, a 2-3 cm long downward cut is made through the bark and slightly in to the wood of the stock. Then, a second cut of about 2.5 cm is made so that it bisects the first cut at an angle of 30-45°. In this way, the chip of wood is removed from the stock. The bud chip is then slipped in the place of rootstock from where chip is removed (Figure 34.3). However, it may be assured that the cambium of the chip bud should have direct contact with the cambium of the stock. It is then tightly wrapped with some wrapping material like polyethylene strip, leaving the bud uncovered. The bud may sprout after 3-4 weeks, and afterwards, the wrapping material may be removed, allowing the bud to swell properly. When the bud starts growing, the stock may be cut close above the union.

I-budding

In 'I' budding, the bud patch is also cut in the form of a square or rectangle as in patch budding. Then, two transverse cuts are given on the smooth surface of the stock. These cuts are then joined at their centres by a vertical cut to produce the 'I' figure. These two flaps of the bark are then lifted and the bud path is placed beneath these flaps. The patch is tied with an alkathene tape or some other material. In tying 'I' budding, care should be taken that the bud patch should not bulge outside (Figure 34.3). 'I' budding is more successful in plants having thicker bark than bud-stick.

Flute Budding

In flute budding, the bark patch from the stock is removed in such a way that it almost completely encircles the stock except a narrow bark connection between the upper and lower cuts on the stock. A similar patch of bark is removed from budstick containing a healthy bud. The shield containing the bud is then inserted in the vacant area of the stock. The shield should fit tightly on the stock. It is then wrapped with a suitable wrapping material, leaving the bud uncovered. The other procedure of flute budding operation is the same as in patch budding. Because of presence of a narrow connecting strip of bark on the stock, it remains alive even if the bud fails to sprout.

Forkert Budding

In forkert budding, the stock is prepared by giving two vertical cuts and a transverse cut above the vertical cuts to join them. The bark is removed carefully along these cuts so that the flap of bark hangs down. The scion is prepared in a fashion similar to patch budding, having the size similar to the cuts made on the stock. The scion is then slipped into the exposed portion of stock and the flap is drawn over the inserted bud patch. This is then tied with a suitable wrapping material. After successful growth of bud, the portion of stock above the union is removed carefully.

Skin Budding

Skin budding can be done in any season when plants are not in active growth and skin does not slip easily. It has been found to be very useful in citrus where it can be done at any time of the year. In skin budding, first of all, an oblong piece of 2.0 cm long bark is removed from the stock. The cut in the stock is given in such a way that it does not injure the wood. A bud of a desired cultivar is removed from a shoot of current season's growth. The bud is then placed on the stock containing a thin layer of bark alongwith it. It is then tied firmly with some wrapping material like polyethylene strip and sealed with melted wax. The stock is looped off above the bud union after the healing has taken place and bud has started taking active growth.

Micro-budding

Micro-budding is used successfully for propagating citrus in Australia. It is similar to 'T' budding except that the shield (bud piece) utilized in it is thin and tiny. Like 'T' budding, the micro-budding is also not done under aseptic conditions. For micro-budding operation, the petiole is cut off just above the bud and then bud is removed from the bud stick by a flap cut just underneath the bud. Thus, only the bud itself and a small piece of wood underneath the bud are utilized in micro-budding, In stock, an inverted 'T' cut is made as is done in the inverted 'T' budding and the tiny shield containing the bud is inserted in it and later tied with a thin plastic tape. The tape may be removed soon after (15-20 days) the healing has taken place.

Top Budding

Like grafting, the old, declining and thrifty orchards of inferior cultivars can also be rejuvenated through top working, following one or the other method of budding. Thus, when budding is used to rejuvenate such orchards, the process is called as *top budding*. It can also be done on young vigorous plants having plenty of shoots. In older trees, the branches are headed back a year before the actual operation of budding to induce vigorous sprouts. Budding with any method can be done on the selected shoots with a desirable cultivar. The number of buds to be utilized for top working depends on the size of the tree but usually 10-15 buds per tree are sufficient. 'T' budding is usually practiced in thin barked species and patch budding in thick barked species. Further, it can be done in spring or rainy season, depending on the climatic conditions of the locality and fruit crop to be rejuvenated. After budding, all other unbudded branches should be removed at the base. Similarly, the shoots arising in summer should also be removed regularly.

BUD-WOOD SELECTION

For any successful budding method, the importance of a healthy and good budwood is always realized. In fact, the success of budding operation mainly depends on the bud-wood. Thus, performance of an orchard lies on the selection of bud-wood and its subsequent budding on the rootstock. Therefore, a propagator should always prefer to collect bud-wood from vigorous, healthy, productive, true-to-type mother plants, which are giving consistent yield of high quality fruits. It is important to mention here that viral diseases are considered as potent factor, which are mainly transmitted through infected bud-wood. In many advanced countries, a due attention is given for bud-wood certification programmes to avoid the spread of diseases through bud-wood. It is thought that the spread of citrus decline in India is mainly because of absence of any budwood certification programme, which has ultimately resulted in the spread of viral diseases of citrus in different parts of India.

SELECTION AND PREPARATION OF A BUD-STICK

Well-matured round twigs of current season's growth are first selected from the mother plants for the collection of buds. The leaves of the bud-stick are trimmed, leaving only a short piece of the leaf stalk or petiole attached to the bud. It helps in handling of a bud-stick. The bud-sticks should be collected just before they are to be used. If the bud-sticks are to be sent to distant places, these should be kept in wet newspaper, packed with moist sphagnum moss grass and kept in a cool and shady place. Usually, the plumpy healthy buds from middle portion of shoot are most desirable and give best results and therefore, small, angular and immature wood of terminal portion should not be selected as these do not give desirable results (Figure 34.4). However, in certain species such as sweet cherry, buds on the lower portion of the shoot are flower buds, which of course, should be avoided.

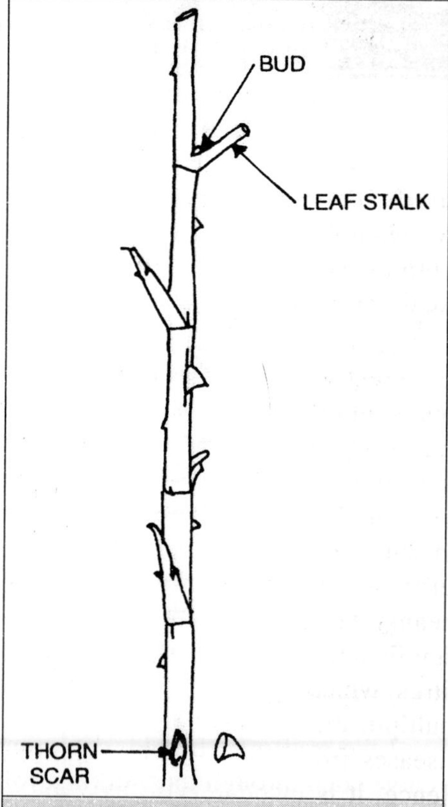

Fig. 34.4 : Prepared bud-stick

• • •

CHAPTER 35

Bud-Wood Certification in Citrus

Vegetatively propagated horticultural plants may have high rate of infection by viruses, viroids and phytoplasma, which are transferred from the one plant to another during propagation. To ensure disease-free planting material, bud-wood certification is mandatory. A brief description of bud-wood certification programme and the process of virus and virus-like diseases detection and their elimination are discussed in this chapter.

Vegetatively propagated plant has two plant-parts the rootstock and the scion. Scion or bud-wood is collected from mother plants. Hence, mother plants are procured from authentic source and should be of known pedigree records in respect to health, vigour, bearing, yield and good quality fruits. Several pathogens, especially virus and virus-like diseases, are transmitted through the bud-wood or scion during the propagation. These pathogens cause systemic diseases and proliferate in vegetative parts of the plant. Hence, it is almost entirely difficult to eliminate such infections in field situations. The infected plants in the orchard serve as secondary source of spread to the other plants through vectors and mechanical means. The uncontrolled movement of non-certified planting material has contributed significantly to the spread of several virus and virus-like diseases in fruit crops particularly in citrus, which is responsible for decline. Unscientific methods of bud-wood utilization in the multiplication of planting material is mainly responsible for the spread of virus and virus-like diseases from one region to another within a country and from one country to another. Hence, it is mandatory to have bud-wood certification programme in the production and supply of quality planting material for the growers.

The bud-wood certification can restrict or exclude the entry transmissible pathogens into the production environment and lower the inoculums potential of transmissible pathogens. Citrus is one of the most affected fruit crops by these transmissible pathogens. Hence, it is necessary to have orchards planted with disease-free nursery stock to realize actual potential of yield and quality of a cultivar. Bud-wood certification programme is an essential component in nursery raising to get disease-free planting material, which is currently lacking in most of the Indian nurseries. The basic requirements of such a programme are insect-free propagation facility, diagnostic laboratory, foundation block, bud-wood increase block, regulatory agency, and well trained scientific and skilled man-power.

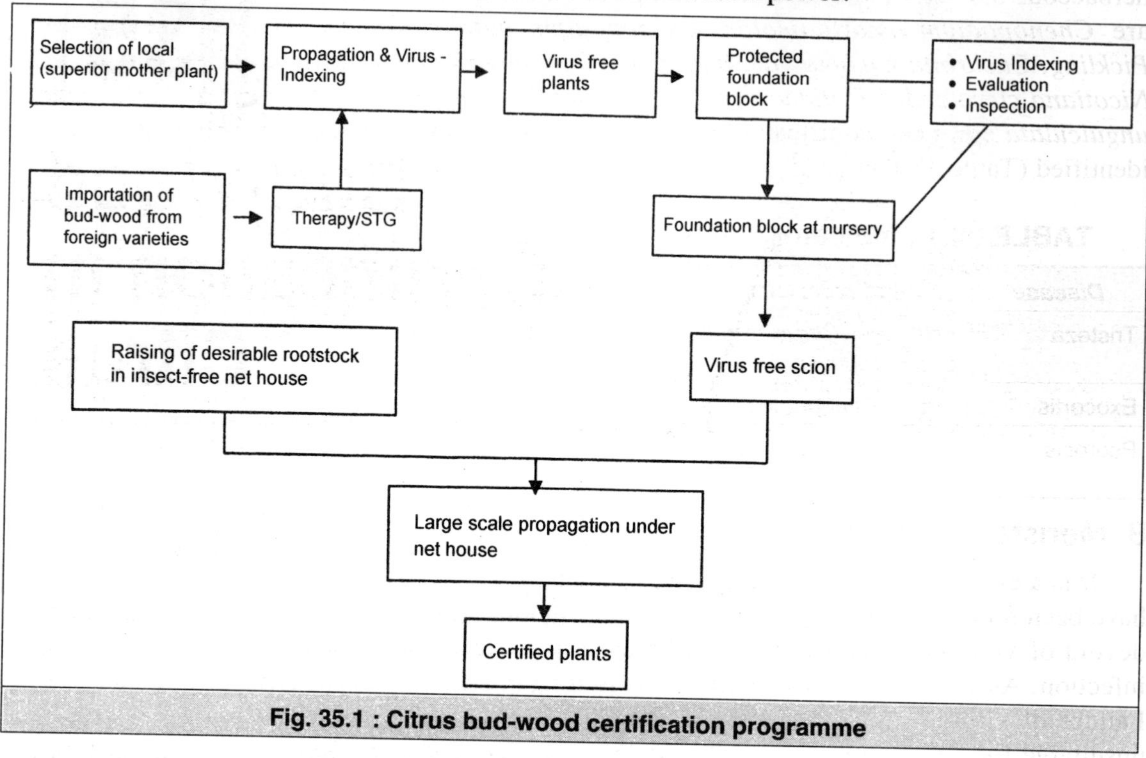

Fig. 35.1 : Citrus bud-wood certification programme

Bud-wood certification programme emphasizes that only indexed and disease-free plants should be used as bud source. Such disease-free plants should be marked and bud-stick should be carefully taken to avoid infection. Bud-wood should always be selected from fairly well matured wood of current season's growth or next to last growth. Round twigs having longitudinal white streaks on the bark with swollen buds are preferred in bud-stick selection. Budding knife should be disinfected with alcohol or sodium hypochlorite solution before use in budding operation. Bud-wood should be kept in moist sphagnum moss grass and gunny bag to avoid desiccation.

METHODS FOR DETECTION OF VIRUSES

The various strategies to get virus-free plants, which are an essential part of bud-wood certification (Figure 35.1) have been discussed hereunder :

A. Biological Indexing or Bio-diagnosis

Biological indexing is an oldest method for determining the presence of viral diseases in the plants. It is achieved by visual inspection of the plants followed by budding or grafting of a scion from a suspected plant onto other susceptible (indicator) plants identified for each virus. The indicator plant produces the typical visual symptoms such as gumming around the bud, stunted and deformed growth etc., which indicates the presence of virus on the source plant. Biological indexing or assays are time, space and labour consuming. Some of the herbaceous indicator plants used for biological indexing of mechanically transmitted viruses are *Chenopodium amaranticolor, Chenopodium quinoa, Cucumis sativus* var. *Boston Pickling, Cucurbita maxima var. Buttercup, Gomphrena globosa, Nicotiana benthamiana, Nicotiana cleveland, Nicotiana occidentals, Nicotiana tabacum, Phaselous vulgaris, Vigna unguiculata* etc. For identification of viruses in citrus, several indicators plants have been identified (Table 35.1) :

TABLE 35.1 : Indicator plants for virus and virus-like diseases in citrus

Disease	*Indicator plant*	*Disease*	*Indicator plant*
Tristeza	Acid lime, *Citrus excelsa, C. macrophylla*	Xyloporosis	Sweet lime, Orlando tangelo
Exocortis	Citron, Rangpur lime	Vein enation	Acid loime, sour orange, rough lemon
Psorosis	Sweet orange, lemon, acid lime	Greening	Sweet orange, grapefruit

B. Meristem Culture

It is a established fact that apical meristematic tissues of stems and roots of some plants have been found to be free of viruses. Root and shoot apices of virus infected plants are often devoid of virus particles and this can be used as the basis for *in vitro* eradication of virus infection. Aseptic culture of meristems excised from stem tips produce virus-free plantlets. Failure of virus to invade meristems may be due to high concentration of auxin that is unsuitable for replication of several viruses or because the meristems are fast growing and viruses are not able to move with the same speed leaving these tissues free from infection. Sometimes elimination of viruses is not possible, then thermotherapy treatment is given before meristem tip culture. Thermotherapy can be done by soaking dormant plant parts (tubers, cuttings, bud-wood) in water at 50°C for a period ranging from several minutes to couple of hours, or by exposing growing plant parts to an air temperature of 35 to 40°C for some weeks.

C. Electron Microscopy (EM)

Electron microscopy is the most convenient method of detection of viruses on the basis of virus morphology. It has the ability to detect viruses within a few minutes in crude plant sap, which ultimately helps in facilitating the selection of healthy planting material to some extent. However, this test should always be supported by other tests. Serology with EM in tests like immunosorbent electron microscopy (IEM) and immunogold labelling will increase the sensitivity of virus detection manyfolds. Electron microscopy, however, is still the best technique to detect phytoplasmas and fastidious bacteria in plants by viewing ultra thin sections. IEM is a powerful tool for detecting specific virus even in mixed infections in decoction tests and also through trapping tests when the concentration of the virus is very low.

D. Serological Methods

The serological techniques are indirect methods of virus detection. Earlier, it was performed using precipitin tests and gel diffusion tests, which have now been replaced by enzyme linked immunosorbent assay (ELISA), a methodology developed in late 1970's by Clark and Adams. Direct antigen coating (DAC) ELISA, double antibody sandwitch (DAS) ELISA and indirect Triple Antibody Sandwitch (TAS) ELISA are more advanced variant of ELISA for detection of viruses. Out of these, DAS ELISA by using monoclonal (CTV) and polyclonal (CTV, RS, mosaic) antibodies are more common for testing the present of viruses in plant sample. 1-10 ng/ml is the detection limit for serological methods.

E. Dot Blot Immuno Binding Assay (DIBA)

It is another protein based serological detection method based on immuno-blotting of nitrocellulose membrane. It is particularly useful when a facility for serological assay is not available at the site of propagation. The dotted membrane can be mailed to referral laboratory for virus testing. The protocol for this test is based on ELISA but nitrocellulose membrane is used instead of polystyrene plates.

F. Immuno Fluorescence (IF)

This method is used for diagnosis of the pathogens restricted in phloem tissues such as fastidious bacteria and phytopiasmas.

G. Molecular Techniques

Recombinant DNA (r DNA) and genome characterization technology have enabled to diagnose presence of viruses in plants using viral nucleic acid. Usually, two basic approaches of diagnostics namely nucleic acid hybridization and nucleic acid amplification are used. Nucleic acid hybridization is a useful and sensitive tool for virus and viroid detection in crops like potato and citrus and also for detecting citrus greening bacterium. However, the detection process is slow as compared to nucleic acid amplification technique using Polymerase Chain Reaction (PCR), where simultaneous detection of presence of more than one virus (mixed infections) is possible. Nucleic acid amplification of DNA is done *in vitro* by using 4 dNTPs and two oligonucleotides (primers) flanking the segment of DNA to be amplified by the Polymerase Chain Reaction (PCR). A thermostable *Tag* DNA polymerase enzyme catalyzes the reaction. PCR can not be applied directly for single stranded RNA viruses and therefore, the RNA is first reverse transcribed into c-DNA using a reverse transcriptase (RT) enzyme. Immunocapture PCR (IC-PCR) in which virus is trapped by antibodies coated on the wall of microfuge tube and then nucleic acid of virus is released for c-DNA synthesis and subsequent PCR. The RT-PCR has also been useful for identification of plant viruses. Less than 0.001 ng/ml RNA is detection limit using hybridization technique while less than 0.0001 ng/ml RNA is the limit for PCR techniques.

Serological and molecular techniques for detecting virus and virus-like diseases in citrus have been given in Table 35.2.

TABLE 35.2 : Serological and molecular techniques for detecting viruses and virus-like diseases in citrus

Disease	*Detection Technique*	*Disease*	*Detection Technique*
Tristeza	Inclusion bodies, ELISA, electron microscopy, fluorescent antibody, single strand conformation polymorphism (SSCP)	Exocortis	Electrophoresis, Dot biot hybridization, return polyacrylamide gel electrophoresis (R-PAGE) and PCR
Tristeza (seedling yellows)	ds RNA, get electrophoresis, florescent marker, serology, monoclonal antibodies MCA 13	Greening	Electron microscopy, ELISA using monoclonal antibodies, Nucleic acid hybridization
Ring spot	Electron microscopy or ISEM, ELISA using monoclonal antibodies	Citrus mosaic	ELISA, electron microscopy

• • •

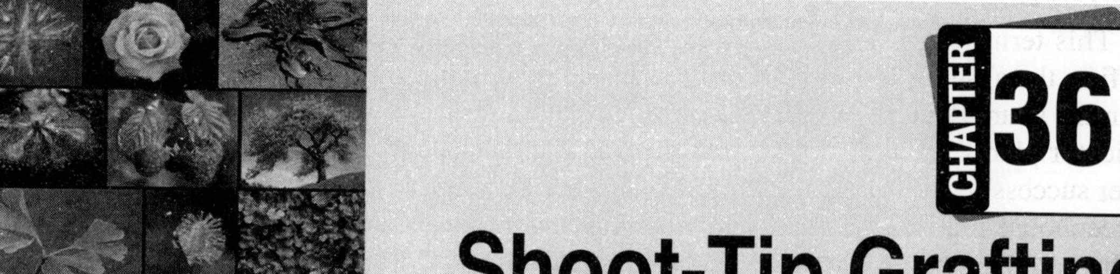

CHAPTER 36

Shoot-Tip Grafting and Micro-budding in Citrus

There are several viruses, which singly or together as complex lead to death of the citrus plants. The principle method of spread of viruses is through use of bud-wood for propagation or excessive activity of insect vectors. Hence, realizing the seriousness of this problem, meristem culture technique was standardized, which was later used in shoot tip grafting on an indicator rootstock seedling for confirming the presence or absence of viruses. Realizing the principle means of spread of most citrus viruses and virus-like diseases in India, several programmes were initiated at PAU, Ludhiana, IARI, New Delhi, North Bengal University, Cooch Bihar, AAU, Jorhat, CHES, Chethali, to clean the infected citrus plants from viruses. However, a comprehensive citrus improvement programme through *in vitro* shoot-tip grafting (STG) was started at National Research Centre on Citrus, Nagpur in 2001 to clean the selected mother trees of Nagpur mandarin to produce and distribute certified disease-free and quality bud grafts on a large scale to the citrus growers/nurserymen.

SHOOT-TIP GRAFTING (STG)

STG is a technique based on cleaning of the stock using *in vitro* culture of apical meristem, which are suppose to contain little or no virus(es). It has been used successfully to produce virus-free citrus plants. This technique involves the precise excision of a minute (0.14 to 0.17 mm) shoot-tip from a young shoot of a plant, which may be contaminated with a graft-transmissible pathogens, and its successful grafting on a decapitated rootstock that is grown on a sterile agar-solidified medium. The grafted rootstock is then placed in a test tube containing liquid growing medium and kept under light for several weeks. Thereafter, emergence of leaves from shoot-tip grafted scion takes place in the test tube. These are then transplanted into pots filled with soil media and maintained in a greenhouse. It is not until 1 to 2 years after STG that the plants are again tested to confirm whether these plants are free from viruses or not. This is achieved by using virus-indexing employing different methods as listed hereunder :

A. Biological Indexing

This term refers to the preferred method of testing STG or other citrus source trees to confirm the absence or presence of viruses and other pathogens. In biological indexing, tissue from the source tree is inoculated or grafted onto a series of indicator rootstock plants. These indicator plants show symptoms for specific pathogens under specific temperature regimes. After successful inoculation, indicator plants are cut back and placed in a chamber or room of a greenhouse that has the specific temperature range for symptom development of expression of the pathogen being tested. The plants are checked for disease symptoms over a period of months. If no symptoms (over a period of months) are seen, then the plants are declared free of the pathogens and mass multiplication can be started thereafter. The major diseases so tested in citrus are CTV, citrus tatter leaf virus (CTLV) and citrus psorosis virus.

B. Laboratory and Serological Tests

The main laboratory test performed for viral testing is the enzyme-linked immuno-sorbent assay (ELISA) for CTV. ELISA is a serological test based on an antigen-antibody reaction that is routinely used for the detection of pathogens. There are numerous variations to the ELISA. However, the one we generally perform is to use a polystyrene plate to trap potential CTV virus particles in plant sap using two antibodies thus trapping the virus referred to as the double antibody sandwich method. Furthermore, use of polyclonal antibody for detection of all known CTV strains is also in vogue, however, a monoclonal antibody is sometimes used to detect certain severe strains of CTV.

In contrast to biological indexing, ELISA for CTV is a quick method to test or absence the presence of the virus, as hundreds of samples can be tested in a short period of time. However, ELISA and other laboratory tests cannot replace the comprehensiveness of biological indexing, which allows the detection of most of the graft-transmissible citrus pathogens. The ELISA plate after testing show the presence of CTV and the wells that are yellow coloured are positive for the virus (CTV). The healthy stock are then shifted to glasshouse and maintained as foundation block.

C. Foundation Block

A foundation block is a collection of virus-free trees that have been biologically indexed for a number of pathogens. The trees from a foundation block become the source trees for bud-stick if they are true-to-type after horticultural evaluation and are also virus-free. They are periodically re-indexed to confirm their virus-free status. Foundation block trees are on a number of different rootstocks and are spaced far apart to prevent possible mechanical transmission of pathogens. Buds from these trees are cut and budded onto rootstocks in an increase block.

D. Increase Block

An increase block is created from buds cut from a foundation block. These trees are spaced closer in nursery setup to increase the number of buds. Bud-wood from these trees are cut and distributed to nurserymen and growers. These trees can be used up to four years, provided an annual citrus tristeza virus test is negative. Increase block trees are planted in a cyclical rotation, to ensure an ample bud-wood supply, as older trees are eventually pushed out and replaced.

E. STG-based Citrus Quarantine

The use of STG is also being adopted worldwide for the import of germplasm from different sources both within and outside the country. A quarantine programme based on tissue culture procedure *In vitro* bud-stick culture and STG is playing a lead role in Spanish and Japanese citrus industry. It is technically superior and safer compared to the traditional quarantine procedures. The technique consists of culturing *in vitro* imported bud-sticks at a constant 32°C day and 30°C night temperature for induction of new flushes in plants from which shoot-tips are excised and micro-grafted *in vitro*. This technique helps as a post entry quarantine check for preventing entry of any virus and virus-like diseases along with the introduced bud-wood material.

Merits of Shoot-tip Grafting Technique

- It is one of the potent methods to obtain virus or viroid-free plants from sick or infected stocks.
- It can be used to rejuvenate old and declining stocks.
- It helps in the study of graft incompatibility in specific stionic combinations based on histological, biochemical and physiological aspects.
- It is used in quarantine procedure for testing the presence or absence of viruses in the exotic material.

This technique has been successfully demonstrated in over 60 species including dahlia carnation, orchids, strawberry citrus, potato, sweet potato, cassava, apple, *Prunus* sp. etc.

Procedure of Shoot-tip Grafting

Prior to employing STG, the rootstock seedlings are raised *in vitro* using blank medium or growth regulators supplemented culture medium. The cultures are incubated in dark in culture room. The proliferated cultures raised from meristems followed by indexing are used as shoot-tip source. The two to three-week-old seedlings are decapitated and a wedge (for apple) or inverted 'T' (for citrus) cut is given aseptically in a laminar air-flow cabinet using a dissecting microscope. The shoot-tips are then excised using a single or double sided slant cut to precisely obtain the apical dome with few leaf primodia. The scion bud is then placed in the slit keeping the cambial contact in mind. The grafted plants are then transferred onto a filter paper bridge with tap root passing through the slit. After few weeks, the swelling of buds and further growth suggest the success of graft union. The developing plant is transferred to hardening unit to improve the *ex vitro* survival. The growing plants are then maintained under glasshouse/insect proof condition for almost one year before they are used for planting in prepared field.

Micro-budding

Micro-budding is a novel technique developed by scientists of Texas and AM University, USA to produce virus-free grafted citrus plants in large number. The procedure is based on budding rootstocks that are only few week-old, using small scion buds, and no taping is done to keep the bud in place. This technique enables the production of budded plants much faster and at a cheaper cost. Furthermore, it occupies less space compared to other conventional systems. The procedure works on several scion-rootstock combinations throughout the year, under greenhouse conditions. Micro-budded plants have been used in biological indexing for tristeza, tatter leaf, and psorosis viruses and exocortis viroid. This technique was used to produce a large number of trees for a commercial trial in an ultra-high density fashion. Standardized micro-budding, a new propagation technique in citrus on just 5-6-month-old rough lemon or Rangpur lime rootstocks, saves grower's time, money, space and enables rapid and year-round multiplication of planting stock.

Procedure of Micro-budding in Citrus

It is a simple procedure in which budding in performed *in vivo* unlike *in vitro* as in STG. The virus-free indexed and certified plants are maintained under insect-proof nethouse or glasshouse. It has to be undertaken in the following manner :

Prior to collecting bud-sticks from such plant, the shoot-tips are decapitated. After three days, the bud-sticks are collected in a beaker containing distilled water to maintain the sap. The previously raised five-month-old seedling in insect-proof net are decapitated at 15 cm from base. Thereafter, a wedge cut is given with a sterile scalpel blade. Similarly, on scion-stick, the petiole is cut, leaving few mm length and 0.4 to 0.5 mm scion-bud are scooped out. The underneath wood portion is removed and the same is placed in the wedge with the bud facing upwards. The bud point is then covered with a micro-tip to keep the bud and rootstock tissue intact and also to maintain a good moisture level. The swelling of bud after 10 to 14 days suggests the careful removal if the micro-tip. The sprout is then allowed to grow. These are regularly sprayed with some plant protection chemicals to avoid insect vectors. The plants so raised are ready for field transfer after 6 months of initial operation.

• • •

CHAPTER 37

Rejuvenation of Old Orchards

Orchards of most of the fruit plants show steady decline in fruit yield after fruitful life. This productive life varies from fruit-to-fruit. For example, life of herbaceous fruit plants such as banana, strawberry etc., is 1 or 2 years, for perennial types, it may be 15-20 years (*e.g.* guava, citrus etc.,), 20-30 years (*e.g.* apple), or 40-50 years (*e.g.*, mango). However, the declining, old and unproductive orchards can be rejuvenated and made productive by top-working. Top working is chiefly employed for rejuvenating old and thrifty plants, changing inferior seedling trees with superior cultivars and replacing the damaged limbs of the fruit plants. Top working can be used in converting the old and declining orchards of mango, cashewnut and ber. Procedure for top working of these fruit plants is given below :

SEQUENTIAL PROCEDURE FOR REJUVENATION

(*i*) Selection of the Plants to be Rejuvenated

The plants, which are giving low yields of inferior quality even under proper management, should be selected. However, there is no limit to the age of the plants, provided they are healthy and vigorous.

(*ii*) Selection of Mother Plants for Bud/Scion Wood

The plants of known history should be selected as mother plants. The plants giving outstanding yield of superior quality should be preferred. The mother plants should be free from major insect-pests and diseases. After the selection of mother plants, these should be maintained properly by following adequate management practices.

(*iii*) Heading Back of the Plants to be Rejuvenated

For top working, the plants to be rejuvenated are headed back. The height at which the plants to be headed back, depends on the age of the plant. In well protected orchards, where stray animals (cattle, goat and sheep etc.) are unable to cause any damage, the small plants in pre-bearing age may be cut at a height of nearly 4 to 5 feet from the ground level. In old and vigorous fruiting plants, it is desirable to cut the branches at places where the plant is 6 to 9 inches in diameter. Thus, the height at which the branches should be cut, would depend on the size of the plant. On the fruiting plants, it is necessary to retain some lateral branches near to the ground to serve as nurse limbs and to provide shade to the trunk. These branches can be removed in the following years when enough crowns would develop by the new scion variety or varieties used for top working.

(*iv*) Methods for Top Working

Different methods can be employed while top working the fruit plants. However, the success is better with crown grafting, side-veneer grafting and budding. These methods can be used singly or in combination.

(*v*) Selection and Preparation of Scion/Bud-wood

For crown and side veneer grafting, shoots of 6 to 9 inches in length are preferred. These should be taken from the terminal mature shoots of current year's growth. If the length of the current year's growth is not sufficient, shoots of previous year's growth may be used. However, for budding, only current year's growth should be utilized. Such shoots are normally available on young vigorous plants. The new growth can also be induced in old and devitalized plants by pruning one or more branches a few months before the actual date of budding/grafting. Leaves from the selected shoots should be removed by secateurs about 7 to 10 days before the actual operation of budding/grafting, leaving the petioles attached to the shoots. During this period, the petioles that are left behind will drop down, thus completely defoliated shoots are ready for use. Only those shoots are used for grafting in which the apical buds have swollen but have not actually started growing. If the apical buds are found in an active state of growth, the shoots must be discarded. Such shoots can, however, be used for taking buds for budding purpose.

(*vi*) Time for Budding and Grafting

The crown grafting and side-veneer grafting can be done during the raining season, when there is a plenty of humidity in the atmosphere and the weather is cloudy. Budding is successful when the maximum temperature varies between 90° F and 100° F. Thus, spring in plains and the summer months in the hills are the best suited for budding. In plains, the month of June has also been found quite suitable if there are good pre-monsoon showers.

(*vii*) Operations of Budding and Grafting

Many budding and grafting techniques are employed for top working of fruit crops but crown grafting, side veneer grafting and 'T' budding give better success.

Crown grafting must be done immediately after the trunk or branch is cut. With the point of a strong knife, two vertical parallel cuts at a distance of 1/2 inch to 1 inch starting from the cut surface downwards to a length of 3 to 4 inches, should be made so that bark is completely cut. By inserting the ebony end of a budding knife or some other pointed material between the bark and the wood, the bark should be separated from the wood and kept back in its position. Two slanting cuts are then made in the bud wood, one nearly 3 to 4 inches in length and the other nearly 1/2. to 1 inch, to give it a wedge shape. It is important that the longer cut of 3 to 4 inches is given in one stroke with a sharp strong knife. The wedge is then inserted between the bark and the wood, the longer surface pointing inwards. Depending on the thickness of the trunk or branch, 2,3 or more grafts can be made in this way. These grafts should be tied firmly but not very tightly with a string used for stitching gunny bags. The cut surfaces should be covered with grafting wax (prepared by boiling together three part resin, two parts beewax and one part coconut or linseed oil). Grafting wax should be applied when it is just warm and not very hot.

In side-veneer grafting, two parallel vertical cuts in the bark are joined by a third horizontal cut on the upper side. To facilitate insertion of ebony end of the knife or other pointed material for lifting the bark in thick branches and trunks, it is necessary to remove a triangular patch of the bark with the horizontal cut acting as a base. This triangular patch can be put back in its position after inserting the wedge shaped portion of the bud-wood between the bark and the wood.

Budding is done on new shoots that grow from the pollarded trunks and branches. Budding should be done quite close to the trunk or branches and started as soon as the shoots are mature. Many shoots grow from the pollarded branches but budding has to be done only on some selected ones, which should be properly spaced. Remaining shoots should be rubbed off as soon as they appear.

Shield shaped buds without wood are removed from the defoliated shoots and are inserted between the bark and the wood by making only one vertical cut on the bark by loosening the bark with the ebony end of the budding knife. Budding tape prepared by passing strips of cloth through a hot mixture of the melted paraffin and resin (mixed in the ratio of 3:1) is used for tying the buds. To get best results, bud-sticks should be used soon after they are removed from the mother plants. They can, however, be stored for 4 to 5 days in damp gunny pieces or in polythene film.

(*viii*) After Care of the Top Worked Tree

Crown and side grafts normally grow very vigorously. These grafts should be protected from being blown away by the storms by giving supports. The string used for tying should be cut and removed when it is observed that the string is girdling the graft union. During summer months, a shade must be provided on the southern-western side to protect the bark from scalding. Watch should be kept for the stem borer, presence of which can be detected by noticing saw dust like powder on the ground.

Grubs should be traced by peeling off the bark. These can be easily removed and killed. If holes are found in the wood, it means that the grubs have entered the wood. Diesel oil or petrol mixture can be used for destroying them. When thick branches are cut, it takes many years for the new bark to cover the cut surface. The central portion of the wood starts rotting, making trunks and branches hollow, and thus shortening the life of the plants. To prevent this, the cut surfaces should be covered during raining season with a plaster, prepared by mixing 3 parts of sand and 1 part of the cement. Similarly, if the cut portions are plastered with Bordeaux mixture, it also helps in preventing the infection of the branches.

• • •

CHAPTER 38

Apomixis and Polyembryony

In general, the seed embryos are produced as a result of fertilization between the ovule and the pollen grains following meiosis. However, in some plant species, the embryo is not produced as a result of meiosis and fertilization, but from the diploid cells of the seed or from the cells of the embryo sac or surrounding nucellus such seed embryos do not under go meiosis but develop to form a zygote similar to the same genetic make up of female parent. This phenomenon of asexual reproduction in place of sexual reproduction is called as *'apomixis'*. The seedling plants raised through such seeds are called as *apomictic seedlings*. The genetic make up of apomictic seedlings is similar to the mother plant and hence propagation by means of apomictic seedlings is similar to vegetative propagation. Plants that produce only apomictic embryos are called *obligate apomicts* and the ones, which produce both apomictic and sexual embryos, are called as *facultative apomicts*.

TYPES OF APOMIXIS

Recurrent Apomixis

In recurrent apomixis, the embryo sac develops from the diploid egg mother cell or from some other diploid cells of the embryo sac with fertilization. Consequently, the egg has the normal diploid number of chromosomes similar to mother plant. The embryo subsequently develops from the egg nucleus without fertilization. This type of apomixis is quite common in *Parthenium,* raspberry, *Malus* species (apple), *Allium* (onion), dandelion and *Poa* etc. In some species, this event occurs without the stimulus of pollination, whereas in others, pollination appears to be necessary for the development of a viable embryo.

Non-recurrent Apomixis

In this type, the embryo develops directly from the haploid egg cell (haploid parthenogenesis) or some other haploid cells of embryo sac (haploid apogamy) without fertilization and as a result, the embryo developed is also haploid. This type of apomixis occurs very rarely and is primarily of genetic interest. The plants produced are haploid and sterile. It is common in *Solanum nigrum* and *Lilium* species etc.

Adventitious Embryony

Adventitious embryony is also called as *nucellar embryony*. In this type of apomixis, the embryo does not develop from the cells of the embryo sac but from a cell or a group of cells of nucellus or integuments. The cells of integument or nucellus are diploid and hence the resultant embryo is also diploid. Such embryos usually develop outside the embryo sac in addition to the regular embryo. This phenomenon is quite common in citrus, where pollination and fertilization takes place in a usual manner and sexual as well as apomictic (nucellar) embryos also develop.

Vegetative Apomixis

In this type, vegetative buds or bulbils are produced in the inflorescence in place of flowers. These buds or bulbils may sprout into new plants while they are still attached to the mother plants. This type of apomixis is quite common in *Allium, Agave, Poa* and *Diascorea* etc.

POLYEMBRYONY

The phenomenon in which two or more embryos are produced in a seed is called as *polyembryony*. This condition may result from many reasons. One of the most common reasons being the nucellar embryony, as in citrus. Sometimes more than one nucleus develops within the embryo sac, which may lead to the development of more than one embryo. Cleavage of pro-embryo during early stages of development may be the other reason for development of multiple embryos as in case of conifers. Among fruit crops, polyembryony is quite common in mango and citrus (Figure 38.1).

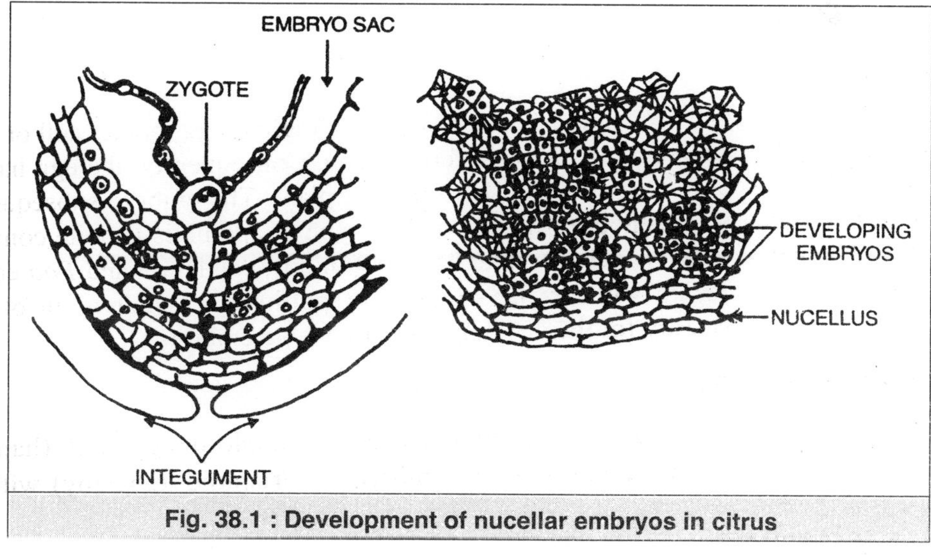

Fig. 38.1 : Development of nucellar embryos in citrus

SIGNIFICANCE OF APOMIXIS

Apomixis occurs in many plant species in nature. The ability of plants to reproduce apomictically has great significance in horticulture because of the following reasons :

- It provides an opportunity for producing genetically uniform population.
- Apomictic seedlings are useful for producing uniform rootstocks in apple (*e.g. Malus toringoides, M. sikkimensis M. sargentii*) and citrus.
- Apomictic seedlings are highly uniform and quite healthy than sexually produced seedlings.
- Apomictic seedling are helpful for identification, as trifoliate leaf character in citrus.
- Apomictic seedlings are suitable for screening viruses, which are not otherwise transmitted through seeds.

• • •

CHAPTER 39

Propagation by Layering

Layering is one of the oldest techniques used by the nurserymen to propagate many horticultural plants. Layering is a method of plant propagation, which is designed to induce plant stems to produce roots, while they are still attached to the parent plant. After proper rooting, stem is detached from the parent plant and planted in the nursery or field where it becomes a new plant, growing on its own roots. It is a reliable means of plant propagation and is adopted for species, which are difficult or impossible to root from the cuttings or other vegetative means. Layering does not lend itself to the production of large number of plants and therefore nurserymen use it only for difficult-to-root plant species. It is a natural means of propagation in black raspberries and trailing blackberries or it may be induced artificially in many plant species.

ADVANTAGES OF LAYERING

Layering has certain advantages over other methods of propagation as under :

- It is easy to perform and does not require much facility.
- It is an effective means of propagating species that usually do not root easily by cuttings as in guava, rubber plant, kumquat, filberts and litchi etc.
- It is the best method of propagation of plants, which reproduce naturally by layering *e.g.*, trailing blackberries and black raspberries.
- It does not require precise control on water, relative humidity or temperature, which is required for other methods of propagation.
- It is possible to produce large-sized plant with layering within a short time.
- Layering is useful for producing relatively a smaller number of plants of good size with minimum propagation facilities.

DISADVANTAGES OF LAYERING

Layering has certain disadvantages as under :

- It is not possible to produce large number of plants within a short time with layering techniques.
- It is a costlier technique in areas where labour availability is a problem.
- The plants produced through layering have usually small and brittle roots, as a result, survival rate of layered plants in the nursery may be less.
- In layering, the beneficial effects of rootstocks on the scion cultivar can't be exploited.
- The mortality rate in layers (particularly air layers) is usually high.

In general, better rooting in the layers can be obtained by ringing or wounding, etiolation (absence of light) and/or use of rooting hormones (IBA, NAA) and by providing favourable environmental conditions (especially temperature and humidity).

FACTORS AFFECTING THE SUCCESS BY LAYERING

As mentioned above, the roots in layers are brittle and small, which hinders their establishment in the field. Many factors affect the root initiation and their development in the layers. Some of the important factors are discussed below :

Nutrition : In layering, the stem remains attached to the plant during the process of rooting and the nutrients and water are continuously supplied through the intact xylem to the layers. Better nutrition and water supply by the mother plant help in better rooting in layers and poor supply affect rooting process adversely.

Stem treatments : Various manipulations like bending of the shoot, notching, girdling, wounding etc., are made in the stem, which cause interruption in the downward translocation of organic materials like carbohydrates, auxins and other root promoting factors from the leaves and the growing shoot-tips. These materials are accumulated near the point of treatment and help in better rooting of the layers. In general, rooting takes place easily and profusely if the stem is etiolated, wounded or girdled before layering.

Etiolation treatment : Exclusion of light (etiolation) from the part of stem where roots are to develop has strong influence on rooting of the layers. The greatest stimuli to root induction results when the newly developing shoots are continuously covered with the rooting medium as in case of mound/stooling and trench layering. In etiolation treatment, the base of the layered shoot (approximately 2.5 cm) is covered with black polyethylene or soil. It is never exposed to light and is allowed to develop roots by covering with the rooting medium.

Physiological condition of the mother plant : The physiological condition of the stock plant at a certain time of the year has dramatic influence on the rooting of layers. Thus, the time at which layering is done, affect the rooting process in the layers. In different types of layering, different timings are associated with the movement of carbohydrates and other substances, which has strong influence on rooting in layers. In general, rooting is better in layers if layering in done at the end of the seasonal cycle of the growth, when there is accumulation of plenty of carbohydrates and other growth promoters, which help in better root initiation and their further development in the layers.

Rejuvenation of the stock plant : In some techniques of layering as in stooling or trench layering, heading-back/cutting-back of the old plants is necessary for encouraging new shoot growth. These new shoots usually root better as compared to the older shoots. Therefore, it is always advisable to follow layering in the juvenile shoots rather than in the mature shoots.

Treatment with growth regulators : Usually, treatment of the wounded part of stem with auxins or other growth promoters helps in better rooting in the layers. Among different growth regulators, Indole butyric acid (IBA) and naphthalene acetic acid (NAA) are commonly used as rooting hormones in layering. The root promoting compounds are applied as a powder or as a paste in lanolin or as solution in 50 per cent alcohol to the girdled/portion of the layers. The concentration of the rooting hormone usually depends on the plant species, type of layering, age of the shoot and the environmental conditions. The root promotion in the layers, however, depends on moisture supply, aeration and temperature in the rooting zone.

The environmental conditions : The environmental conditions, specially temperature and relative humidity exert greatest influence on rooting of layers. For example, profuse rooting takes place in air layers if the humidity is high. Similarly, rooting is better in stooling if mounds are kept moist throughout.

TYPES OF LAYERING

Different types of layering techniques are used for propagating different horticultural plants depending upon the plant species, facilities and the environmental conditions. A brief description of these techniques is given hereunder :

(*i*) Tip layering : In tip layering, the tips of the shoots are bent to the ground and the rooting takes place near the tip of current season's shoot. It is the simplest form of layering, which often occurs naturally and is a natural method of propagation of blackberries, raspberries and dewberries. The stem of these plants completes its life in two years. During first year, vegetative growth takes place, while in the 2nd year, fruiting takes place. After harvesting of the fruits, the plants are heavily pruned, which give rise to a number of lateral shoots. The tips of these shoots are buried 5 to 10 cm deep in the soil (Figure 39.1). Rooting in

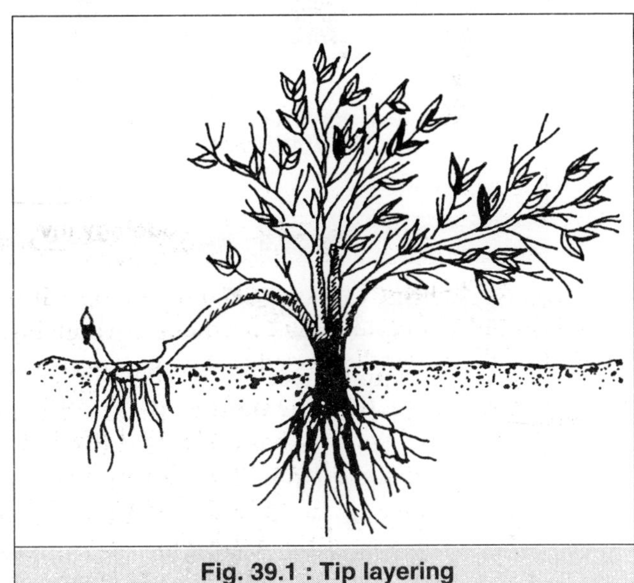

Fig. 39.1 : Tip layering

the buried shoots takes place within a month or so. The new plants (layers) may be detached and transplanted in the nursery during spring. Currants, gooseberries and rambling roses can also be propagated by tip layering easily.

***(ii)* Simple layering :** It is the easiest method of plant propagation used by the nurserymen and plant propagators to propagate many horticultural plants. It is similar to tip layering except that some manipulations are done to the buried shoot for better rooting. In actual practice of tip layering, the flexible shoots of a plant are bent downwards over to ground in early-spring or in rainy season. A second bend is made in the shoot, a short distance from the tip, which is covered with soil and held in place with a wire or a wood stake. This portion is sometimes injured or wounded either by notching, wiring or girdling to stimulate rooting (Figure 39.2). Rooting in shoots takes place within 2-3 months, however, the layered shoots should be examined carefully before severing and if insufficiently rooted, these should be left in the field for sometime. Plants propagated through tip layering include magnolias, rhododendron, citrus and filberts. The best time for simple layering is early-spring.

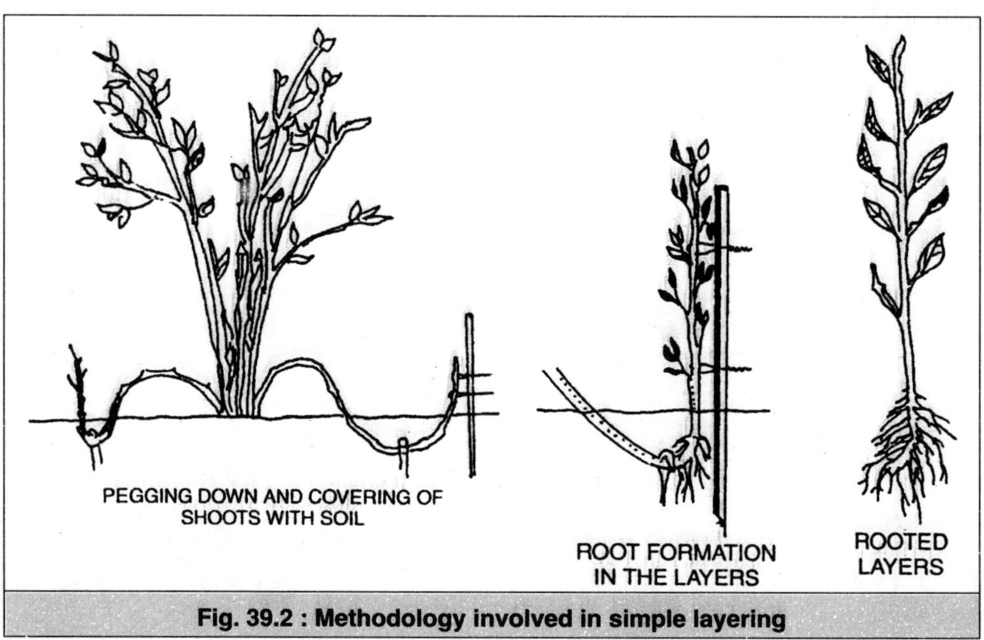

Fig. 39.2 : Methodology involved in simple layering

***(iii)* Serpentine or compound layering :** It is also an easy plant propagation method to perform but is only suitable for plants producing slender, long and flexible shoots. It is a modification of simple layering in which one-year-old branch is alternatively covered and exposed along its length (Figure 39.3). The stem is girdled at different points in the underground part. However, the exposed portion of the stem should have atleast one bud to develop a new shoot. After rooting, the sections are cut and planted in the field. In this way, many new plants can be prepared from one branch. Muscadine grape is commercially propagated by this method. Serpentine layering is also used for propagation of ornamental plants like honeysuckle, *Wisteria chinensis,* Jasmine and *Clematis* etc.

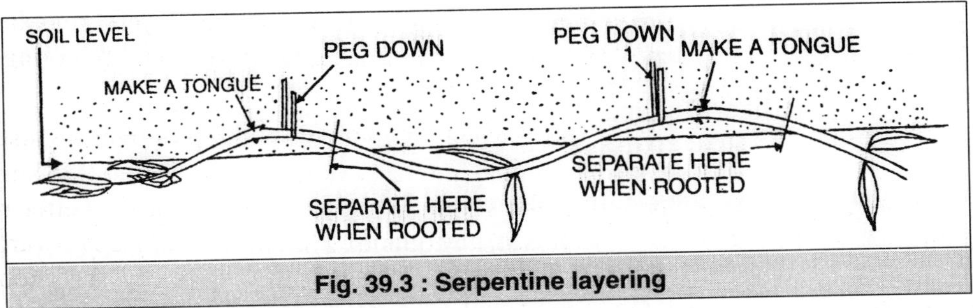

Fig. 39.3 : Serpentine layering

***(iv)* Continuous or trench layering :** Continuous layering is a popular method of propagation of rootstocks of some fruit plants. In this method, it is important to establish a permanent row of plants to be propagated in the nursery. The mother plants are planted at the base of a trench at an angle of 45° in rows, spaced 90 cm apart. The long and flexible stems of these plants are pegged down on the ground to form a continuous line of layered plants (Figure 39.4). The young shoots that arise from these plants are gradually mounded up to a depth of 15-20 cm in autumn, winter or at the end of the growing season, depending on the

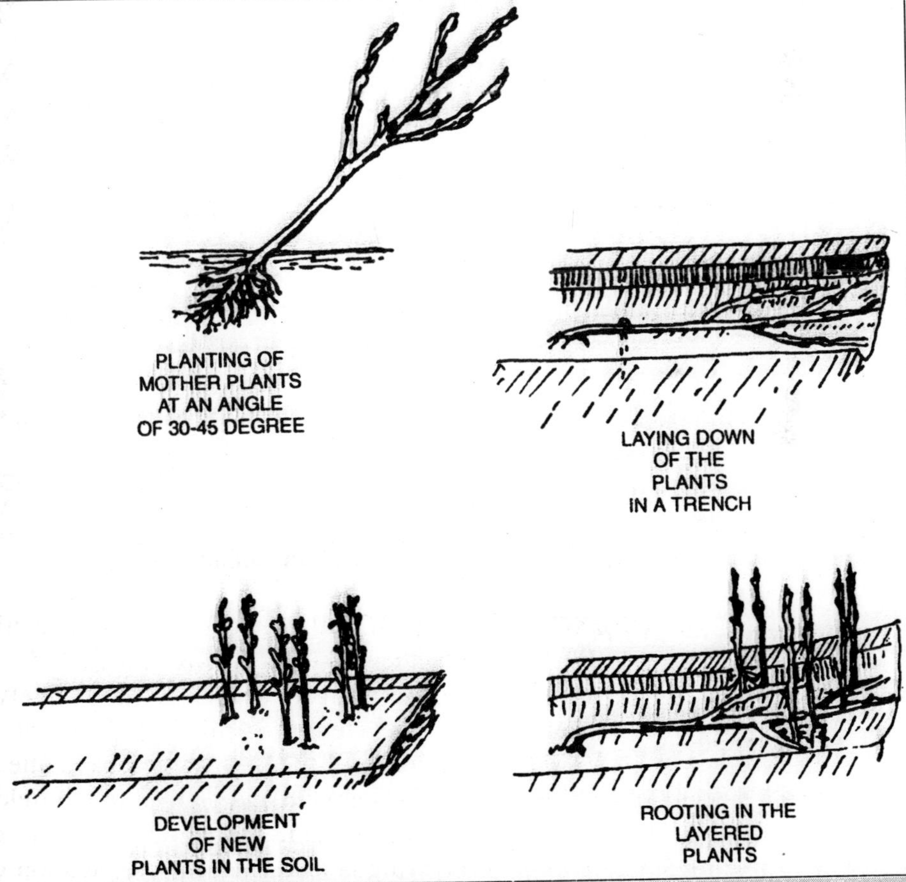

PLANTING OF MOTHER PLANTS AT AN ANGLE OF 30-45 DEGREE

LAYING DOWN OF THE PLANTS IN A TRENCH

DEVELOPMENT OF NEW PLANTS IN THE SOIL

ROOTING IN THE LAYERED PLANTS

Fig. 39.4 : Different steps of trench layering

plant to be propagated. It is the most suitable method of propagation in woody plants, which produce long vines and are difficult to propagate by other methods of propagation. Vigorous rootstocks of apple like M_{16} and M_{25} and walnut are usually be propagated by trench layering.

(*v*) Herbaceous layering : In herbaceous layering, very young shoots of plants (especially carnation) are used for producing new plants. The layering of border carnations is usually regarded as a skilled operation but it is not really difficult. In this layering, first of all, sandy compost is placed around the plant to be layered. Young shoots are selected and their lower leaves are removed. Now, a slit near the base about halfway through the stem and in an upward direction is made. The stem is then bent so as to open the slit, and at this point is pegged down firmly into the soil and covered with an inch or so with compost. The layers will root by autumn and then be severed and potted or planted in the nursery.

(*vi*) Air layering : Air layering is a specialized form of layering, which was originally introduced in China and now widely used for propagation of many horticultural plants. It is popularly known as '*gootte*' or '*marcottage*'. In all forms of layering, rooting is induced by bringing the stem in contact with the soil or other moist medium but air layering is quite distinct from others. In this method of layering, the moist medium is brought to the stem, the latter being left in its original position *i.e,.* roots are formed on stem well above the ground without using soil at all. It is an useful method for propagating rare and valuable plants, which are difficult or impossible to propagate through cuttings and where ordinary methods of layering are not practicable. Thus, plants with rigid branches borne high above the ground, which could not be brought down to low level, are highly suitable for air layering.

Clean young shoots are best for air layering. They are prepared by making an upward cut about 5 cm long at or about the centre of the shoot. The shoot is then girdled by removing a 2 cm wide ring of bark. The upper part of the wound is applied with IBA paste made in lanolin. The wound is covered with moist sphagnum moss in a way to provide complete cover to it. In the past, there had been difficulty in air layering because of drying of the moss grass but now this problem appears to have been largely overcome by the use of very thin polyethylene film as a wrapping material. This film is wrapped around the moss grass in such a manner that it does not leave any opening, which would allow evaporation of moisture from the moss grass. The advantage of this film is that it allows transfer of air but practically no moisture can escape through it. Air layering may be done in the spring or rainy season. Rooting often occurs within a month or so under all favourable conditions. The rooted layers may be severed from the mother plant and planted in the nursery under shaded place until they are established in the field. A wide range of plants like litchi, lime and lemons, jackfruit, guava, fig, croton, maple, rubber plant, dog-wood, hibiscus, wisteria, cashewnut etc., are propagated by air layering.

(*vii*) Mound layering : It is commonly called as 'stooling'. The term stooling was first coined by Lynch in 1942. It is somewhat similar to continuous layering but in this method, the stems/shoots of desired plant are not pegged down. In the basic method of stooling, the mother plant is headed back to 15 to 20 cm above ground level during dormant season. The new sprouts will arise within 2 months (Figure 39.5). These sprouts are then girdled near the

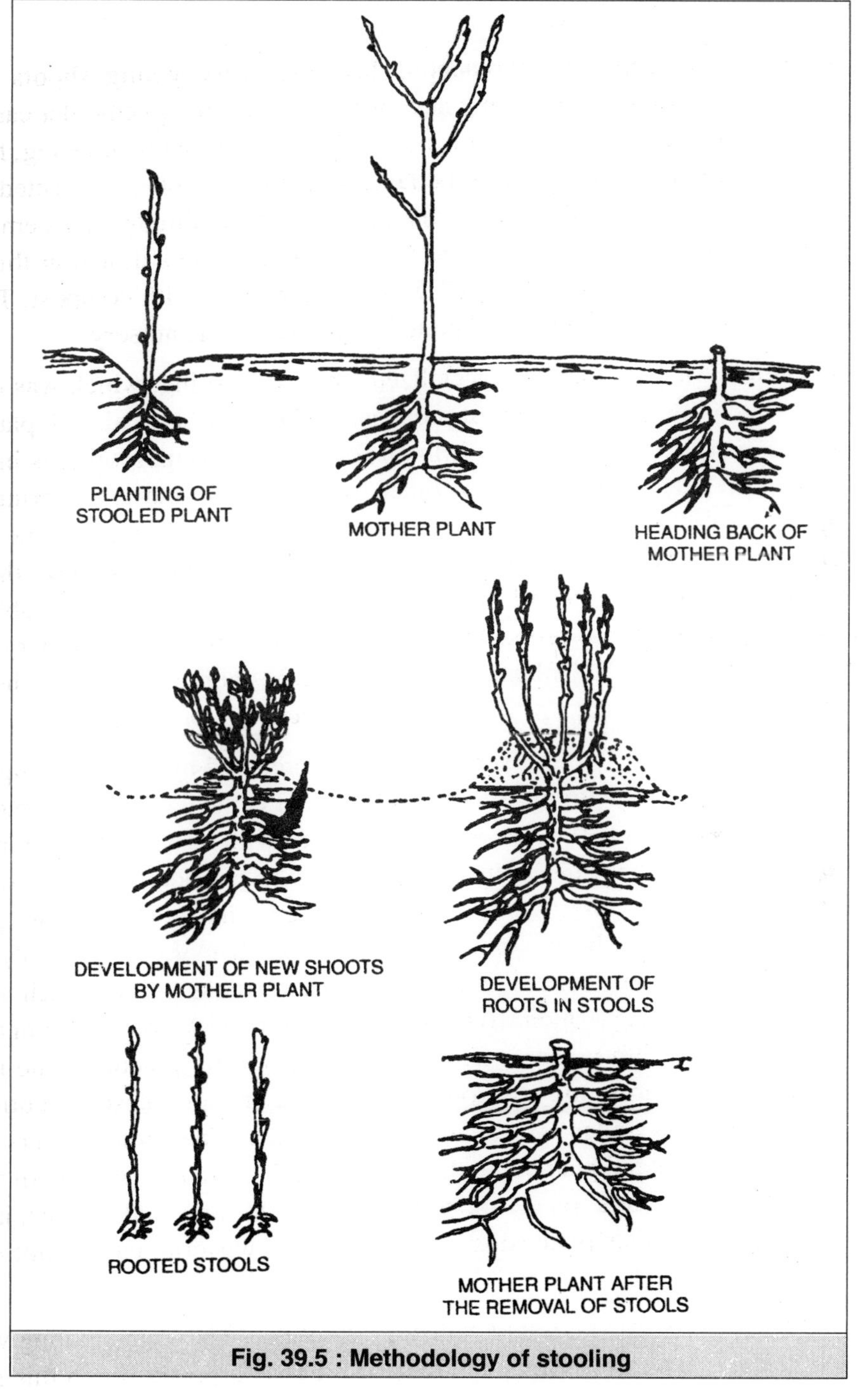

Fig. 39.5 : Methodology of stooling

base and rooting hormone (IBA), made in lanolin paste is applied to the upper portion of the ring. The concentration of rooting hormone varies from plant-to-plant but in general 3,000-5,000 ppm is most commonly used. After application of IBA paste, these shoots are left as such for 2-3 days for proper absorption of rooting hormone (IBA). After this, these shoots are covered with a moist soil. Care must be taken to keep the soil heaps moist all the times, which facilitates rooting in the stools. The roots in shoots may emerge within 30 to 40 days. However, the rooted stools should be severed from the mother plants only after 60 to 70 days and then planted in the nursery. Apple, guava, quince, currants and gooseberries are commercially propagated through stooling. This method is also advocated in other fruits like plum, cherry, hazelnut, pecannut, mango, jackfruit and litchi etc.

• • •

CHAPTER 40

Propagation by Specialized Vegetative Structures

Some horticultural plants develop certain natural specialized structures or organs, which can also be used for vegetative propagation. Propagation through specialized structures is probably the simplest and easiest of all the methods of propagation and certainly the one of the most successful for a beginner. These modified plant parts may be stem or root or some other part. Description on some of the specialized organs and methods of propagation some important horticultural plants has been given hereunder :

Tubers

Tubers are formed from the swollen underground stems or roots and are used by the plant for storing food. The commonest example of tuber is the potato. A tuber has all parts of a typical stem except that it is much swollen. It has nodes (eyes), internodes, lateral and terminal buds. The eyes (nodes) of a tuber are spirally arranged, consisting of one or more small buds, subtended by a leaf scar. Usually, one bud in the eye sprouts and suppresses the growth of others, which shows the phenomenon of apical dominance. Usually, the removal of first sprout from the stem permits other buds to sprout. Jerusalem artichoke, caladium and dahlia are other examples of tuber.

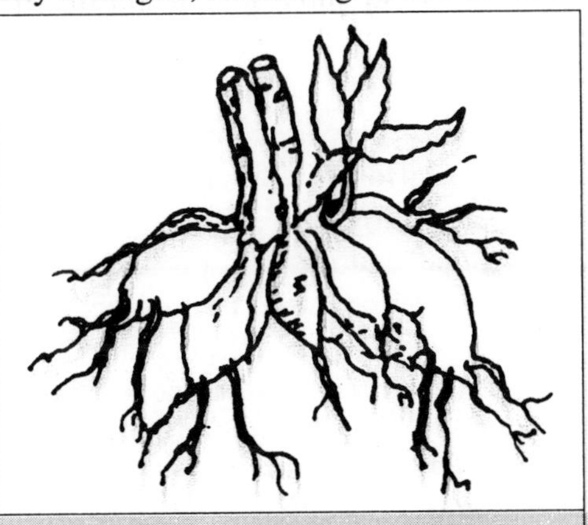

Fig. 40.1 : Tubers of dahlia

Methods of propagation : Propagation by tubers is usually done either by planting whole tuber (Figure 40.1) or by cutting the tuber into pieces. While propagating through pieces, it must be ensured that each piece should have atleast one bud or eye. The weight of a tuber piece should vary between 28 to 56 g to provide sufficient food for the new plants. The pieces should be cut with a sharp knife and kept at a warm temperature (20°C) and high relative humidity (90 per cent) for 2-3 days prior to planting for rapid healing. These pieces may also be treated with suitable fungicide to avoid *Rhizoctonia* rot and further spread of other diseases.

When whole tuber is used for propagation, it may be kept in mind that the tuber may sprout under favourable conditions but it usually does not sprout because of dormancy. The period of dormancy may vary from variety-to-variety, which can be overcome by dipping the tubers in thiourea, ethrel or KNO_3 solution for sometime. Sprouting in tubers takes place only after the dormancy is over.

Bulbs

A bulb is a complicated structure, which consists of a short thickened stem bearing roots on the underside and thick fleshy leaves on the upper side. Bulbs are usually produced by monocotyledonary plants in which these are modified for storage of food and reproduction. Most of the bulbs have bulb scales. The outer bulb scales are fleshy, containing reserve food materials, whereas the inner scales are leaf-like and don't store food. In the centre of bulb, there is either a vegetative meristem or an unexpanded flowering shoot. The meristems more often produce a miniature bulb, called *bulblet,* grown fully to the offset. In some species, the bulblets are produced in the leaf axils, called *bulbils* (when in aerial shoots) and stem bulblets (when produced in the underground organs).

Bulbs are of two types, i.e., *Tunicate* and *non-tunicate.* Tunicate bulbs have dry and membranous outer bulb scales (*e.g.,* onion, tuberose, daffodil and tulip). These membranous coverings provide protection from drying and mechanical injury and give a solid structure to the bulb. In non-tunicate bulbs, the dry coverings are loosely arranged like petals of a flower overlapping each other. Drying and rough handling usually injure non-tunicate bulbs. Garlic and lilies are examples of non-tunicate bulbs.

Methods of propagation : The bulbous plants are propagated through offsets, bulb cuttings, scaling, scooping, scoring, stem cuttings, leaf cuttings, leaf buds and bulbils.

(*a*) **Offsets :** Many bulbous plants like tulip, daffodils, lily etc., are easily propagated through offsets (Figure 40.2). Small bulbets are produced by the mother bulbs in the axil of leaves, which on development, are called as *offsets*. These offsets are detached from the mother bulb and planted in soil for further multiplication. Offsets usually take lesser time to flower as compared to seeds, scales or bulbils.

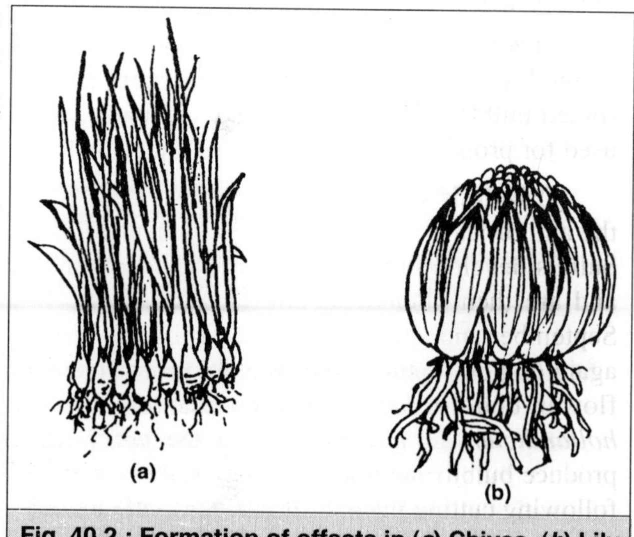

Fig. 40.2 : Formation of offsets in (*a*) Chives, (*b*) Lily

(*b*) **Bulb cutting :** In bulb cutting, the fully mature bulb is cut into small segments, containing a piece of basal plate and 3-4 scales. These segments are planted vertically in a suitable medium. New bulblets will develop from these segments under favourable conditions, which can be transplanted in the nursery. Bulb cutting method is used for propagation of *Lycoris, Scilla, Narcissus, Haemanthus* and *Hymenocallis* etc.

(*c*) **Scaling :** Scaling is an useful method for propagating lilies in late-fall or in mid-winter. In this method, individual bulb scales are separated from the mother bulb and placed 5-6 cm deep vertically in a suitable growing medium (*e.g*., sand or vermiculite). The scales usually develop 3-5 bulblets after 4-5 weeks of planting. Treatment of scales with NAA (1 ppm) is useful for stimulating bulblet formation. Further, the scales are quite delicate and should be handled carefully. Similarly, for better results, damaged, dried or infected scales should not be used for propagation purpose.

(*d*) **Scooping and scoring :** Scooping and scoring methods are used chiefly for propagating hyacinth and scilla. In scooping, the entire basal plate of the mature bulb is scooped with a special sharp and curved scalpel to destroy the main shoot. In scoring, however, 3 straight cuts are given through the basal plate and growing point but the complete plate is not removed as in scooping. In scooping, adventitious bulblets develop from the base of the exposed bulb scales, while in scoring, growing points in the axil of bulb scales grow into the bulblet. After scoring and scooping, the bulbs are placed about 2.5 cm deep in dry sand or open trays at about 21°C for callusing. Then, these are incubated at 25-30°C in dark or diffused light for 2 weeks or so for bulblet formation. The mother bulb alongwith bulblets are planted 10 cm deep in the nursery during fall. The bulblets produce sufficient foliage by next spring and mother bulbs degenerate by the first summer. The bulblets of considerable size should be dug out and re-planted. They normally take 3-4 years for flowering. With scooping, a propagator can get 50-60 bulblets and with scoring about 20-30 bulblets per bulb.

(*e*) **Stam cutting :** In stem cutting, the stem of the bulb is cut off in small pieces and planted in a shallow trench. After sometimes, the bulblets would appear on the stems. The rooted bulblets are then removed and planted in the nursery beds. Stem cuttings are mostly used for propagating lilies.

(*f*) **Bulbils :** Lilies are also propagated through bulbils, which are produced naturally in the axil of leaves of some species such as *Lilium bulbiferum, L. tigrium, L. sargenti* etc. The bulbils develop in the early part of the season and fall. These are harvested before they fall and are then planted in mid-October in the field about 10 cm deep. They are dug next September and again re-planted 12-13 cm deep and 10-12 cm apart in single rows. They are again dug out at the end of second year. To enhance bulbil formation, disbudding soon after flower bud formation is recommended. Some species of lilies like *L. candidum, L. hollandicum, L. maculatum, L. testacum, L. longiflorum* and *L. chalcedonicum* etc. don't produce bulbils naturally. However, these can be induced artificially by flower bud pinching following cutting the half of the stem after a week.

(*g*) **Micropropagation :** Micropropagation is the present day hi-tech propagation technique for large-scale and faster multiplication of the bulbs. The micropropagation is chiefly advantageous and valuable to multiply newly released cultivars and to produce virus-free stock of them.

Corms

A corm is a short, solid, much swollen underground stem, with one or several buds near the top, a tuft of leaves at the upper side and a ring of thick fibrous roots around the base. In contrast to bulb, a corm is a solid structure with distinct nodes and internodes. Most of the storage tissues contain parenchymatous cells. The dry thin membranous scaly leaves usually cover the corm called *tunic*. These coverings protect the corm from injury and water loss. At the apex of corm, is a shoot, which develops into a flowering shoot or leaves. Corms produce two types of roots, fibrous roots from the base of the mother corm and fleshy contractile roots from the new corms. Gladiolus (Figure 40.3), crocus and water chestnut are examples of some typical cormous plants.

Methods of propagation : Cormous plants can be propagated through new corms, comels or corm segments or micropropagation.

(*a*) Corms and Cormlets : The production of corms and cormlets is a varietal character. Usually, a mother corm produces 2-3 new corms and 15-20 cormlets. The corms are planted in winter, which flower in spring and plants are left in the field for a month or so following blooming. After digging the plants, new corms and cormlets are separated and corms are sorted and graded according to size. The corms are then cured at 35°C for few hours and stored at 5°C and 70-80 per cent relative humidity in a well-ventilated room. These corms may be planted next year. Cormels are miniature corms that develop between the old and the new corms. One or two year's growth is necessary for the corm to flower. Shallow planting

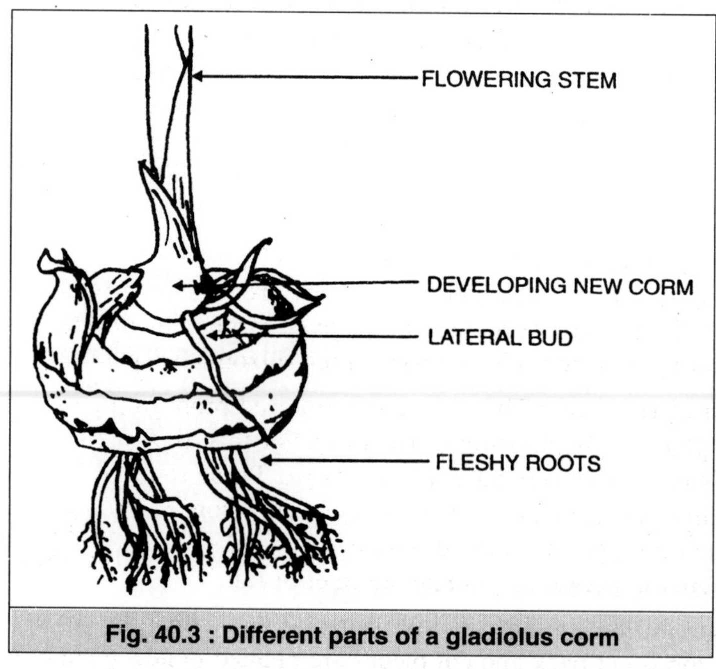

Fig. 40.3 : Different parts of a gladiolus corm

enhances cormel production, while deep planting reduces their production. The cormels are separated from mother corm and stored at 5°C in a well-aerated room up to planting. In general, a corm flowers within a year of planting, whereas it takes, 2-3 growing seasons for a cormel to flower.

(*b*) Division of corms : Division of corm is also sometimes used as a method of propagation of cormous plants. While making segments, it must be ensured that atleast one eye should be present in one piece of a corm. These pieces are planted vertically, 5 cm deep in soil, which develop in a full plant soon.

(*c*) Micropropagation : Micropropagation technique has also been developed for large-scale multiplication of cormous plants. Micropropagation protocols are now standardized and commercially used in gladiolus.

Rhizomes

A rhizome is a modified stem structure in which the main axis of the plant grows horizontally at or just below the soil surface. The stem is segmented having nodes and internodes. A leaf like sheath is attached to each node, which on expansion, becomes the foliage. Usually, the roots develop in the vicinity of nodes. The flowering stems, called as *culms,* are either produced from the rhizome tip or from the lateral branches. The banana is a typical example of rhizome. Other rhizomatous plants are iris, ginger, bamboo and sugarcane. Most of them are monocotyledonary but a dicot, blueberry (*Vaccinium angustifolium*) has also been classified as rhizome.

On the basis of growth and structure, rhizomes are of three types *viz., pachymorphs leptomorphs* and *mesomorphs*. The pachymorph rhizome is a thick, fleshy and short structure with many-branched clump. It grows horizontally with roots arising from the lower side. It is usually determinate type *i.e.,* each clump terminates into a flowering stalk. The pachymorph rhizome is found in iris, ginger and turmeric. The leptomorph rhizome is slender with long internodes. They don't produce clumps but spread extensively in a given area. These are indeterminate type *i.e.,* grows continuously in length from the terminal apex. It is found in lily of the valley (*Convallaria majalis*). The mesomorphs are intermediate forms between pachymorphs and leptomorphs.

Methods of propagation : Many horticultural plants like banana, turmeric, ginger, canna, iris, lily of the valley and water hyacinth are propagated through rhizomes. Even some ferns and grasses have rhizomes and are propagated through them. The whole rhizome, piece of a clump or cutting may be used for propagating rhizomatous plants.

(*a*) Division of rhizome : The most common used propagating material for propagation of rhizomatous plants is the division of rhizome. For this, rhizome is cut into sections. Each section should have atleast one lateral bud or eye. These sections are planted in a suitable medium. Adventitious roots and new shoots develop from the nodes and ultimately in to a plant. Propagation through divisions should be done either at the beginning of spring or at or near the end of growth period *i.e.,* in late-summer or fall.

(*b*) Other methods : In pachymorph rhizome, the culms are cut off at the point of the attachment, the top is cut back and cut pieces are planted in new plantations. In leptomorphs, a single lateral 'off shoot' is removed from the rhizome and planted in the nursery. In lily of valley rhizome, the tip of the flowering bud, called a 'pip' is removed and planted in the nursery. Similarly, rhizomes producing turf grasses are cut in to different sections (sprigs) and individual sprig is planted for further multiplication. However, in large rhizome bearing plants, such as bamboos, the aerial shoot or culm is used as cutting. The cuttings/sections containing 3-4 nodes are planted vertically in the soil. The new branches arise from the nodes, which produce new plant.

Bulbils

In some species, bulblets may develop in the leaf axils either on the underground portion or on the aerial portion of bulb (stem). These aerial bulblets are called *bulbils*. Some species of lilium like *Lilium bulbiferum* and *L. sulphureun* produce bulbils and are propagated through bulbils (Figure 40.4).

Fig. 40.4 : Lily plant showing bulbil production

Tuberous Roots and Stems

In some herbaceous plants, there is a massive enlargement of the secondary roots. Their internal and external structure is similar to roots but nodes and internodes are present. The buds are produced only on the crown or the stem end (proximal) and the fibrous roots are present at the distal end, thus polarity is reverse to tubers. Typical examples of tuberous roots are dahlia (Figure 40.5*a*) and sweet potato. Tuberous roots are biennial. Tuberous stems are produced by excessive enlargement of the hypocotyl of seedling plant. They have vertical orientation with one or more vegetative buds on the upper end. The fibrous roots on such stems are produced at their base.

The typical examples of tuberous stem are cyclamen and tuberous begonia (Figure 40.5*b*). These are perennial and continue to enlarge laterally every year. Tuberous stem and roots producing plants (*e.g.* sweet potato and dahlia) are propagated by dividing crown, stem and leaf cuttings. Treatment of cuttings with some rooting hormone like IBA, helps in better rooting of the cuttings prepared from roots and stems.

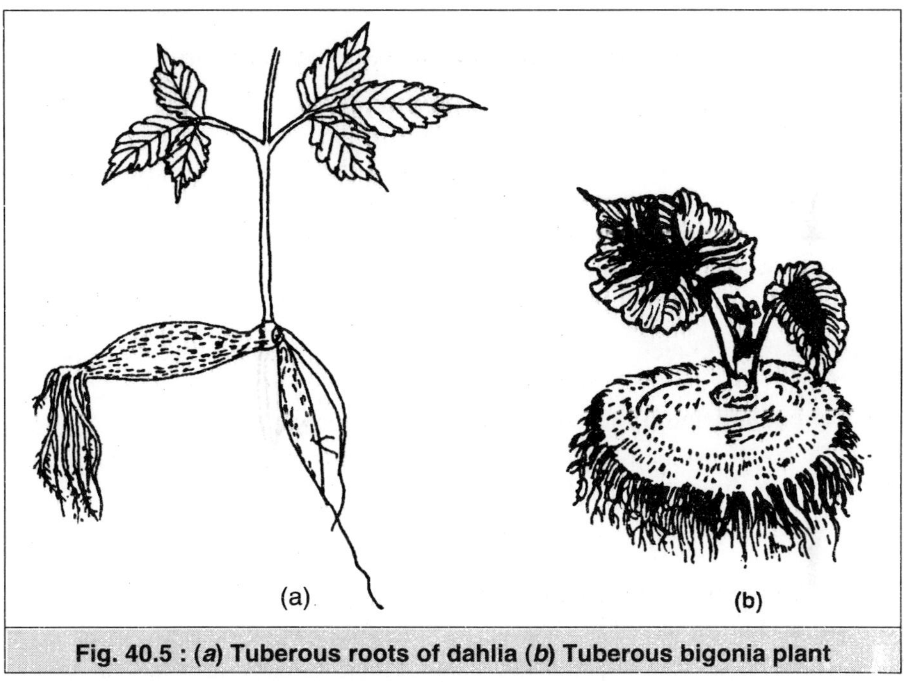

Fig. 40.5 : (*a*) Tuberous roots of dahlia (*b*) Tuberous bigonia plant

Runners

It is a specialized long slender stem, which is produced from the leaf axil at the crown of a plant and prostate horizontally along the ground. The roots appear at one of the nodes having contact with soil. After root formation in the new plant, the contact with the mother plant is automatically detached and the new plant can be separated and planted in the field or nursery. Strawberry is a typical example, which is commercially propagated through runners. *Doob grass* (*Cynodon dactylon*), *Oxalis corniculata*, fern and spider plant are some other examples of plants propagated through runners (Figure 40.6). Nowadays large-scale multiplication of runners (*e.g.*, strawberry) is possible through micropropagation.

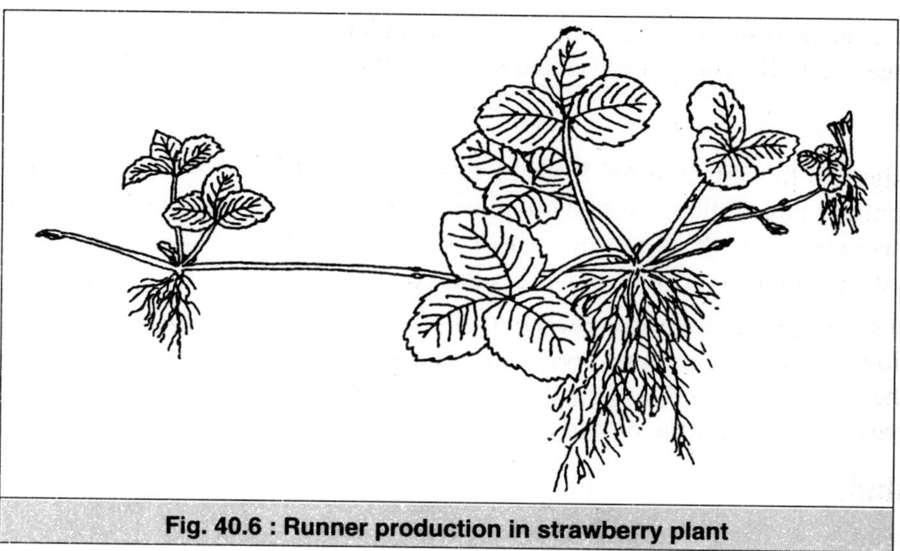

Fig. 40.6 : Runner production in strawberry plant

Suckers

A sucker is a shoot, which arises on a plant below the ground. It may also arise from an adventitious bud on a root. However, in practice, shoots arising from the vicinity of crown of the plant are also referred to as *suckers*. The capacity of a plant to form suckers varies from species-to-species and even from variety-to-variety.

Pineapple is usually propagated through suckers. The 'Giant Kew' variety is a shy suckering type, whereas 'Queen' produces profuse suckers. In banana, two types of suckers are produced –water suckers and sword sucker. Water suckers are broad leaved, whereas, sword suckers are pointed and have sword shape. For propagation purpose, sword suckers are preferred over the water suckers. Suckers are separated from the mother plant and planted either in the nursery or directly in the field for rooting.

Offsets

An offset is a young plant produced along the side of the parent plant, which can easily be detached from it. The term offset is generally applied to a shortened, thickened stem of rosette like appearance. It is also applied to lateral shoots arising on the stems of monocotyledonous fruit plants like date palm, pineapple and banana. Similarly, some cacti and succulents produce a cluster of small plants around the base and these offsets merely need separation. Bromeliads produce offsets around the old flowering shoot and some bulbs also produce offsets freely.

Many plants are propagated by means of offsets. For propagation, these offsets are removed by cutting them close to the main stem with a sharp knife. If offset is well rooted (as in cacti and succulents), it can be potted immediately and if roots are not developed in sufficient amount, these offsets may be placed for rooting in a suitable medium. Offset development in plants can be stimulated by heading back the crown of the plant. Usually, the natural production of offsets by the plants is very low, for which many techniques have been developed but the most common is the micropropagation. Fruit plant like date palm, pineapple, banana, some succulents and cacti, and bulbous plants like tulip, daffodil, iris, lily, hyacinth and amaryllis can be propagated through offsets.

Crown

It designates the above ground part of a plant from which new shoots are produced. In perennial plants and shrubs, the transition zones between the stem and the root, whereas in herbaceous perennials, the part from which new shoots arise is basically the crown. Some plants produce extensive crowns and others only one.

Because of simplicity and reliability, division of crown is an important method of propagation of many herbaceous perennials. In this method, the crown is divided in spring with the help of a sharp knife. The cut pieces are planted in pots or nursery bed immediately after preparation. Sansevieria, day-lily and daisy are propagated through division of crown.

Pseudobulbs

A pseudobulb is an enlarged fleshy stem with several nodes. It is produced by some horticultural plants especially for the storage of food. The structure of pseudobulb varies from species-to-species. The pseudobulbs are usually propagated through offshoots, division or from black and green bulbs produced by them. In few orchids like *Dendrobium* the offshoots are developed at the nodes of long slender pseudobulbs, from which roots develop. The rooted shoots arc cut from thc mother bulb and planted in the nursery. Some orchids like *Cattleya, Laelia, Miltonia* are multiplied commercially through the division of pseudobulbs. The bulb is cut into different sections by a sharp knife during dormant season. Each section is sown in a nursery and new growth begins at nodes. Sometimes black bulbs *i.e.,* bulbs without foliage as in *Cymbidium* orchid, are used for propagation and sometimes the green bulbs (with foliage) are used.

Tuberacles

Some plants like *Begonia* and *Dioscoria batatas* produce small aerial tubers in the axil of leaves called *tuberacles*. These tuberacles are removed in fall and stored for 3-4 months in winter and planted in spring. Shoot and root emergence takes place within a short time but tuber formation is induced under short-day conditions.

Slips

Slips are the shoots just arising below the crown but above the ground. Pineapple is commercially propagated through slips.

Stolons

Stolon is a modified stem, which grows horizontally above the ground. These are found in mint and *doob* grass and sometimes involved in tuberization also. The stolon is a naturally occurring layering in which individual layer can be cut from the parent plant and planted in the nursery.

• • •

CHAPTER 41

Micropropagation : Principles and Practices

Micropropagation refers to the production of plants from a very small plant part, tissues or cells, grown aseptically in a test tube or containers under controlled nutritional, environmental and aseptic conditions. Tissue culture or *in vitro* culture are two broadly used terms for micropropagation, which basically include aseptic culture of various plant parts. All the biological principles of micropropagation technique are based on the phenomenon of *totipotency* of a cell, which is defined as the capacity of a plant cell to regenerate into a full-fledged plant having different organs. German plant Physiologist, Haberlandt (1902) is known as the father of tissue culture technique, who coined the term *totipotency*.

With the advancement in science and technology, micropropagation technique has been standardized for many horticultural plants, and it is now widely used for their propagation at a large scale.

ADVANTAGES OF MICROPROPAGATION

Micropropagation is gaining popularity day-by-day because of the following advantages over conventional propagation techniques :

1. Large-scale multiplication in a limited time and space is possible.
2. Micropropagated plants are usually free from viruses.
3. Under *in vitro* conditions, the plants can be produced throughout the year.
4. Micropropagation is highly beneficial in dioecious fruit plant species (*e.g.*, date palm and papaya), where large-scale production of female plants is desirable.
5. It helps in reducing the breeding cycle.
6. Production of homozygous plants is possible under *in vitro* conditions.

7. It is highly beneficial in plants in which vegetative propagation is not possible or the propagation rate is very slow (*e.g.* date palm, papaya, coconut etc.).
8. Tissues culture systems produce secondary metabolites, which are used by many pharmaceutical industries for production of several medicines.
9. Long-term storage of the material in lesser space is possible under *in vitro* conditions.
10. *In vitro* systems have the potential for long-term transportation or shipment of propagation material.
11. Production of seedless varieties (*e.g.*, grape, peach) is possible by utilizing embryo rescue technique.
12. Distant hybridization is possible by following micropropagation techniques.

LIMITATIONS OF MICROPROPAGATION

Micropropagation has certain limitations also. These are as under :

1. The facilities required are very costly.
2. Technical skill is required to carry out the different micropropagation procedures.
3. Pathogens once appeared in the system, they also multiply at a very faster rate in a short time.
4. Plants having high levels of phenols (mango, date palm, coconut etc.) usually do not respond to micropropagation techniques.
5 Establishment of laboratory-raised plants in the field is a very difficult task.

PRINCIPLES AND PRACTICES OF MICROPROPAGATION

In raising plants through micropropagation, many principles and practices are involved, which have been discussed briefly hereunder :

Titipotency and Plant Regeneration

All *in vitro* methods of plant propagation rely on the phenomenon of totipotency of plant cells. Totipotency may be defined as the capacity of plant cell, tissue or organ to have the capacity to regenerate in to a full plant. As stated earlier, this term was given a German Plant Physiologist, Haberlandt in 1902, who is considered as the father of tissue culture systems. This can be achieved by excising the explant and placing it on culture medium, which involves the following consecutive events : (*i*) Dedifferentiation of source tissue or organ, as a result of which, all physiological mechanism activate in the cell, which leads to cell division, (*ii*). Active cell division in the entire cut surface of the explant, resulting in proliferation of the callus tissue, (*iii*). Organisation of defined meristems, resulting in the formation of shoot or root meristems or both, (*iv*). Regeneration and differentiation of new organs through organogenesis or somatic embryogenesis. Both organogenesis or somatic embryogenesis may be direct, coming from the explant and not involving callus formation, or indirectly from the callus tissue. However, it is emphasized that the entire micropropagation process must be carefully controlled and monitored, otherwise, it may result in generation of plantlets that are not true-to-type.

Axillary Bud Proliferation

Axillary bud proliferation under *in vitro* conditions is considered as a convenient route for micropropagation, because it does not include a callus stage and hence it is considered safer for the preservation of clonal characteristics. It has been observed by many biotechnologists that a very large number of axillary shoot buds grow extensively when shoot-tips, or even smaller apical meristems are cultured in an appropriate culture medium, containing higher concentration of cytokinins. Both the excision of the explant and the cytokinin rich medium activate the bud, leading to proliferation of many side shoots, which can later be separated for further culture.

Organogenesis : Development of Organs

Organogenesis from explants results in the formation of both shoot and roots. Formation of these organs takes place simultaneously, but either shoots or roots are formed first. However, in general, shoots are formed first than the roots. Regeneration of the other organs occur after the development of shoot and roots, either in the same medium or after sub-culturing in another medium, supplemented with different concentrations of salts under different environmental conditions. Organogenesis is controlled by the type of explant, composition of culture medium (specially the concentration of growth regulators), and the environmental conditions during culture. Sometimes, organogenesis brings about the formation of modified shoots and roots, usually the storage organs like bulbs and tubers. In such cases, the shoot or the root meristems are formed, but instead of further growth and elongation, mini-bulbs, mini-tubers or mini-corms are developed. Such propagules can also develop from axillary buds, especially when basal plates of bulbs, corm or tuber are used as explants.

Somatic Embryogenesis : Development of Embryos

Somatic embryogenesis is different from organogenesis in that the regeneration and organization are bipolar *i.e.*, the shoot and root meristems in the explants are formed simultaneously from proembryonic masses. Differentiation and organization of somatic embryos take place directly in the explant or from the callus, depending on the type of explant and composition of culture medium, and the subsequent sub-culturing. In somatic embryogenesis, usually, a two-stage culture is involved. First, induction of pro-embryonic masses, which is usually favoured by 2,4-D or other auxins and second, by the transfer of the pro-embryonic masses to a differentiation medium in which the somatic embryos fully differentiate in to organs. In general, somatic embryogenesis is more often favoured by culture in an agitated liquid medium. In the process of embryogenesis, first globular embryos are formed, which later change to ear shape in which shoot and roots can be distinguished, and finally to torpedo stage in which shoot and roots have vascular connections. However, it is very much sure that for development of somatic embryos, coordinated and synchronized approach is required.

BASIC REQUIREMENTS FOR MICROPROPAGATION

The basic requirements for successful micropropagation are :

(*i*) A well equipped laboratory. (*ii*) Asepsis.
(*iii*) Culture medium, and (*iv*) Controlled culture environment.

Culture Laboratory

A well-equipped laboratory is required for micropropagation techniques. It should have sufficient working area for performing cultures and their subsequent handling. The laboratory should have all necessary components like washing facilities, large sinks, sterilizing equipments like autoclave, incubation hood etc, in a proper order. It should have constant supply of water and electricity. Provision of generator should also be there. Scalpels, forceps and needles are the most commonly used accessories in a tissue culture laboratory. A large amount of glassware (conical flasks, culture tubes, funnels, beakers, pipettes and petri-dishes) are used in such labs. For temperature control, an automatic thermostat with timer should be installed in the laboratory. For various operations, distilled, double-distilled and sterile water should always be available in the laboratory.

Asepsis

A high degree of aseptic conditions are absolutely necessary for the success in micropropagation. To start with, the plant part to be cultured (explant) should be treated with a suitable sterilizing agent (*e.g.*, mercuric chloride, hydrogen peroxide, sodium hypochlorite, silver nitrate, bromine water or ethanol etc.) before culture. The concentration of these sterilization agents is different for various plant species. In general, the choice of the chemical and duration of treatment depends upon the sensitivity of the plant material to be sterilized (Table 41.1). After treating with sterilizing agent, the explant should be washed with sterile water for 4-5 times for removing the sterilizing agents. The culture medium and culture instruments are sterilized by autoclaving at a pressure of 15 lbs/inch^2 and at a temperature of 121°C for 15 minutes in an autoclave. The thermolabile substances (*e.g.*, vitamins and hormones) are sterilized by filtering their solution through glass or micro-filters. All the conical flasks and test tubes are plugged with non-absorbent cotton and then covered with aluminium foil. After autoclaving, all the material should be kept in an incubation hood or chamber till use.

TABLE 41.1 : Chemicals used for sterilizing explants in micropropagation

Name of the sterilent	*Effective concentration*	*Duration of treatment (minutes)*
Mercuric chloride	0.1 to 1 per cent	2 to 10
Calcium hypochlorite	8 to 10 per cent	5 to 20
Sodium hypochlorite	2 to 3 per cent	5 to 20
Hydrogen peroxide	10 to 12 per cent	2 to 15
Absolute alcohol	95 per cent	3 to 5
Bromine water	1 per cent	5 to 25
Silver nitrate	15 per cent	2 to 15

Culture medium

The nutrient requirement of various kinds of tissues is met by a culture medium. A culture medium contains both organic and inorganic salts (major, minor salts, vitamins and growth regulators). The concentration, however, depends on the different media. The commonly used media in micropropagation are MS (Murashige and Skoog, 1962), B_5 (Gamborg *et al.*, 1968), WPM (Woody Plant Medium), White's (1963) and Nitsch and Nitsch (1956) medium. One medium suitable for a particular plant species may not be of any use to the other plant species. Out of the above mentioned media, MS medium and B_5 medium are commonly and widely used in micropropagation of horticultural crops. The major constituents of these media are given in Table 41.2 and Table 41.3, respectively.

TABLE 41.2 : Composition and preparation of the stock solutions for MS basal medium

Stock solution	*Ingredients*	*Weight of the ingredients (mg)*	*Volume of the water used*	*Volume of stock solution for marking 500 ml of medium (ml)*	*Final concentration of ingredients in the medium ($mg1^{-1}$)*
		INORGANIC SALTS			
(I) MACRONUTRIENTS					
A	NH_4NO_3	33,00			1,650
	KNO_3	38,000			1,900
	$MgSO_4 \cdot 7H_2O$	7,400	1,000	25	370
	KH_3PO_4	3,400			170
B	$CaCl_2.2H_2O$	8,800	500	12.5	440
(II) MICRONUTRIENTS					
C	$MnSO_4 \cdot 4H_2O$	2,230			22.3
	$ZnSO_4 \cdot 7H_2O$	860			8.6
	H_3BO_3	620	1,000	5	6.2
	KI	83			0.83
	$NaMoO_4.2H_2O$	25			0.25
D	$CuSO_4.5H_2O$	2.5			0.025
	$COCl_2\text{-}6H_2O$	2.5			0.025
E	$Na_2EDTA\ 2H_2O$	1,492			3.73
	$FeSO_4 \cdot 7H_2O$	1,112	200	2.5	27.8

ORGANIC SALTS					
F	Glycine	40			2.0
	Nicotinic acid	10			0.5
	Pyridoxine HCl	10	200	5	0.5
	Thiamine HCl	2			0.1
	Myo-inositol	2.000			100
G	Sucrose	30,000			30,000

TABLE 41.3 : Composition and preparation of the stock solutions for B_5 basal medium

Stock solution	*Ingredients*	*Weight of the ingredients (mg)*	*Volume of the water used*	*Volume of stock solution for making 500 ml of medium (ml)*	*Final concentration of ingredients in the medium (mgl^{-1})*
INORGANIC SALTS					
(I) MACRONUTRIENTS					
A	$(NH_4)_2$ SO_4	2,680			134
	KNO_3	50,000			2.500
	$MgSO_4$. $7H_2O$	5,000	1,000	25	250
	NaH_2PO_4. H_2O	3,000			150
B	$CaCl_2$.$2H_2O$	3,000	500	12.5	150
(II) MICRONUTRIENTS					
C	$MnSO_4$. $4H_2O$	1,000			10
	$ZnSO_4$.$7H_2O$	200			2
	H_3BO_3	300	1,000	5	3
	KI	75			0.75
	$NaMoO_4$.$2H_2O$	2.5			0.025
D	$CuSO_4$.$5H_2O$	2.5	100	0.5	0.025
E	Seq. 330 Fe	1,120	200	5	28
ORGANIC SALTS					
F	Myo-inositol	2,000			100
	Nicotinic acid	20			1
	Pyridoxine HCl	10	200	5	0.5
	Thiamine HCl	2			0.1
	Myo-inositol	2,000			100
G	Sucrose	20,000			20,000

Preparation of culture medium First of all, the stock solutions of major and minor salts, vitamins and growth regulators are prepared separately for future use and stored at 4°C. For preparation of medium, the required quantity of agar is weighed, dissolved in half the required double distilled water by heating at 80–100°C. The other components (salts) are also dissolved in double distilled water and added to the agar. Then, the required amount of sucrose is added to this mixture and after making the final volume, the pH of the medium (5.7 to 5.8) is adjusted with a dilute HCl or NaOH. The medium is then boiled and poured into flasks or test tubes. The flasks or test tubes are plugged with non-absorbent cotton and covered by an aluminium foil. The test tubes/flasks containing medium are then autoclaved at a pressure of 15lbs/square inch and at a temperature of 121°C for 15 minutes. After removing the flasks from autoclave, these are kept in incubation room for 4-5 days. Afterwards, these can be utilized for culture establishment.

Controlled Culture Environment

The culture environment (temperature and light) of the laboratory has great influence on the success of the explant under *in vitro* conditions. Light is needed to regulate the processes like root initiation, formation of shoots, chlorophyll synthesis and subsequent growth of shoots. Usually, a light intensity of 300-400 Klux is required in a tissue culture laboratory. Duration of light is another factor influencing plant regeneration. Normally, 16 hours light and 8 hours dark photoperiod is necessary or required in such laboratories. Further, the quality of light is equally important for organogenesis. Blue light favours shoot initiation, whereas red light does not influence it. However, red light stimulates root initiation and blue light has no effect on root formation. It is, therefore suggested to provide balance exposure of both red and blue light for getting balanced root and shoot formation. It is also essential to maintain an optimum temperature throughout in a tissue culture laboratory. In general, a temperature of 27 ± 2°C is required for woody perennials including most of the fruit crops. However, for crops like gladiolus and carnation, a temperature of 25 ± 2°C is sufficient for culture initiation and subsequent development of the plantlets.

STAGES OF MICROPROPAGATION

Micropropagation is an integrated process in which cells, tissues, or organs of the selected plants are isolated, surface sterilized, and incubated in growth promoting, sterile medium and environment to produce a large number of plants. In this, atleast five stages are involved. A brief description of these stages has been described briefly hereunder :

Stage 0 : Selection of Mother Plant for Explant Isolation

The success of micropropagation is largely dependent on the quality of the source/mother plant. Thus, mother plant from which explant has to be excised, should be :

(*a*) A certified and true-to-type representative of the desired species and or a cultivar.

(*b*) Healthy and free from insect-pest and diseases.

(*c*) Quite vigorous.

Desired explant should be excised from the selected mother plants, surface-sterilized and prepared which can range in size from 0.1 mm to large corms of banana and dates, to seeds or fragile stem and leaves.

Stage 1: Explant Establishment in Culture Medium

During this stage (lasting from one week to 2-3 months or even longer), the surface sterilized and prepared explant is cultured in a suitable culture medium for tissue activation and multiplication. This stage is usually carried out on agar based media, but liquid media can also be tried. The selection of basal media and growth regulators at this stage is of special significance, and may vary according to plant and tissue type and to the desired multiplication method. There are always chances of microbial contamination at this stage and hence one has to keep vigilant eye, which can be avoided to some extent by specific indexing.

Stage 2 : Proliferation and Rapid Multiplication

Explants that have successfully established in the culture medium are later transferred aseptically to stage 2 for proliferation *i.e.* for generation of numerous plantlets. For this, repeated sub-culturing is done to encourage more proliferation. The types of regeneration and proliferation largely depend on the combination of growth regulators. In general, higher concentrations of cytokinins stimulate multiplication of axillary or adventitious shoots but higher concentrations of auxins are necessary for callus proliferation and somatic embryogenesis. The combined and balanced adjustment of growth regulators, basal medium composition and environmental conditions are important factors, which affect the proliferation rate. The duration of this stage is unlimited and largely depends on the choice of propagator, but it usually lasts for several months to 1-2 years.

Stage 3 : Plantlet Establishment and Rooting

After repeated sub-culturing and screening, the desirable plantlets are transferred to the final *in vitro* stage. In this stage, the selected plantlets are forced for root formation, which can be achieved by media modification and modifying the concentration of the growth regulators. Usually indole butyric acid (IBA) is used as rooting hormone in micropropagation systems and its different concentrations are used for rooting *in vitro*. The plantlets are forced for the development of shoots and roots and for stimulation of photosynthesis and other physiological activities. For this, media modifications are desirable. For example, the concentration of cytokinins is reduced or completely eliminated, sugars are reduced, auxins are increased and proportion of gelling agents is increased. In addition, the environmental conditions (especially light intensity is increased) in the laboratory are also modified.

Stage 4 : Acclimatization or Hardening

Healthy plantlets produced under laboratory conditions are very delicate and are usually incapable of being established in natural field conditions because of entirely different environment in the field. Hence, these should be acclimatized or hardened for 4-8 weeks by modifying the environmental conditions in the laboratory itself. For this, they are subjected to low light, high temperature and high humidity conditions in the greenhouse. Once the stomata start functioning and new and functional roots develop, photosynthetic activity is increased; as a result, plants become autotrophic. Later, light intensity is raised and the temperature and humidity are regulated to natural growing conditions. These conditions help the plantlets in getting acclimatized to the natural conditions, which help them in getting established in the field easily.

MICROPROPAGATION AND TISSUE CULTURE SYSTEMS

Based on the plant part to be cultured, different *in vitro* techniques have been employed in micropropagation of different plant species (Figure 41.1).Some of the commonly used systems have been described briefly hereunder :

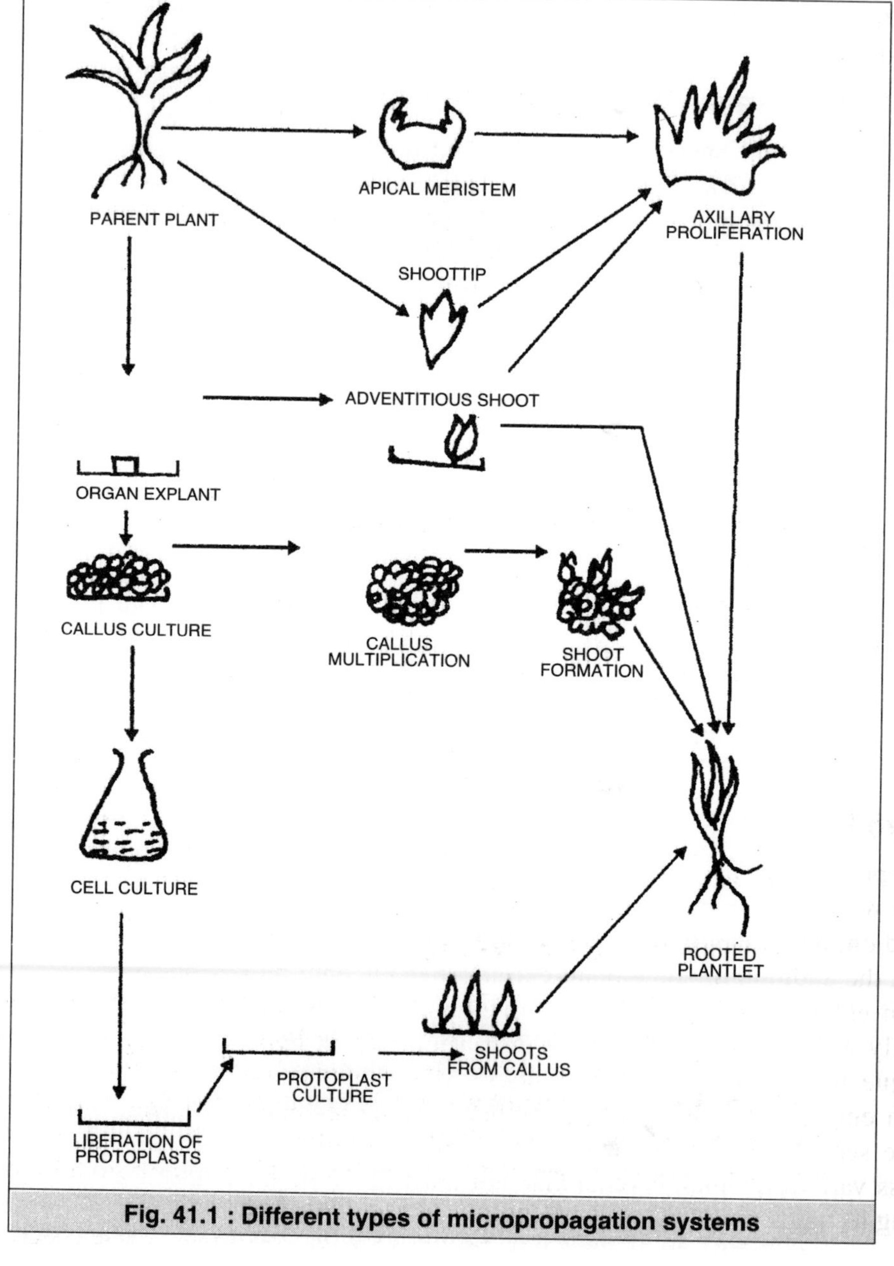

Fig. 41.1 : Different types of micropropagation systems

Meristem tip Culture

This technique is widely used in horticultural plants like potato, dahlia, carnation and orchids. In this method, the meristem tip consisting of one or two pairs of leaf primordia are cultured in a culture medium. After a few weeks, the plantlets are re-generated and after hardening of the plantlets, these are transplanted in the soil under natural environmental conditions. Meristem tip cultured plants give rise to polyploid plants instead of diploid plants. Moreover, meristem tip culture is very useful for the elimination of viruses from infected plant material. Rapid multiplication of the plants, which are otherwise not easily propagated by vegetative means, is also possible through meristem tip culture. Plants produced are free from pathogens and can be stored for longer period and in a smaller space. The different steps involved in meristem tip culture are given in Figure 41.2.

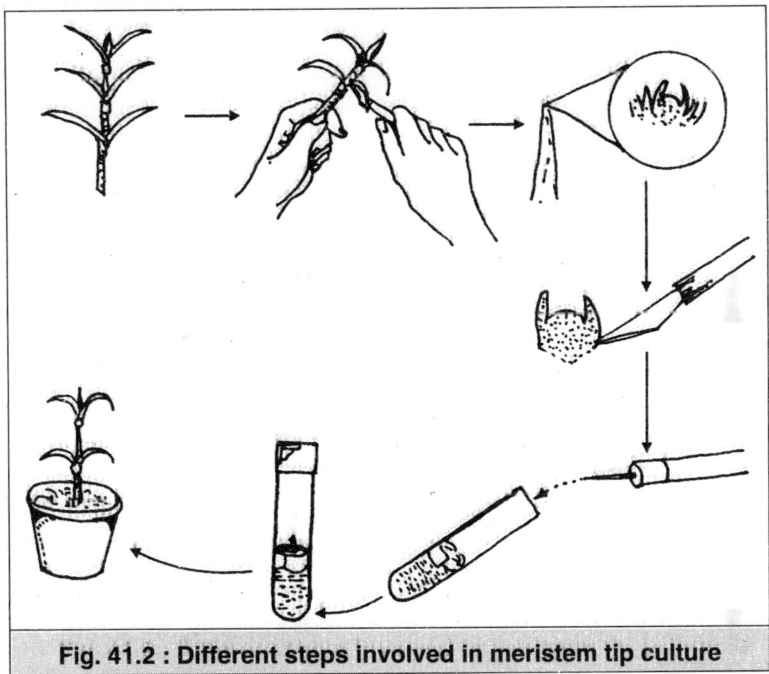

Fig. 41.2 : Different steps involved in meristem tip culture

Embryo Culture

Aseptic culture of embryo has been standardized in many plant species. The embryo is rescued while it is immature and before the degeneration is initiated. The excised embryo is cultured on the semi-solid or solid culture medium as the case may be. However, in embryo culture, the addition of some plant extracts like coconut milk or yeast extract in the culture medium enhances the growth and development of the embryos. Embryo techniques are primarily useful in case of making distant crosses in hybrid production. It is also useful technique to overcome seed dormancy. Many embryological problems can be studied through embryo culture. In grape, embryo rescue technique has been used worldwide to produce seedless varieties by excising immature embryo in crosses between Seedless × Seedless varieties, which is otherwise not feasible through a conventional propagation or a breeding technique.

Callus Culture

A callus is an unorganized mass of parenchymatous cells. Callus culture has been induced artificially in a wide range of plant species using some explants. Callus formation is basically a natural phenomenon essential for the healing of wounds, graft and bud union and rooting of cuttings. In callus culture, the stimulation of cell division and multiplication is required and thus culture medium is accordingly supplemented with necessary growth regulators. For callus culture, auxins are required frequently followed by cytokinins. In callus culture, organogenesis occurs in two stages. In first stage, formation of meristems takes place and in the second stage, active growth of stem buds and roots takes place. As the nutritional requirement of these two stages is different, hence transferring of first stage to the second stage with a culture medium having different supplementations (especially hormones) is a must.

Anther Culture

Anther culture is used to produce haploid plants. Haploid plants are very useful in genetical and biochemical studies and for producing homozygous lines (within six months) of F_1 hybrids which otherwise may take six generations or more with conventional methods of breeding. Guha and Maheshwari (1959) reported anther culture for the first time in Datura. In this technique, anthers are excised and cultured onto required and standardized culture medium. The stage, at which the anther is excised, depends on different species. However, in most cases, it is excised at or near microscopic mitosis, when the pollen grains pass from uninucleate to the binucleate condition. As a result of cell division, multicellular masses are developed, which may develop into embroids or callus. Plantlets are developed from callus by organogenesis. Haploid plants are treated with colchicine for chromosome doubling and for producing the true breeding diploid plants.

Cell Culture

Cell suspension cultures are obtained by homogenizing a piece of callus into liquid medium and shaking with shaker until medium becomes cloudy with suspended cells. Due to recent advances in Plant Biotechnology, it is now possible to derive full plant from a single cell. This technique is primarily useful to plant breeders because now it is possible to induce desirable variability in an individual cell, which can be regenerated into a full plant.

Protoplast Culture

In protoplast culture, the rigid cell walls are first removed either mechanically or enzymatically to expose the protoplast. The exposed protoplast is then cultured onto a suitable culture medium. Individual protoplast of the plants has been regenerated into full plant in *Nicotiana*, *Brassica* and *Asparagus* etc. In plant breeding systems, protoplast culture has many advantages: (*i*) a large homogenous population (10^6 to 10^7) of plants can be obtained from a small sample (1 g leaf), (*ii*) fusion of two protoplasts of two different plant species is possible, which may bring a greater variability in plants.

Ovule Culture

In this method, unfertilized ovules are excised and cultured into medium. This technique has potential application to produce hybrid seeds in wide genetic crosses, which is rather difficult through conventional means. It is also efficient method to obtain plants in self-incompatible species. The technique of growing ovules in sterile culture is not a difficult job. However, one should have sound knowledge of the flowering and fertilization timing for isolation of the ovule at a proper time.

Micropropagation techniques have been successfully adopted in many horticultural crops. Many flowering crops like gladiolus, carnation and orchids etc., are being commercially propagated through micropropagation. Among fruit crops, strawberry was the first fruit to be propagated commercially through micropropagation. Micropropagation techniques have been standardized for many temperate, tropical and sub-tropical fruit crops. In India, tissue culture technique has been perfected in banana, grape and papaya (Table 41.4). In banana, shoot-tip

TABLE 41.4 : Micropropagation technology developed for some horticultural crops in India

Crop	*Institution, which standardized technology*
	Fruit crops
Annona	IIHR, Bangalore; BARC, Mumbai; NCL, Pune
Banana	IIHR, Bangalore; TNAU, Coimbatore; NCL, Pune
Citrus	NBRI, Lucknow; NRC Citrus, Nagpur; University of Jodhpur, Jodhpur
Grape	IARI, New Delhi
Guava	GBPUAT, Pantnagar
Papaya	IARI, New Delhi
Pineapple	BARC, Mumbai
Strawberry	TERI, New Delhi
	Vegetable crops
Brinjal	BARC, Mumbai
Cabbage	BARC, Mumbai
Chillies	IIHR, Bangalore
Onion	IIHR, Bangalore
Peas	PAU, Ludhiana
Potato	CPRI, Shimla
Tomato	BARC, Mumbai
	Ornamental crops
Bougainvillea	IIHR, Bangalore; NBRI, Lucknow
Carnation	IARI, New Delhi; UHF, Solan; Delhi University, Delhi
Chrysanthemum	NBRI, Lucknow; NCL, Pune
Gladiolus	NBRI, Lucknow; IARI, New Delhi; IIHR, Bangalore
Lily	NBRI, Lucknow
Orchids	IIRH, Bangalore; NBRI, Lucknow
Rose	NCL, Pune; IARI, New Delhi

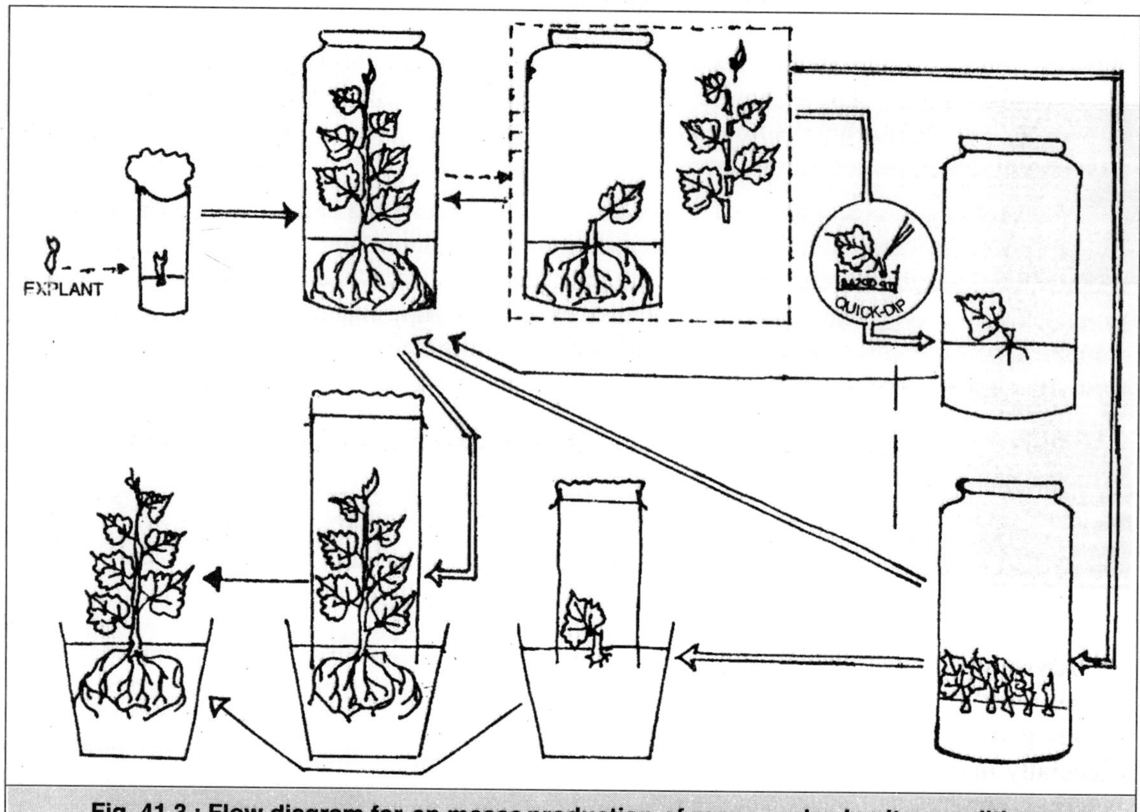

Fig. 41.3 : Flow diagram for *en masse* production of grape under *in vitro* conditions

excised from rhizomes of sword suckers are suitable explants explants and MS medium supplemented with sucrose (3 per cent), gelite (0.25 per cent), is the best. Shoot tips and two nodal micro-cuttings are highly suitable explants for faster and disease free production of grape plants through tissue culture (Figure 41.3). Salt tolerant rootstock of grape (*e.g.* Salt Creek and St George) has great demand and *in vitro* propagation has been successfully used for raising these rootstocks successfully. Shoot tip culture technique has been demonstrated in papaya to produce female plants in a desired ratio. Similarly, pineapple is now being commercially propagated through micropropagation. In difficult-to-propagate fruit species like date palm and coconut, *in vitro* techniques have been perfected only recently. Scientists in India and abroad, are trying hard for propagation of mango and coconut through tissue culture but the success is limited.

• • •

CHAPTER 42

Major Propagation Methods of Fruit Plants

About seventy-eight fruit plants are cultivated in India. These are propagated both by sexual (seed) and asexual (vegetative) methods. As stated earlier. vegetative propagation avoids genetic variation, whereas, the sexually propagated plants show great variability, especially in the cross-pollinated plants. Vegetative propagation includes budding, grafting, layering, cuttage, and some specialized organs. In each propagation method, various other methods are adopted for propagation. One method of propagation may or not be suitable for all types of fruit crops in the same or different locality.

Some of the most commercially accepted propagation methods for different fruit crops are described hereunder :

Almond

The chief method of almond propagation is 'T' budding, which is usually done in autumn (fall). It can also be done in June and September-October.

Aonla

Of the various methods of vegetative propagation 'T' budding is the most practical and commercial method for *aonla* propagation. In this method, one-year-old seedlings with 1 cm girth are selected. The methodology for 'T' budding is the same as described for citrus but the right time of budding in almond is early-June. The older trees of inferior *aonla* type can also be rejuvenated and easily changed into superior type by top-working with 'T' budding. For this, the inferior trees are headed back to a height of 1.2 m from the ground during March and the shoots so emerged are shield-budded in early-June.

Apple

Tongue grafting is the most successful method of apple propagation. In tongue grafting, a 5-mm slanting cut is made tapering into the bark on the lower end of the scion-wood. In the middle of this slanting cut, another cut is made upward to form a tongue. A similar slanting cut of identical length is made on the rootstock pointing upwards. Then, a similar tongue is also made in the rootstock, corresponding to the exact location of the tongue made in the scion-wood. The scion tongue is inserted into the rootstock tongue in a way that the two cuts fit each other firmly. If the thickness of the scion-wood is different to the rootstock, the tongue should be cut in a way that on inserting it in the rootstock, the cambium layer of both the rootstock and scion make as much contact as possible. After the scion-wood is interlocked with rootstock, it is tied with a thread and waxed or wrapped with an alkathene strip so as to make it air-tight. The height of the scion graft on stock should normally be 15 cm above the ground. February-March is best time for tongue grafting in apple.

Apricot

Apricot is commercially propagated by 'T' budding. 'T' budding is commercially done is autumn, which gives maximum success. However, budding can also be done in spring and summer.

Arecanut

Arecanut is mainly propagated by seed, called as 'seed nuts'. For details, please read Chapter 44.

Bael

Bael is usually propagated through seeds. Seeds are extracted from mature, fully ripe fruits during December-January and immediately sown in the nursery. If environmental temperature is very low, the nursery should be covered with an alkathene sheet. The seeds can also be sown in poly bags. 'T' budding is also practical in some areas.

Banana

Suckers are commonly used as planting material in banana. Suckers are of two types, sword suckers and water suckers. Sword suckers have a well-developed base with narrow sword-shaped blades at the early stages, whereas water suckers have broad leaves. For planting, sword suckers should be used because water suckers do not produce healthy banana clumps. Similarly, banana produces bigger and heavy bunches if sword suckers are used as a planting material.

Ber

Ber is commonly propagated by 'T' budding. The method of 'T' budding in *ber* is the same as in citrus. However, the best time for budding is late-June.

Blackberry

Trailing type blackberries are propagated by tip layering, whereas upright types are raised through suckers, which are more often produced by them.

Blueberry

It is commercially propagated by hardwood cuttings. The cuttings of 10-15 cm long size are prepared from one-year-old dormant wood, which are planted in winters.

Cape Gooseberry

The perennial types are propagated either by softwood grafting or layering, while annual types are usually raised through seeds.

Carambola

It is preferably propagated by air layering or 'T' budding because there is a wide variation in plants if raised through seeds. Air layering is done in monsoon season and 'T' budding either in March or September-October, when the bark slips easily.

Cashewnut

Until 1980's, the biggest hurdle in deriving the benefits of new cultivars in cashewnut was the lack of reliable vegetative propagation method. However, the Agri. Res. Station, Ullal has developed softwood grafting technique for its propagation, which is gaining popularity day-by-day. In softwood grafting, the rootstock should be 20 to 25 days old with good growth. Tall and lanky seedlings should be avoided. The scion should be 8-10 cm long and of pencil thickness with bulging tips. The colour of scion should be turning from green-to-brown and at the top, 4-5 leaves should be dark green indicating proper maturity. For pre-curing, leaf blades are clipped off, 7-10 days before grafting. When petioles dry and drop off, the scion can be detached from the mother tree and used for grafting. While grafting, atleast 1-2 leaves are left at the bottom and grafting is done in the softwood portion above these leaves. Longitudinal cut of 3-4 cm is made in the stock and scion wedge is inserted in the rootstock and tied with alkathene strip of 75-gauge thickness. The grafts are kept in shade for a week and later transferred to an open place.

For rejuvenation of old and unthrifty plants, top-working with softwood grafting can be done by grafting scions of superior cultivars. The basic method is the same as described above. The trees for top-working should be beheaded during November-December or during April-May for forcing of new growth. 2-3 months old shoots should be used for top-working. Its propagation has been discussed in details in Chapter 44.

Cherry

Cherries are propagated commercially by 'T' budding, which is done autumn. These can also be raised through seeds but seeds require stratification at 3-4°C for 3-4 months before sowing.

Citrus

Citrus is mainly propagated by 'T' budding and air layering.

'T' budding is the most common method of vegetative propagation of sweet oranges, mandarins, grapefruit and pummelo. In this method, a 'T' shape cut is made on the stock at a height of about 20-25 cm above the ground level. The bark is loosened near the cuts and a bud after removing wood, is inserted and tied with alkathene tape. The successful bud may start sprouting after 10-15 days of budding and then the stock is cut step-by-step. The budded plants become ready for planting in the orchard in about 6-8 months after budding. Basically, budding can be done at any of the year when there is active sap flow in the plants. However,

February-March and September-October are the best months. For successful budding in citrus, the following points should be kept in mind.

- Use rootstock of pencil thickness.
- Select buds from 10-12 months old non-angular branches.
- For high success, budding should be done in February-March.

Air layering is mainly done for the propagation of lemon and limes. It is also called as 'goottee' or 'marcottage'. In this method, the aerial branches are girdled, covered with moist moss and tied with a polyethylene strip. Application of 5,000 ppm IBA (Indolebutyric acid) in lanolin paste is beneficial for rooting in the air layers. The rooted layers are ready for planting after a month or so. The best time for air layering is the monsoon season.

Coconut

Coconut is commonly propagated by seeds. The productivity of the palm depends on the quality of the seedling and therefore every care must be taken to select best nuts for sowing. The seed nuts should be collected from healthy and vigorous mother plants, which have a known history of giving consistent yield of quality nuts. Usually, twelve months old nuts having a minimum weight of 750 g are good for planting and better germination. The seed nuts are usually collected during February-to-May. These seed nuts should be stored for a month or so before sowing. Different methods of planting can be used but vertical planting is preferred in India. Its propagation has been discussed in details in Chapter 44.

Cranberry

Cranberries are propagated by cuttings made from runners and one-year branches.

Currants

Currants are mostly propagated by hardwood cuttings. Cuttings of 20-25 cm length are prepared from pruned wood in dormant season and planted in the field/nursery in early-spring.

Custard Apple

It is usually propagated by seeds but cleft grafting and 'T' budding are useful methods to raise it commercially.

Date Palm

In India, date palm is raised through seeds, as a result of which, the seedling plants have greater variability. Such plants come into bearing vary late and their sex can't be judged easily in the early years. Thus, date palm should be raised through cuttings or offshoots (suckers), which arise from the base of the main trunk. Suckers, which are produced near or below the ground level usually root better than those produced above the ground level. To promote rooting in suckers, the base of the offshoot should be in contact with moist soil for atleast a year before separation. It can be done by mounding or by putting soil boxes on the offshoots arising high on the ground. Offshoots of 12-15 Kg weight and 25-30 cm diameter usually root better. The most suitable time to remove and plant offshoots is spring (February-March) and autumn (August-September).

Durian

Durian is usually propagated by seeds. Its seeds have aril and if it is not removed before sowing, these may have poor germination. The seeds are extracted in September and before sowing, the aril should be removed by thorough washing with tap water and sowing can be done in October. Durian can also be propagated by budding and air layering.

Fig

The fig is propagated by hardwood cuttings. Hardwood cuttings of 25-30 cm length are prepared from 2-3 year old shoots in January-February from the pruned wood. These may be treated with indole butyric acid (2,500 ppm) and planted in nursery or field in January-February.

Guava

At present, stooling is the best method of guava propagation, which was standardized at IARI, New Delhi. Stooling is a very easy, simple and less cumbersome method of guava propagation. In this method, plants of desired variety are headed back to the ground in January-February and shoots are allowed to develop. When the shoots attain a size of 20-25 cm, these are girdled at the base by removing a 2-2$\frac{1}{2}$ cm ring of bark. Then 2,500 ppm indole butyric acid (IBA) in lanolin is applied to the upper portion of the girdled shoots. These shoots are left open for 2-3 days after IBA application and then covered with soil to encourage rooting in the stools. After 6-8 weeks, the roots are fully developed in the shoots. These rooted shoots can be separated from the mother plant and planted in the nursery. Before separating the stools, a light irrigation should be given if soil is dry. After separating the shoots, level the field after applying ample quantity of manures and fertilizers. Stooling can be done twice a year and in this way, one can get 30-40 stools per mother plant per year.

Grape

Grape is commercially propagated through hardwood cuttings. The cuttings of 25-30 cm length can be prepared from the pruned wood. The cuttings are usually prepared in mid-January under north indian conditions, while these can be prepared in March-April and September-October in south India. The cuttings should be of pencil thickness (7-8 mm in diameter). Further, the cut at the base of the cutting should be made perpendicular, just below the bud, while the upper cut should be made slanting and about 1.5 cm above the apical bud. It avoids injury to the apical bud and facilitates orientation of cuttings at the planting time. After preparation, the cuttings should be planted in the nursery bed or at a permanent location in the field. However, these can also be stored in polyethylene bags in a refrigerator until planting.

Jackfruit

The most common method of jackfruit propagation is by seeds, though it can be raised by air layering also. Seeds should be sown immediately after extraction because they loose viability very soon. Pre-soaking of seeds in NAA (25 ppm) for 24 hours or GA_3 (25 ppm) for 48 hours is useful for better germination.

Jamun

The most common method of *jamun* propagation is by seeds. Seeds should be sown fresh because they loose viability if stored. In north India, the seeds are sown in monsoon season. Inarching can also be used for its propagation.

Karonda

Karonda is chiefly propagated by seeds. The seeds are extracted in winters and sown immediately in nursery because they loose viability very soon.

Kiwifruit

Commercially, kiwifruit is propagated either by whip grafting or 'T' budding. The best time for budding and grafting is spring season. A new indigenous method of rooting of cuttings has been standardized at IARI, Regional Research Station, Shimla, only recently. In this method, the hardwood cuttings of kiwifruit are planted in a bottom heat chamber to facilitate rooting in them.

Litchi

Litchi is commercially propagated by air layering (*goottee*). In air layering, upright, 1-2 cm in diameter and 30-60 cm long healthy branches from well-developed trees are selected, A 2-2.5 cm ring of bark is removed from the selected branches. Then, 5,000 ppm IBA (Indole-butyric acid) made in lanolin paste is applied to the upper cut of the ring. This cut is then wrapped with polyethylene sheet containing moist sphagnum moss grass. The rooting in these layers takes place in $1\text{-}1\frac{1}{2}$ month and the rooted layers can be separated from mother plant and planted in the nursery. The best time for *goottee* in litchi is spring and early-summer. However, gootee in rainy reason also gives high success.

Longan

Longan is usually propagated by seeds. It can, however be propagated by air layering. The best time for layering is between October-to-February.

Loquat

For commercial propagation, air layering is most commonly followed for the propagation of *loquat*. In this, 3 months old shoots are ringed and layered. Basic procedure of air layering is the same as described for litchi.

Macadamia Nut

It is mostly propagated by side-veneer grafting. The procedure is the same as followed for mango propagation.

Makhana (Gorgon Nut)

The commercial method of propagation is by seeds. The seeds should be first sprouted before sowing. For sprouting, water is usually sprinkled on the gunny bags containing seeds. Seeds sowing is done in ponds with the help of a boat.

Mango

Mango in India is commercially propagated by side-veneer grafting and epicotyl grafting. Side-veneer grafting technique was standardized at IARI (Pusa) New Delhi-12. This is the improvement over inarching, which has been the traditional method of mango propagation in India.

Following points should be kept in mind while using side-veneer grafting in mango :

- Scion should be selected from 4-5 month old shoots of the desired variety.
- Defoliate the scion, 6-7 days before actual grafting process.
- February-March and July-August months are best for side-veneer grafting.

Side-veneer grafting has the following advantages :

1. It is less cumbersome, laborious and time-consuming method of propagation than inarching.
2. Seedlings of 2-3 years age can also be utilized for grafting.
3. Two or more varieties can be grafted on a single seedling stock.
4. Scions of desired variety can easily be sent to distant places.
5. Old/declining orchards of inferior variety can be converted in to bearing orchards of superior or desirable variety.
6. It is easy to establish *in situ* orchards with side-veneer grafting technique.

Mango is also propagated by epicotyl grafting, which is also known as *stone grafting*. It gives high success in localities having high humidity, as in Konkan region of India. In this method, the stones are placed in a sand bed and covered with 5.0-7.5 cm thick layer of leaf mould for germination. 10-15 days old seedlings are beheaded about 5.0 cm above the stone and a wedge shaped scion of desired variety is inserted in the vertical split of the beheaded stock. Polyethylene tape (200-gauge thickness) is used for tying the grafts. Immediately thereafter, the grafts are planted in polyethylene bags filled with soil and farmyard manure (FYM) mixture (1:1). The grafts are watered frequently and kept under semi-shade conditions to avoid the affects of direct sun and rain. The successful stone grafts are shifted to open space when their leaves have developed. The best time for stone grafting is monsoon season.

In the recent times, softwood grafting method has also been standardized for propagation of mango. It gives good success than side-veneer grating in certain localities. The procedure of softwood grafting has been described in cashewnut.

Mangosteen

It is mainly propagated by seeds. The seeds are not zygotic thus produce true-to-type seedlings. Seeds, however, loose viability soon and therefore these should be sown immediately after extraction.

Mulberry

Mulberry is commercially raised by hardwood cuttings. Cuttings of 25-30 cm in length are prepared during spring (February-March) or rainy season (July-August) from one-year-old wood. These may be treated with root promoting hormone (IBA, 2000 ppm) and planted in the nursery soon after their preparation.

Naranjilla

It is usually propagated by seeds and hardwood cuttings of 20 to 40 cm length.

Olive

Olive is commercially propagated by hardwood cuttings. The cuttings of 15-20 cm length are prepared from one-year-old shoots during December-January and planted soon in the nursery or field.

Papaya

Papaya is mainly propagated by seeds. However, the mucilase of seeds should be removed after extraction as it hinders germination process adversely. Micropropagation techniques have also been standardized by various institutes for speedy multiplication of papaya.

Passion Fruit

Cleft grafting is followed for commercial propagation of passion fruit. One-year-old seedlings, which are of 30-40 in height are cut off about 15-20 cm above ground and spilt-down the middle for about 1-2 cm to accommodate the scion. The scion consists of a small piece of vine wood, about 8-10 cm long with its larger leaves removed. The lower end is cut to a fine wedge to fit neatly into the cleft made in the stock. The scion is completely bound with narrow thin PVC plastic tape. Grafted plants are ready for transplanting in the field after about 6-8 weeks of grafting.

Pear

The pear is commercially propagated by 'T' budding and tongue grafting. 'T' budding is done in April-May or August-September when bark starts splitting and the cambium tissues responsible for union are in active state. Rest of the procedure for 'T' budding is the same as described for other fruit crops. Tongue grafting is also employed successfully when scion and stock are of equal thickness. It is done in December-January.

Pecannut

Patch budding is commercially used for propagation of pecans. Patch budding is done in March under Indian conditions.

Pepino

It is commercially propagated by herbaceous cutting. Cuttings of 10-25 cm are usually considered as the best.

Persimmon

Persimmon is commercially propagated by whip grafting. The best time for whip grafting is September.

Phalsa

Phalsa is mainly propagated by seeds. Seeds are extracted during May or June in north India and are sown in the nursery. Cuttings can also be used for its propagation but rooting in them is very difficult.

Pineapple

Pineapple is usually propagated vegetatively through slips, suckers, crowns and stump bits but slips are the best planting material followed by suckers. Usually, slips of 350-400 g and suckers of 400-500 g are recommended for planting.

Plum

Plum is mainly propagated by 'T' budding. The best time for 'T' budding is early-spring. However, it can be done in fall. The basic procedure for 'T' budding is the same as described for other fruit crops.

Pomegranate

The common method of propagation of pomegranate is the hardwood cuttings. The cuttings of 20-25 cm are prepared from fully mature one-year-old wood. These can be prepared in January from deciduous species and in July-August from evergreen species. Dipping of cuttings in 50 ppm IBA for 10-15 minutes gives high rooting in the cuttings and subsequent better survival.

Rambutan

Normally patch or forkert budding methods are used for the propagation of rambutan. Patch budding is done in March.

Raspberry

Red raspberries are commercially propagated by root suckers and cuttings. The roots are cut into small pieces (5-7 cm long) in early-spring and planted in the nursery. The rooted cuttings can be transplanted in the permanent locations in the next spring.

Sapota

Air layering (gootee) is commonly followed method of sapota propagation in India. In this method, one-to-two-year-old, 40-60 cm long mature branches of pencil thickness are selected. A ring of 2.5-3 cm bark is removed carefully from the selected branches. Over this, moist rooting medium, such as sphagnum moss or vermiculite is applied and then the layers are wrapped with a plastic or polyethylene film. In about 3 months, roots emerge from the upper ring of the layer. To detach layers from the mother plant, one has to be very careful because the mortality rate of air layers is very high in the field. When the roots have fully developed, first a 'V' shaped notch is made in the layered branch, 10-15 cm below the ring. It is further deepened after 15 days and then finally, layers are removed from the mother plant and planted in the nursery.

Strawberry

Strawberry is commercially propagated by runners. Runners can be detached from the mother plant in July and planted in the nursery. These plants are ready for planting in a field within a month or so. The best time for planting in hills is July and October end in the plains. Strawberry is commercially propagated through micropropagation is different parts of the world. In India also, some institutes are raising strawberry plants at a commercial scale.

Tamarillo

It can be propagated by seeds and cuttings as well. Plants raised by seeds are tall and are suitable for cultivation in protected environment. The plants raised by cuttings are shruby and suitable for open fields.

Tamarind

It manly propagated by seeds.

Walnut

Patch and 'T' budding are the best methods of walnut propagation. The best time for budding is fall. However, it can be done in spring with little success.

Water Chestnut (*Singhara*)

It is commercially propagated by seeds. Seeds are sown during December in ponds, which are transplanted in rainy season.

• • •

CHAPTER 43

Propagation Methods of Vegetable Crops

In our country, about 70 different types of vegetables are grown. After China, we are the largest producer of vegetables in the world. As per information available, an average Indian consumes about 92 g of vegetables/day, whereas, according to ICMR reports, one must include 295 g vegetables in his/her diet/day. It is, therefore, necessary that vegetable production and consumption has to be increased manifolds. In the first few five year plans of India, vegetable production and development programmes did not get priority, but in the 8th and 9th five year plans, emphasis was given on research and development programmes of vegetables. As a result, a lot of progress has been made on the development of a number of varieties and hybrids, and production technologies of many vegetable crops. In this chapter, emphasis has been given on propagation aspects of vegetables. Vegetables are propagated both by seed and vegetative methods.

SEED PROPAGATION

Most vegetables, which are propagated by seeds (*e.g.* cucurbits, beans, peas, radish, turnip, carrot, leafy vegetables and okra etc.) are required to be sown directly in the fields whereas, in others like, tomato, brinjal, capsicum, chilies, cabbage, cauliflower, knol-knol, lettuce, brussel's sprouts and onion, seeds are first sown in the nursery beds where seedlings are raised and then transplanted in the fields. Proper nursery management for raising seedlings and transplanting them in the field are important operations in vegetable production. In general, it takes 10-15 days for the seeds to germinate. The seedlings are planted in the fields when these attain a age of 1-1.5 month. The process of germination, factors affecting seed germination and principles involved in seed propagation have been discussed in detail in the previous chapters. However, brief information on seed rate, sowing time, transplanting time and planting distance for important vegetables, which are commercially propagated by seeds, is given in Table 43.1.

TABLE 43.1 : A list of some important vegetables propagated by seeds

Common name	*Scientific name*	*Family*	*Seed rate/ha*	*Sowing time*	*Transplanting time*	*Planting distance*
Tomato	*Lycopersicon esculentum* Mill.	Solanaceae	400-500 g	Autumn-winter crop : June-July, Spring-summer crop : November. In hills : March-April	4-6 weeks after seed sowing	Autumn-winter crop : 75 × 60 cm, Spring-summer crop : 75 × 45 cm
Brinjal	*Solanum melongena* L.	Solanaceae	375-500 g	Autumn-winter crop : June, Spring-summer crop : November, Rainy season crop : March. In hills : April	4-6 weeks after seed sowing	Bushy non-spreading type : 50-60 × 50-60 cm, Spreading type : 90 × 60 cm
Chilli	*Capsicum annum* L.	Solanaceae	1.5 -2 kg	Autumn-winter crop : June-July, Spring-summer crop : November. In hills : March-April	4-8 weeks old seedlings	North India : 45 × 45 cm, South India : 55 × 55 cm
Cauliflower	*Brassica oleracea* L. var. *botrytis*	Cruciferae	For early crop-600-750 g For late crop-375-400 g	Early sowing : May-June, Mid season sowing : July August, Late sowing : Sept.- October	4-8 weeks old seedlings	60 × 45 cm
Cabbage	*B. oleracea L. var capitata*	Cruciferae	For early crop-500 g, For late crop-375 g	Early sowing : August, Mid season sowing : September, Late sowing: October	4-6 weeks after seed sowing	60 × 45 cm
Cucumber	*Cucumis sativus* L.	Cruciferae	1.5 kg	Summer season crop : January Rainy season crop : June-July In hills : April	Direct sown	1.5-2.5 m × 60-90 cm
Muskmelon	*Cucumis melo* L.	Cucurbitaceae	2-2.5 kg	November-March, In hills : April-May	Direct sown	Channels : 3-3.5 m Hills : 0.9-1.2 m
Water melon	*Citrullus lanatus* Thunb.	Cucurbitaceae	2-2.5 kg	November-March	Direct sown	Channels : 3-3.5 m Hills : 0.9-1.2 m

Bottle gourd	*Langenaria siceraria*	Cucurbitaceae	4-5 kg	Summer season crop: January-February, Rainy season crop : June-July	Direct sown	Channels : 2-2.5 m Hills : 0.9-1.2 m
Bitter gourd	*Momordica charantia* L.	Cucurbitaceae	5-6 kg	Summer season crop : January- March, Rainy season crop : June-July. In hills : March-June	Direct sown	Channels : 2-2.5 m Hills : 0.6-1.2 m
French bean	*Phaseolus vulgaris* L.	Leguminosae	Bush variety : 85-90 kg Pole variety : 25-30 kg	First sowing July-Sept. Second sowing: Jan.-Feb. In hills : March-May	Direct sown	Bush type : 30 × 5 cm, Pole type : 90 × 7.5 cm
Cowpea	*Vigna sinensis* Savi	Leguminosae	12.5-20 kg	North India : Feb-March, South India : Dec-Jan. Rainy season crop : June- July	Direct sown	45-60 × 10-15 cm
Cluster bean	*Cyamopsis tetragonoloba* L. Taub.	Leguminosae	20-40 kg	February-March and June-July	Direct sown	45-60 × 15-20 cm
Moth bean	*Vigna acontifolia* (Jacq.). Marrchal	Leguminosae	10-15 kg	June-July	Direct sown	30-45 × 15-20 cm
Broad bean	*Vicia faba* L.	Leguminosae	70-100 kg	September-October and February-March	Direct sown	45 × 10-15 cm
Winged bean	*Psophocarpus tetragonolobus* L.	Leguminosae	8-10 kg	June-July	Direct sown	Tall variety : 90 × 90 cm, Dwarf variety : 45 × 45 cm
Pea	*Pisum sativum* L.	Leguminosae	Early variety : 100-200 kg. Main season variety : 80-90 kg	Early variety : September Main season variety : October-November	Direct sown	30-45 × 5-10 cm

Soybean	*Glycine max* L. Merril.	Leguminosae	*Kharif* crop : 20-30 kg. Spring crop : 30- 80 kg	*Kharif* season crop : June-July Spring season crop : February-March	Direct sown	*Kharif* crop : 45-60 × 15 cm Spring crop : 30-45 × 15 cm
Okra	*Abelmoschus esculentus* L.	Malvaceae	Spring-summer crop : 20 kg Rainy season crop : 8-10 kg	Spring summer crop : February-March, Rainy season crop : June-July. In hills : May-June	Direct sown	Spring-summer crop : 45 × 15 cm Rainy season crop : 60 × 30 cm
Amaranthus	*Amaranthus* spp.	Amaranthaceae	Direct sowing 2-2.5 kg Transplanting : 1-1.5 kg	North India : Throughout the year except May-June South India : Throughout the year	Direct sown as well as trans-planting at an age of 4 week	20 × 15 cm
Spinach	*Spinicia oleracea* L.	Chenopo-dicaceae	35-45 kg	Plains : Sept-Oct., Mid hills : March-June, High hills: July-Sept.	Direct sown	30 × 10 cm
Fenugreek	*Trigonella foenum graecum* L.	Leguminosae	Common : 25 kg, Kasuri : 20 kg	September onwards	Direct sown	20-25 × 5 cm
Lettuce	*Lactuca sativa* L.	Asteraceae	400-500 g	September-October	5-6 weeks old seedling	45 × 30-45 cm
Radish	*Raphanus sativus* L.	Cruciferae	9-12 kg	Plains : August-December Hills : April-June	Direct sown	45 × 6-8 cm
Turnip	*Brassica rapa* L. var. *glabrakitamura*	Cruciferae	3-4 kg	Plains : September-Dec. Mid Hills : July-Oct. High hills : July-Sept.	Direct sown	45 × 15 cm
Carrot	*Daucus carota* L.	Apiaceae (Umbelliferae)	8-10 kg	North India : Aug-Dec. South India : Jan. Feb., June-July, Oct-Nov.	Direct sown	45 × 10 cm

SOIL PREPARATION

For raising nursery of vegetables, first suitable soil should be selected. The soil should be fine, moist and firm to provide better germination and excellent medium for seedling growth. Enough well decomposed farmyard manure (FYM) should be mixed thoroughly in the soil. After repeated ploughing, the nursery beds are made. The length of nursery bed should be 150 cm or more and width should not be more than 120 cm. This width facilitates weeding and watering without trampling the bed. The bed is raised about 15 cm above the ground level for proper drainage. The level of bed surface is also made slightly raised in the centre with a little slope on the two sides.

Some vegetables are raised in pots also. For pots, usually, the soil mixture should have two parts of garden soil, one part of sand and one part of leaf mould.

Seed Sowing

The common practice is to broadcast seeds in the nursery bed but line sowing is preferred so as to check proper germination and to facilitate weeding, hoeing and plant protection operations. The rows are usually kept about 5 cm part. The size of the nursery bed to raise seedlings to plant a hectare of tomato, brinjal and capsicum would be 50 × 1.2 m, 54 × 1.2 m and 180 × 1.2 m, respectively. In case of cole crops such as cauliflower, cabbage and knol khol etc., a nursery bed of 115 × 1.2 m would be sufficient to raise seedlings for one-hectare field. Instead of making a single large bed, it is better if smaller beds are made. Before sowing small seeds, these should be mixed with sand, and after sowing, be covered with a thin layer of sand or soil. This covering should be still lighter in heavy soils. After this, a light irrigation is given with a sprinkler.

After Care of the Seedlings

Watering of the beds is done uniformly and gently so as to avoid a packing of the soil or washing away of the soil. It is always advisable to make the nursery beds near the source of irrigation water. In the beginning, if the sun is too warm, the mid-day sun may be avoided by leaves, twigs or thatch. During summer, the seedlings may be protected from hot wind and sunshine. When the seedlings are more than 2.5 cm tall, too much of shade and water make them yellow, succulent, lanky and susceptible to insect-pests and diseases, mainly damping off, especially in humid and warm weather. In such case, the affected plants should be removed immediately. A week before transplanting, the seedlings may be exposed to full sunshine. In addition, the number of irrigations should be reduced so that the seedlings become hardy to bear the shock of transplanting.

Transplanting of the Seedlings

If seedlings are allowed to grow too tall, they become weak and they may start flowering in the beds itself. Thus, transplanting should be done as soon as seedlings are about 4-8 weeks old, 10-15 cm long and have formed about 3-4 true leaves. The bed must be watered 24 hours before uprooting the seedlings for transplanting. Transplanting should always be done in the evening so that plants may establish themselves in the cool weather at night and may recover from the shock of transplanting before the sunrise nexday.

Transplanting should be done as early as possible after carefully removing the plants without damaging the roots. The soil of the field should be prepared thoroughly by repeated ploughing before transplanting. Each seedling is placed vertically in the centre of the hole made in the field and the soil near the roots is pressed down with the fingers to make the soil firm and to exclude air pockets. This may be followed by immediate but light irrigation. During transplanting, care should be taken to protect the seedlings against wilting by frequently sprinkling of water on them and by covering their root zone by moist soil or the leaves. Because the seedlings are very delicate, regular watering is absolutely necessary after transplanting. Seedlings not doing well may be removed and replaced by the new ones. Any attack of insect-pests or diseases must be controlled immediately. Sometimes, seedlings are transplanted twice. As soon as they begin to crowd after the first transplanting, they are taken out and transplanted again in the field, spacing them further apart for proper growth and development. For example, cauliflower seedlings are sometimes transplanted twice to get stockier seedlings. Studies have shown that the hardening of cauliflower plants by repeated transplanting markedly improves the growth of plants in all respects, which helps in better curd yield. However, double transplanting may damage seedlings and result in delayed maturity. It is needless to say that it would increase the cost of production with whatever advantage it may have and that is why this system is not commonly practiced.

VEGETATIVE PROPAGATION

Some vegetables are propagated commercially through vegetative means also. A brief description on propagation of some important vegetables by vegetative methods is given in Table 43.2.

TABLE 43.2 : Vegetable crops raised by vegetative methods of propagation

Common name	*Scientific name*	*Family*	*Commercial propagation method*
Asparagus	*Asparagus officinalis*	Liliaceae	Stem cuttings
Basella	*Basella alba*	Basellaceae	Seeds; stem cuttings
Chow-chow	*Scheium edule*	Cucurbitaceae	Whole fruit
Colocasia	*Colocasia esculenta*	Araceae	Corm or cormel
Drum stick	*Moringa oleifera*	Moringaceae	Seeds and stem cuttings
Elephant's foot	*Amorphophallus campanulatus*	Araceae	Offsets of corms
Garlic	*Allium sativum*	Alliaceae	Cloves and bulbils
Globe artichoke	*Cynara scolymus*	Asteraceae	Suckers or offshoots
Horse radish	*Aromoracia rusticana*	Cruciferae	Root cuttings
Ivy gourd (*Kundru or tondli*)	*Coccinea grandis*	Cucurbitaceae	Stem cuttings
Jerusalem artichoke	*Helianthus tuberosus*	Asteraceae	Sets of tubers with 2-3 eyes

Kakrol	*Momordica dioca*	Cucurbitaceae	Tuberous roots
Multiplier onion	*Allium cepa* var. *aggregatum*	Alliaceae	Bulblets
Onion	*Allium cepa*	Alliaceae	Seeds and bulbs
Pointed gourd (*Parwal*)	*Trinchosanthes dioca*	Cucurbitaceae	Vine cuttings
Potato	*Solanum tuberosum*	Solanaceae	Division of tubers; Seed potato weighing 40 to 50 g
Rhubarb	*Rheum rhaponticum*	Polygonaceae	Division of crown
Sweet potato	*Ipomaea batatas*	Convolvulaceae	Vine cuttings with atleast 4 nodes (about 40,000 to 50,000 cuttings are required/ha).
Tapioca	*Manihot esculenta*	Euphorbiaceae	Stem cuttings
Tree onion	*Allium cepa* var. *viviparum*	Alliaceae	Bulbils
Yam	*Dioscorea alata*	Dioscoreaceae	Division of tubers

• • •

Propagation of Plantation Crops

In general, it is very difficult to define plantation crops and rather very difficult to distinguish them from other horticultural crops. However, a few researchers refer plantation crops to those crops, which are cultivated on an extensive scale in a large continuous area, owned and managed by an individual or a company. Thus, the crops like tea, coffee, rubber, cocoa, coconut, areca nut, oil palm, cashew etc form the group of plantation crops. These crops are considered as high value commercial crops of economic importance, which play a pivotal role in the economy of our country. A brief description on the propagation aspects of some of the important plantation crops is given in this chapter.

Tea (*Camellia sinensis*)

Tea is an evergreen shrub, belonging to Camelliaceae family, the newly emerged leaves of which has the value of commerce. Tea can be propagated both by seeds and by cuttings. However, seed propagation not practiced now-a-days, but new varieties are first raised by seeds only. Moreover, rootstocks for grafting are also raised by seeds. Before sowing, the seeds are soaked in water for overnight and the seeds, which sink in water, are used for sowing. Germination of seeds occurs in about 25-30 days and the seedlings are ready for transplanting in about 10-12 months. Among different methods of vegetative propagation, single nodal cuttings have given very good success. For this, first of all healthy mother bushes are selected and to get juvenile shoots, the mother plants should be pruned. Juvenile shoots should be excised in the morning hours and 3 cm long cutting each with a healthy mother leaf and axillary bud should be prepared. Cuttings from the top tender and bottom brown wood should be avoided because these cutting give very poor success. The cuttings are then planted in polythene sleeves (30 × 10 cm) of 150 gauge thickness, filled with suitable growing medium. The sleeves are then covered with polyethylene sheets over the 9.5 feet wire arches and the sides are tugged well to preserve moisture. In general, rooting in cuttings takes place within 8-10 weeks, which can be transplanted after 6-8 months. Some researchers have got good success with cleft grafting, particularly when one has to take the advantage of drought tolerant rootstocks.

Coffee (*Coffea arabica*)

Coffee is bushy short-day plant, belonging to family Rubiaceae, the roasted berries of which have commercial value. It is commercially propagated by seeds. For this, healthy mother plants are selected. Healthy and 3rd to 4th mature ripe fruits of normal size are harvested. The fruits are soaked in water atleast for 24 hours. Next morning, floated fruits are discarded and the rest are pulped to remove seeds. The extracted seeds are dried for 4-5 days. These are sown in raised beds with flat side facing the soil at a distance of 1.5 to 2.5 cm and then covered with a fine layer of sand, followed by paddy straw. The beds are irrigated lightly daily and protected from direct sunlight by constructing overhead pandal. Seeds germinate in about 30-40 days and the seedlings can be transplanted in the secondary nursery beds or in polythene bags (30 × 10 cm) in February or March, when they are in button or topee stage, containing suitable growing medium. The seedlings mature in about 2 months. During this period, the seedlings should be provided with adequate nutrition, water and shade. Finally, the seedlings (16 to 18 months old) may be transplanted in the field at the desired or recommended spacing. Seedlings must be provided with cross stakes to prevent uprooting by wind.

Rubber (*Hevea brasiliensis*)

Rubber of commercial importance is obtained from many plants but rubber plants *(H. brasiliensis)* belonging to family Euphorbiaceae is considered as the most important commercial source. Rubber is propagated both by seeds and budding.

Propagation by seeds : Seeds are used to raise rootstock seedlings, which are available from July-to-September. After harvesting, seeds should sown immediately because they loose viability soon after extraction. Seeds must be sown in raised beds of 1 m width, 3 cm height with any convenient length, in a single layer, touching one another and pressed firmly. A thin layer of dry grass may be put on the beds to avoid falling of direct sunlight. Regular and light irrigation is required for higher success. In general, seeds start germinating within 7 to 10 days. The sprouted seeds should be picked and planted in the nursery at 30 × 30 cm to raise seedling stumps or at 60 × 90 cm or 60 × 120 cm to raise bud wood nursery or stumped budding or may be directly planted in the field.

Propagation by budding: In rubber, the scions of desired types are maintained in the bud-wood nursery by planting the budded stumps or by budding the clone on the seedlings *in situ* in the nursery. The term, budded stumps refers to the budded plant whose scion shoot is cut very close to the budding zone, leaving few dormant buds in the scion shoots. However, if rootstock is cut as stump and budding is done (usually green budding is done when seedlings are 4-5 months old), then it is called as stumped budding. Two types of budding techniques are followed in rubber. Brown budding is done by using buds from one years old shoot on 10 months old rootstock, whereas, in green budding, both bud-wood and rootstock are tender and green.

Cocoa (*Theobroma cacao*)

Cocoa is a beverage crop, belonging to family Sterculaceae, growing primarily in humid tropics. It is propagated commercially by seeds and rarely vegetatively from buds and cuttings.

For seed propagation, those plants are selected, which bear 80 or more pods and each pod have atleast 400 g weight. Fresh beans from the selected pods should be used for sowing because cacao beans lose their viability very soon after extraction. Before sowing, the beans should be rubbed with dry sand or wood ash to remove mucilage. These are sown either in polythene bags of 25 × 10 cm size with 150 gauge thickness, or in raised beds with their pointed ends upwards. Seeds start germinating in 15 to 20 days and the seedlings are ready for transplanting after 3 to 4 months.

Cacao can be propagated by cuttings, soft-wood grafting, but the success is less and hence these methods of vegetative propagation have not yet been employed for large scale multiplication on commercial basis.

Cashewnut (*Anacardium occidentale*)

Cashew is a low spread evergreen tree, belonging to Anacardiaceae family, which was introduced in Indian to Malabar coasts in sixteenth century to cover bare hills for soil conservation. It can be propagated through seeds and vegetative means but seed propagation is seldom practiced because of wider variability in the resultant seedlings raised from seeds. However, in general, seeds are collected during March to May and sown soon after extraction. Germination takes place in a month's time. 6-8 months old seedlings can be transplanted in the field.

Among different methods of vegetative propagation, air layering, cuttage, veneer grafting, side grafting, patch budding etc. have been tried but the success has been limited. However, the techniques of epicotyl grafting and soft-wood grafting has been very successful. By adopting these techniques of vegetative propagation, propagator can get 60-70 per cent success. However, the success in these methods primarily depends upon the climatic factors, more specially humidity. In addition, prevailing temperature, incidence of fungal diseases, number of rainy days and rate of cambinal growth also affect the success. Both these methods have been discussed in detail in the chapter on special grafting methods/ techniques.

Arecanut (*Areca catechu*)

It is a monoecious tall growing plant, belonging to Palmae family, which is primarily grown for its kernels. It is mainly propagated from seed, called as seed nuts. For raising quality planting material, first of all healthy and high yielding mother palms are selected on the basis of weight (35 g and above), vigour, which have the capacity to produce seedlings of better vigour, health and quality. The selected seed nuts are sown 5 cm apart in the beds, having sufficient provision for shade, with their pointed ends upwards. Seed nuts are covered with a thin layer of sand and are watered daily. Germination starts in about 30 days after sowing and when the seedlings have developed 2-3 leaves, these are transplanted to secondary nursery beds at a spacing of 25 × 25 cm or 30 × 30 cm during July-August.

Secondary nursery must be provided with some shade either by constructing pandal or growing banana. When the seedlings are 16-18 months old and have developed 5-6 leaves, these can be transplanted in the fields.

Coconut (*Cocos nucifera*)

Coconut palm is one of the most beautiful and useful palms in the world, belonging to Palmae family. Almost every part of this palm is utilized for one or other purpose and hence it is called as *Kalpavriksha* i.e. the tree of heaven. It is a perennial cross-pollinated palm, primarily propagated through seeds, called as *seed nut*. For this, good mother palms are selected. Following criteria are taken into consideration while selecting mother coconut palm.

1. Mother palms should have spherical or semi-spherical crown. However, the palms with dropping growth habit should not avoided.
2. Mother palm should have 12-15 flowering bunches and 30-40 fully mature leaves.
3. It should be in the age group of 30-50 years. Very young or very old palms should be avoided.
4. Mother palm should have the capacity to produce nuts of 700-800 g weight, which should be spherical or nearly round in shape.
5. Selected mother palms should be healthy *i.e.* the palms should be free from insect-pests and diseases.
6. It should be giving consistent yield of high quality nuts for years together.
7. Mother palms should not be near to farm house, cattle shed or compost heap etc.

For raising new plants, fully mature nuts (11 to 12 months) are harvested and then covered with a 7.5 cm thick layer of sand till planting time. Nuts can be planted both vertically and horizontally but horizontal planting should always be preferred. It takes about 90 to 100 days for nuts to germinate. In general, one year old seedlings are transplanted in the main field.

• • •

CHAPTER 45

Propagation Methods of Condiments, Spices, Medicinal and Aromatic Plants

In India, a number of plantation, spices, medicinal and aromatic plants are grown commercially. In fact, we have a rich wealth of these plants in our country. These plants are raised by different propagation methods. Major methods of their propagation are given below :

A. CONDIMENTS AND SPICES

Numerous condiments and spices are grown in India but only a few are grown on a commercial scale. Common propagation methods of major condiments and spices are given in Table 45.1.

TABLE 45.1 : Common propagation methods of major condiments and spices

Common name	*Scientific name*	*Family*	*Common propagation methods*
Allspice	*Primenta officinalis*	Myrtaceae	Seeds, budding and approach grafting
Black pepper	*Piper nigrum*	Piperaceae	Shoot cuttings
Cardamom (large)	*Ammonium subulatum*	Zingiberaceae	Suckers
Cardamom (small)	*Elettaria cardamomum*	Zingiberaceae	Seeds and suckers
Cinnamon	*Cinnamomum verum*	Lauraceae	Seeds, cuttings or air layering
Clove	*Syzygium aromaticum*	Myrtaceae	Seeds
Ginger	*Zingiber officinale*	Zingiberaceae	Cutting of rhizomes

Peppermint	*Mentha piperita*	Labiateae	Suckers
Saffron	*Crocus sativus*	Iridaceae	Corms with more than 2.5 cm in diameter
Tamarind	*Tamarindus indica*	Leguminosae	Seeds and patch budding
Turmeric	*Curcuma longum*	Zingiberaceae	Rhizomes
Vanilla	*Vanilla planifolia*	Orchidaceae	Rooted cuttings

Seed Spices

A number of seed spices are grown in India, the fruits of which are used either as whole or ground to fine powder for adding fragrant odour or pleasant aromatic taste to the food. Most of the seed spices are propagated by seeds. A list of some important seed spices along with seed rate and best propagation time is given in Table 45.2.

TABLE 45.2 : Common propagation methods of seed spices

Common name	*Scientific name*	*Family*	*Seed rate (kg/ha)*	*Sowing time*
Aniseed	*Pimpinella anisum*	Apiaceae	8-10	February-March
Celery	*Apium graveolens*	Apiaceae	0.5	March-April
Coriander	*Coriandrum sativum*	Apiaceae	10-15	Mid-October for *rabi* season
Cumin	*Cuminum cyminum*	Apiaceae	8-10	March-April
Dill	*Anethum gravelens*	Apiaceae	3-4	October-November
Fennel	*Foeniculum vulgare*	Apiaceae	9-12	Plain : September to November Hills : March
Fenugreek	*Trigonella foenum graecum*	Fabaceae	20-25	Aug-Sept. if to be raised as late *kharif* crop
Mustard	*Brassica nigra*	Brassiceae	6-7	February-March

Herbal Spices

Some herbal spices are also grown in India. In these spices, their leaves are usually used for seasoning of various dishes. In addition, these spices have some medicinal and cosmetic uses. A brief account of some of the important herbs and their propagation methods is given in the Table 45.3.

TABLE 45.3 : Important herbal spices and their propagation methods

Common name	*Scientific name*	*Family*	*Propagation method*
Basil	*Ocimum basilicum*	Labiatae	Seeds
Common mint	*Mentha piperita*	Labiatae	Division of stolon
Horse radish	*Armoracia rusticana*	Cruciferae	Rooted cutting
Marjoram	*Marforama hortensism*	Labiatae	Seeds and stem cuttings
Oregano	*Origanum vulgare*	Labiatae	Seeds and stem cuttings

Rosemary	*Rosemarinus officinalis*	Labiatae	Seeds and rooted cuttings
Sage	*Salvia officinalis*	Labiatae	Stem cuttings
Savoy	*Saturcia hortensis*	Labiatae	Seeds
Tarragon	*Artimisia dracunculus*	Compositae	Root division and stem cuttings
Thyme	*Thymus vulgaris*	Labiatae	Cuttings

B. MEDICINAL AND AROMATIC PLANTS

Our country has a rich wealth of medicinal and aromatic plants. These are maintained at different research centres of Government and public organizations. The commercial methods of propagation of the important medicinal and aromatic plants are given in Table 45.4.

TABLE 45.4 : Methods of propagation of medicinal and aromatic plants

Common name	*Scientific name*	*Major propagation method*
A. MEDICINAL PLANTS		
Belladonna	*Atropa belladonna*	Seeds and stem cuttings
Cinchona	*Cinchona* spp.	Seeds, patch budding and veneer grafting
Dill	*Anethum graveolens*	Seeds
Dioscorea	*Dioscorea spp.*	Tuber and its cuttings
Foxglove	*Digitalis lanata*	Seeds
Ipecac	*Cephaelis ipecacuanha*	Seed and stem cuttings
Isabgol	*Plantago ovata*	Seeds
Liquorice	*Glycyrrhiza glabra*	Mostly by cuttings and rarely by seeds
Poppy	*Papaver somniferum*	Seeds
Rauwolfia	*Rauwolfia serpentina*	Mostly by seed, stem and cuttings of root
Senna	*Cassia agustifolia*	Seeds
B. AROMATIC PLANTS		
Citromella grass	*Cynpogon winteriana*	Rooted slips
Eucalyptus	*Eucalyptus spp.*	Seeds
Mehndi	*Lawsonia inermis*	Stem cuttings
Kewra	*Pandamus odoratissimus*	Stem cuttings
Khus	*Vetiveria zizanioides*	Rooted slips
Lemon grass	*Cympogon flexuousus*	Seeds
Mint	*Mentha arvensis*	Stolons
Palmarosa grass	*Cympogon martint*	Rooted slips
Peppermint	*Mentha piperata*	Stolon
Rose geranium	*Pelargonium graveolens*	Stem cuttings

• • •

CHAPTER 46

Propagation Methods of Floricultural Plants

All plants having ornamental value are considered as the floricultural plants. These may be annuals, biennials, perennials, climbers, grasses, bulbous, herbaceous perennials, shrubs or trees. All these are propagated by seeds or some vegetative method of propagation. A brief description about the propagation of floricultural/ornamental plants is given in this chapter.

A. MAJOR ORNAMENTAL PLANTS

Bougainvillea

Bougainvillea is commonly propagated by hardwood cuttings. The cuttings of 25-30 cm length with 2-2.5 cm diameter are most suitable. These cuttings root better if treated with some growth regulators like IBA (1,000 to 2,000 ppm). The cuttings are prepared in July-August or February-March in north India, during monsoon season in east India but throughout the year under Bangalore or similar climatic conditions. 'T' budding is also done in February-March if cultivars don't respond to cuttings. The rootstocks like Partha and R.R. Pal should be preferred for budding.

Cacti and Succulents

Cacti and succulents are the two most fascinating groups of plants grown indoor by the plant lovers. These are propagated by different methods but for majority of succulents, seed is the major method of propagation, although, it is the slowest method of their propagation. The seeds are sown in spring or early summer in shallow trays or pots. Some succulents are also propagated by cuttings of stem, leaf and offsets. Succulents such as cotyledon, sedum etc., can be multiplied from leaf cuttings also. Kalanchoe is propagated by bud cuttings of the leaves and agave by division of suckers. Now-a-days, different grafting techniques have also been standardized for raising succulents and cacti plants. Though, the method of grafting cacti is very cumbersome, tedious and complex, but comparatively large sized plants are obtained with a single attempt. Among various methods of grafting, cleft method gives the best results.

Canna

The commercial method of propagation of canna is the division of rhizomes. The best season for planting the rhizomes is before rainy season.

Carnation

Marguerite and Chabaud types are easily raised by seeds. In north India, seeds should be sown in July-August but sowing can be extended up to October in Bangalore or similar climatic conditions. In hills, sowing can be done between August-to-October and during March-April. Carnations do propagate easily through terminal cuttings of 10-15 cm size. These can be treated with IBA (50 ppm) before planting in the nursery. However, micropropagation techniques have now been standardized in India and can be used for its faster multiplication.

China Aster

China aster is commonly propagated by seeds. Seeds are sown during August-to-October in areas where rainfall is low. Under mild-climate of Bangalore, seeds can be sown at any time of the year but June is the best time.

Chrysanthemum

Chrysanthemum is mainly propagated by suckers and terminal cuttings. The side suckers are separated from the mother plant in January-February and planted in the nursery. Usually, 5-7 cm long terminal cuttings after dipping in the rooting hormone (IBA 200 ppm) can be planted in the nursery under shade. After rooting, these can be planted in the pots or field. The best time for raising plants by cuttings is July-August.

Climbers

Most of the climbers are propagated by layering or cuttage during rainy season. However, some climbers like *Campsis radicans, C. grandiflora, Clerodendron splendens,* and *Pergularia odoratissima*, *Petrea volubilis* etc., can also be raised through suckers.

Cycads and Palms

Most of the palms and cycads are propagated commercially by seeds and rarely by suckers and division of clumps. The seeds produced by different palms vary in shape and size and may range between the size of a pea and a coconut seed. More often, they grow very slow and sometimes may take even more than a year to germinate. Seeds can be sown at any time of the year but the best time is the spring or rainy season. However, the palms like *Caryota mitis, Geonoma gracilis*, *Wallichia disticha, Licuala peltedla* etc., can also be propagated by suckers.

Dahlia

The best method for raising true-to-type plants of dahlia is its multiplication by cuttings. The cuttings of 7.5-10 cm length are taken from young green shoots, produced from the crown of the tubers The lower leaves of the cuttings should be removed by clipping and cut ends be treated with rooting hormone (IBA, seradix etc.) and planted in the nursery. The best time for preparation of the cuttings is July to mid-September. Dahlia can be propagated by division of the tubers also. The tubers are planted in June under north Indian conditions, during May in Bangalore and during March-April in hills.

Ferns

Ferns are usually propagated from spores. However, several vegetative methods of propagation have also been standardized. Ferns produce thick rhizomes and thus divisions of rhizomes is one of the effective methods for propagating ferns commercially. Certain species also produce bulblets, which can also be used for multiplication.

Flowering Annuals

Almost all the flowering annuals are commercially propagated by seeds. On the basis of climatic requirement, the flowering annuals have been classified as winter annuals (antirrhinum, china aster, ageratum, carnation, calendula, cineraria, larkspar, lupin, nasturtium, nigellia, pansy, petunia, phlox, salvia, sweet pea, stock, viola, verbena etc), summer annuals (cosmos, coreopsis, gaillardia, portulaca, sunflower, zinnia etc.). and the rainy season annuals (balsam, cockscomb, gompherena, marigold, cosmos, portulaca etc.). The sowing of summer annuals starts from January to end of June, depending on the prevailing climatic conditions of the locality. The rainy season annuals are sown with the outbreak of monsoon season *i.e.* in June and July. The winter season annuals are sown in September-October. The seeds of these annuals can be sown in the nursery beds or polyethylene bags or in shallow pans. The time of transplanting of the seedling is generally one month after sowing when they have developed atleast four leaves. The water in the beds should be withheld a day or two before actual date of transplanting for hardening of the seedling. Further, the transplanting should be done either in the early morning or late evening hours. Similarly, preference should be given for transplanting on a cloudy day. All these practices help to reduce the mortality of the seedlings in the field.

Gerbera

The surest and simplest method of raising true-to type plants of gerbera is the division of the mother plants. The best time for raising the plants through division is the monsoon season. However, in hills, it should be done in spring.

Gladious

It is propagated normally by corms and cormlets. Corms remain dormant for about 3 months and do not sprout easily. The best time for planting the corms in hills is February-March and October in plains. Micropropagation techniques have also been standardized for faster multiplication of gladiolus.

Herbaceous Perennials

Herbaceous perennials are mostly raised by seeds. However, some herbaceous perennials like *Angelonia grandiflora* (stem tip cuttings), aster (suckers), geranium (stem cuttings), phlox (stem and root cuttings), *Vinca rosea* (stem cuttings), pyrethrum (division) can be propagated by different vegetative methods also.

Jasmine

Jasmines are commonly multiplied by cuttings of mature wood. However, layering (especially ground layering) is also sometimes followed. The best time for either method is the rainy season. The rooting in the cuttings is easy if these are treated with IBA (1,500-2,000 ppm) and planted in the nursery under semi-shade conditions. For layering, tender shoots should be girdled (removing 2-2.5 cm bark around the shoot) in June-July. IBA paste (1,000 ppm) in lanolin is applied to the upper part of the cut of the ring. High humidity should be maintained to promote rooting process in the layers. Rooting in the layers occurs in about a month or so and the layers can be detached from the mother plant afterwards and planted in the nursery or directly in the field.

Marigold

Seeds are mostly used for raising marigold. The seeds are sown in May to July in the nursery beds or in pans during October-November and February-March in plains, In hills, the best sowing time is March-April.

Orchids

Orchids are raised by seeds, tissue culture and by several other vegetative methods. For the amateur gardeners, it is difficult to raise orchids by seeds or tissue culture technique. For them, division of pseudobulb or rhizome is the best method. Orchids produce millions of seeds but these don't grow because they lack food substance or endosperm. Therefore, seeds are grown in cultural flasks containing sterilized nutrient agar medium (Knudson's Formula C) under aseptic conditions. When seeds germinate, they are transplanted in the pots or nursery beds.

Ornamental Grasses

Most of the ornamental grasses are raised by cuttings, stolon (*doob* grass), division (andropogon) or by seeds (*Trichilaena rosea*).

Rose

Rose is propagated by various vegetative means like cuttings, layering, budding and grafting but 'T' budding is by-and-large, the most commonly used method of propagation in India. 'T' budding can be done in dormant season (October-November) or in February-March. For budding, rootstock of pencil thickness is the most desirable. The commonly recommended rootstocks for rose are *Rosa indica, Rosa bourboniana* (north India) and *Rosa multiflora* (south and east India). Basic technique for budding is the same as described citrus. Layering and cuttage are commonly followed for the multiplication of polyanthas, climbers and ramblers, whereas new hybrids are commonly raised by seeds.

Tuberose

Tuberose is commonly propagated by bulbs. These are planted during rains in the plains and during May-June in the hills. In general, the old roots of the tubers are cut at the base before planting. Large sized bulbs come into flowering within a year of planting, whereas, small tubers take 3 to 4 years to flower.

B. ORNAMENTAL BULBOUS PLANTS

Many plants of horticultural importance are propagated by specialized structures like bulbs, corms, rhizomes and tubers. Bulbous plants are also commercially propagated from such vegetative parts. Apart from these structures, these are also propagated through structures like offsets, bulblets, cormlets, bulbils and scales etc. The details about major bulbous plants, their propagation method, planting time and propagation method is given in Table 46.1.

TABLE 46.1 : Major bulbous plants and their commercial propagation methods

Common name	*Scientific name*	*Specialized structure used for propagation*	*Best planting time*	*Ideal planting depth*
Alocasia	*Alpcasoa* spp.	Rhizomes	Spring season	3-5 cm
Amaryllis	*Hippeastrum* spp.	Bulbs, bulblets	February-March	10-15 cm
Begonia	*Begonia bicolor*	Tuberous roots	February-March	2-3 cm
Belladonna lily	*Amaryllis belladonna*	Bulbs and bulblets	December-January	8-10 cm
Blood flower	*Haemanthus multiflorus*	Bulbs	February-March	7-10 cm
Caladium	*Caladium bicolor*	Tuberous rhizomes	April-May	4-5 cm
Canna	*Canna indica*	Tuberous roots	March-April	12-15 cm
Cap lily	*Crinum asiaticum*	Bulbs	Early summer	8-10 cm
Climbing lily	*Gloriosa superba*	Tubers	March	3-5 cm
Corn lily	*Ixia maculata*	Corms	In hills : February-March In plains : September-October	12-15 cm
Crocus	*Crocus* spp.	Corms	February-March	5-7 cm
Crown Imperial	*Fritillaria imperialis*	Bulbs	September-October	10-12 cm
Cyclamen	*Cyclamen persicum*	Tubers	In plains : October In hills : February-March	5-8 cm
Daffodil	*Narcissus cyclamineus*	Bulbs	In hills : February-March In plains : September-October	8-10 cm
Dahlia	*Dahlia* spp.	Cuttings and tuberous roots	August-September	4-5 cm

Day lily	*Hemerocallis fulba*	Division of roots	February-March	4-5 cm
Fair lily	*Zephranthes candida*	Bulbs	February-March	4-5 cm
Four O'clock	*Mirabilis jalapa*	Tubers	February-March	12-15 cm
Freesia	*Fressia hybrida*	Corms	September-October	5-7 cm
Garland flower	*Hedyehium coccineum*	Division of rhizomes	September-October	5-7 cm
Gladiolus	*Gladiolus hortulanus*	Corms	In hills : February-March In plains : October-November	12-15 cm
Gloxinia	*Gloxinia speciosa*	Rhizomes and stem cuttings	In plains: Dec.-January In hills : May-June	6-8 m
Hyacinth	*Hyacinthus orientalis*	Bulbs	In hills : February-March In plains : September-October	12-15 cm
Iris	*Iris* spp.	Rhizomes	June-July	6-8 cm
Lily	*Lilium auratum*	Bulbs and offsets	In hills : Spring In plains : Autumn	10-15 cm
Lily of nile	*Agapanthus umbellatus*	Offsets and tuberous roots	Spring	3-4 cm
Lily of valley	*Convallaria majalis*	Rhizomes	February-March	4-5 cm
Magic flower	*Achimenes grandiflora*	Tuberous roots	Spring season	5-7 cm
Oxalis	*Oxalis corymbosa*	Bulbs	March-April	7-10 cm
Ranunculus	*Ranunculus asiaticus*	Tubers	In hills : Spring In plains : October	5-8 cm
Scented gladiolus	*Acidanthera bicolor*	Corms	In hills : March In plains : October	7-8 cm
Scilla	*Scilla hispanica*	Bulbs	September-October	5-7 cm
Snake lily	*Arisaema flavum*	Tuberous roots	October-November	5-7 cm
Spider lily	*Hymenocallis littoralis*	Offsets	October-November	5-7 cm
Tuberose	*Polyanthes tuberosa*	Tubers	February-March	4-5 cm
Tulip	*Tulip* spp.	Bulbs	February-March	7-10 cm
Water hyacinth	*Eichornia speciosa*	Rhizomes	Rainy season	Shallow water
Wind flower	*Anemone coronaria*	Tubers	October-November	5-7 cm

C. INDOOR PLANTS

A variety of plants are now used as house plants. These include some shrubs, bulbous plants, ferns, cacti and succulents. The major propagation techniques methods of several such plants are given in Table 46.2.

TABLE 46.2 : Common indoor plants and their commercial propagation methods

Common name	*Scientific name*	*Commercial propagation method*
Aglaonema	*Aglaonema commutatum*	Division of basal shoots, nodal cuttings
Anthurium	*Anthurium magnificum*	Seeds
Aphelandra	*Aphelandra squarrosa*	Half ripened wood cuttings
Aralia	*Aralia filiciplia*	Division of crown
Araucaria	*Araucaria heterophylla*	Seeds
Asparagus	*Asparagus afticans*	Division of crown
Aspidistra	*Aspidistra lurida*	Division
Bird of paradise	*Strelitzia reginae*	Seeds
Calathea	*Calthea insignis*	Root cuttings
Coleus	*Coleus hybridus*	Herbaceous cuttings of slips
Croton	*Cochiaeum* spp.	Air layering
Dieffenbachia	*Dieffenbachia exotica*	Terminal and nodal cuttings
Draecena	*Dracaena deremensis*	Cane pieces
English ivy	*Cissus* spp.	Cuttings
Indian rubber tree	*Ficus elastica*	Air layering
Iron plant	*Aspidistra elator*	Division of crown
Monstera	*Monstera deliciosa*	Air layering
Palms	Different species	Offsets
Pandanus	*Pandanus utilis*	Offsets and division of suckers
Peacock plant	*Calathea* spp.	Division of crown
Philodendron	*Philodendron scandens*	Cuttings
Pitcher plant	*Nepenthes khasiana*	Hardwood cuttings
Podocarpus	*Podocarpus macrophyllus*	Shoot tip grafting
Sago palm	*Cycas revoluta*	Offsets
Sansevieria	*Sansevieria* spp.	Offsets
Spider plant	*Chlorophytum comosum*	Division of crown
Tradescantia	*Tradescantia* spp.	Cuttings
Yucca	*Yucca gloriosa*	Seeds and offsets
Zebrina	*Zebrina pendula*	Hcrbaceous cuttings

D. ORNAMENTAL SHRUBS AND TREES

Shrubs and trees are important group of woody perennial plants. Several shrubs and trees have ornamental value, hence these are used in landscape gardening. These plants are propagated either by means of seeds or vegetative methods or both. The major method of propagation of important ornamental shrubs and trees are given in Table 46.3.

TABLE 46.3 : Important shrubs and trees and their commercial propagation methods

Common name	*Scientific name*	*Commercial propagation technique*
A. ORNAMENTAL SHRUBS		
Acalypha	*Acalyha* spp.	Hardwood cuttings
Allamanda	*Allamanda cathatica*	Air layering
Azalia	*Azalia* spp.	Hardwood cutting and layering
Bottle brush	*Callestemon* spp.	Hardwood cutting and layering
Brya	*Brya ebenus*	Hardwood cutting and layering
Camelia	*Cammelia japonica*	Hardwood cutting and layering
Cestrum	*Cestrum noctrum*	Hardwood cuttings
Clerodendron	*Clerodendron* spp.	Hardwood cuttings
Clematis	*Clematis* spp.	Leaf bud cuttings
Coral tree	*Erythrina indica*	Hardwood cuttings
Hibiscus	*Hibiscus rosa-chinensis*	Hardwood cutting and layering
Hydrangea	*Hydrangea* spp.	Softwood and leaf-bud cuttings
Ipomea	*Ipomea carnea*	Hardwood cuttings
Ixora	*Ixora* spp.	Hardwood cuttings and layering
Jatropha	*Jatropha multifida*	Hardwood cutting and layering
Juniper	*Juniperus* spp.	Hardwood cuttings
Kachnar	*Bauhinia acuminata*	Hardwood cuttings
Kamini	*Murraya exotica*	Hardwood cuttings
Lagerstoemia	*Lagerstoemia indica*	Hardwood cutting and layering
Lantana	*Lantana camara*	Seeds and Hardwood cuttings
Lilac	*Syringa vulgaris*	Hardwood cuttings and layering
Malphigia	*Malphigia glabra*	Hardwood cuttings
Nerium	*Nerium oelifera*	Cuttings and layering
Poinciana	*Poinciana pulcherrima*	Seeds
Poinsettia	*Euphorbia pulcherrima*	Hardwood cuttings

Portlandia	*Porlandia grandiflora*	Hardwood cuttings
Russelia	*Ruselia equisetiformis*	Root cuttings
Tecoma	*Tecoma grandis*	Seeds
Thunbergia	*Thunbergia arecta*	Hardwood cuttings
Woodfordia	*Woodfordia floribunda*	Seeds and cuttings
B. SHADE, ORNAMENTAL AND FLOWERING TREES		
Agasthi	*Sesbania grandiflora*	Seeds
Amherstia	*Amherstia nobilis*	Seeds and layering
Arjun	*Terminalia arjuna*	Seeds
Ashoka tree	*Saraca indica*	Seeds
Auracaria	*Auracaria cooki*	Seeds, cuttings and layering
Barringtonia	*Barringtonia racemosa*	Seeds
Bead tree	*Melia azedirach*	Seeds
Beef wood tree	*Casuarina equisetifolia*	Freshly harvested seeds
Bottle brush	*Callistemon lanceolatus*	Air layering
Burmese rose wood	*Pterocarpus indicus*	Seeds
Cammelia	*Cammelia japonica*	Seeds and cuttings
Caper tree	*Crataeva religiosa*	Seeds and root suckers
Chalta	*Dillenia indica*	Stem cuttings
Champa	*Michelia champaca*	Air layering
Colville's glory	*Colvillea racemosa*	Seeds
Deodar	*Cedrus deodara*	Seeds
Devil's tree	*Alstonia schlolaris*	Seeds
Dilo oil tree	*Calophyllum inophyllum*	Seeds
Divi-divi tree	*Caesalpinia coriaria*	Seeds
Easter tree	*Holarrhena antidysenterica*	Root suckers
Ficus	*Ficus infectoria* *F. bengalensis,* *F.religiosa* and other species	Seeds
Flame of forest	*Butea monosperma*	Seeds
Flame or tulip tree	*Spathodea campanulata*	Seeds and root suckers
Geiger tree	*Cordia sebestena*	Seeds
Golden shower (Sondal)	*Cassia fistula* and other species	Seeds
Green ebony	*Jacaranda mimosaefolia*	Stem cuttings
Indian coral tree	*Erythrina indica*	Hardwood cuttings

Indian cork tree	*Millingtonia hortensis*	Hardwood cuttings
Iron wood tree	*Mesua ferrea*	Seeds
Jeruslem thorn	*Parkinsonia aculeata*	Hardwood cuttings
Jhau	*Casuarina equisetifolia*	Seeds
Kadamba	*Anthocephalus cadamba*	Seeds
Kalanj	*Pongamia glabra*	Seeds
Kalanj champa	*Pterospermum acerifolium*	Seeds
Lily tree	*Magnolia grandiflora*	Layering
Mad tree	*Sterculia diversifolia*	Seeds
Mad tree	*Gliricida maculata*	Hardwood cuttings
Magnolia	*Magnoila grandiflora*	Air layering
Mahogany	*Switenia mahogani*	Seeds
Mahua	*Madhuca indica*	Seeds
Manila tamarind	*Inga dulcis*	Bead tree
Milletia	*Milletia ovalifolia*	Seeds
Monkey bread tree	*Adansonia digitata*	Seeds
Mulsari	*Mimusaps elengi*	Seeds
Neem	*Azadirachta indica*	Seeds
Palas	*Butea frondosa*	Seeds
Pea cock flower (gulmohar)	*Delonix regia*	Seeds
Pines Tree	*Pinus roxburghii*	Seeds
Plumeria	*Plumeris* spp.	Hardwood cuttings
Poplar	*Populus* spp.	Seeds
Pride of India	*Lagerstroemia indica*	Seeds
Purple bauhunia (*Kachnar*)	*Bauhinia purpurea*	Seeds
Rain tree	*Enterolobium saman*	Hardwood cuttings
Rhododendron	*Rhododendron* spp.	Hardwood cuttings
Shisham	*Delbergia sissoo*	Seeds
Silk cotton tree (*Simbal*)	*Bombax malabaricum*	Seeds
Silver oak	*Grevillea robusta*	Seeds
Siris	*Albizzia lebbeck*	Seeds
Sterculia	*Sterculia alata*	Seeds

Tabeuia	*Tabebuia alata*	Seeds
Teak	*Tectoma grandis*	Seeds
The mast tree	*Polyanthia longifolia*	Seeds
Torch tree	*Ixora parviflora*	Seeds
Traveller's tree	*Ravenala madagascariensis*	Seeds
Turmphet flower	*Bignonia megapotamica*	Seeds
Umbrella tree	*Brassaia actinophylla*	Seeds
Weeping willow	*Salix babylonica*	Hardwood cuttings
White cedar (*Morphankhi*)	*Thuja occindentalis*	Seeds
Wild olive (*Putranjiva*)	*Putranjiva roxburghi*	Seeds

• • •

CHAPTER 47

Modern Techniques of Raising Annuals

In horticulture, many plants are propagated from seeds, because it is a quick and economical method of their propagation. However, for successful germination of seeds, proper environmental and cultural conditions should be maintained. Traditionally, flowers and vegetables are produced by germinating seeds in flats and seedbeds. However, modern producers have adopted plug production as the preferred method for transplant production. Plugs transplant easily with a high degree of uniformity. Plugs do not undergo 'transplant shock.' The following are the different methods to produce seedlings of annuals in plug trays.

NURSERY RAISING IN PLUG TRAYS

Raising of healthy, vigorous and disease-free seedlings is very important for growth and development of vegetables and ornamentals. Traditionally, seedlings are raised in flats or nursery beds. Plug trays have completely replaced traditional methods of raising seedlings in western countries. Seeds of many ornamental and vegetable crops are very costly, thus every seed is valuable necessitating plug production. Plug means a single cell transplant. A plug tray consists of 50-800 cavities, which are 2-5 cm deep.

Advantages of Plug Tray Nursery Raising

- This type of nursery is adaptable to different scales of production.
- Healthy, vigorous and disease-free seedlings can be raised in a short period.
- Since the seedlings can be removed without any damage, the transplantation shock is minimized, which increases the chances of better survival of seedlings.
- As seedlings are raised in individual cells and are subjected to uniform conditions, there is a greater uniformity among seedlings.

- Most of the plug production is carried out in protected conditions, as result, the nursery can be raised under adverse conditions.
- The crop can be raised early in the season because of the controlled conditions.
- In plug trays, the media is filled in the tray cells, hence the medium is amenable for sterilization easily.
- The plug tray seedlings can be transported to long distances easily without damage, which is not possible in other systems of nursery raising.
- Several ornamentals and vegetable crops have specific requirement of light and darkness for germination and growth, which can be met by this system.
- The quantity of seed required in this system is less than other methods, as a result, it economizes seed.
- The chances of spread of soil-borne diseases like damping-off is less as crowding of seedlings is avoided because of individual planting in plug trays.
- The insect and disease management schedule can be operated very easily.
- Plug system of nursery raising is amenable to mechanization automation, thereby, it helps in saving labour.

REQUIREMENTS OF PLUG-TRAY NURSERY RAISING

Containers/Plug-Trays

Seedlings are raised in a variety of containers ranging from trays-to-pots. The trays are manufactured from different materials. Plastic/Styrofoam is the most commonly used material for plug trays. Rigid foam cell trays are an alternative to plastic trays. Light weight plastic trays, in contrast, are easy to store and handle in 'off season' but are less durable.

Marigold plugs are gown in 512, 406 and 288 play fiats. Cells shape varies from inverted pyramix round, square and hexagonal. The depth varies from 2-5 cm. The cell number varies from 50-800 in a single tray. The type of cells used varies with the crop species and duration in nursery. The deeper cell trays have larger volume of media and nutrients for root growth. They promote faster growth of seedlings.

Plastic trays have different sizes of cells for raising vegetable seedlings. Two kinds of plastic pro-trays are used, which have cells of 3.75 cm size and the others have cavity of 2.5 cm size. Crops like cucumber, muskmelon, tomato and brinjal require trays of 187 cells of 3.75 cm size, whereas cabbage, cauliflower and chillies require trays of 345 cells of 2.5 cm size. The trays can be placed on thermocol base trays having similar proportion. Flower crops are most often grown in 288, 392 or 406 plug trays.

Propagation medium

Medium for plug-tray nurseries should retain moisture, provide anchorage and supply the nutrients to the growing seedling without fixation. The pH of the medium should be neutral and soluble salts must be low. The medium components generally used in plug-tray planting are sand, peat, vermiculite, coco-peat, perlite and moss. The right mixture of these media depends on crop species. Nowadays, several commercial mixes are also available.

Sowing of Seed and Germination

Three types of seeders are available to plug growers. These are template, needle and drum seeders. The template seeder uses a template with holes that match the location of cells in the tray. This employs vacuum to hold seeds and drop them directly in the cell. Templates with different sized holes are available to handle seeds of different sizes and shapes. It is relatively fast seeder because it sows an entire flat at once. The needle seeder is fully mechanical. Individual needles, under vacuum pressure lift single seeds from a tray and deposit one seed per plug cell. The needle seeder can also sow odd shaped seeds but operations are slower. The drum seeder has a rotating drum that picks up seeds using vacuum from a seed tray and drops one seed per plug cell. This is the fastest seeder but is costly. Drum seeders work best with round, semi-round or pelleted seeds

Irrigation

The moisture content of the medium is very important for success of plug-tray nursery production. Several systems like overhead sprinklers, sub-irrigation systems and manual irrigation are employed for irrigating plugs. Water quality is important for success regardless of the system used. Irrigation frequency depends upon cell size, media and crop species. Over watering increases the susceptibility of seedlings to fungal diseases.

Nutritional Management

Nutrients can be supplied to plugs through various methods. Quantity of nutrient application depends upon crop species and medium. Nutrients are generally supplied by fertigation. Nutrients are also incorporated in the medium. Foliar application of nutrients particularly micronutrients can be done after the plugs have produced true leaves.

Transplanting of the Seedlings

When the seedlings are ready for transplanting, these should be hardened off. Hardening is done by increasing the temperature, reducing the irrigation and shifting the plugs to semi-controlled conditions. Once the plugs are hardened, they can be transported to the final destination.

• • •

Glossary

Adventitious buds : Buds, which occur at sporadic and unexpected places on a vegetative structure.

Adventitious shoots : Shoots arising from an unusual place on a plant.

Agar : A gelatinous substance widely used in solidifying nutrient medium used for micropropagation.

Air layer : Aerial shoot of a plant on which roots are forced to develop by ringing or with other treatment.

Allogamous : Flowers, which are structurally adapted for cross-pollination.

Allogamy : Cross-pollination.

Androecium : Male reproductive part of a flower.

Anemophilous : Flowers, which are structurally adapted for pollination by wind.

Anemophily : Pollination by wind.

Annual : Plant, which grows, flowers, produces seed, and dies within a single growing season or in one year.

Anther : Sac-like structure at the top the stamen, which contains pollen.

Apical : A mass of undifferentiated cells in root or shoot tip, which are responsible for plant growth.

Apomictic species : A plant species, which produces seed from vegetative cells and not through sexual means.

Apomixis : A process of development of diploid embryos without fertilization.

Asexual reproduction : The production of a new plant without involving sex organs.

Autogamous : Flowers, which are structurally or otherwise adapted for self-pollination.

Autogamy : Self-pollination.

Auxins : Plant hormones that accelerate growth by stimulating cell enlargement.

Axil : The angle between leaf-stalk and stem.

Axillary : Any new growth or flower bud arising from an axil is called as axillary.

Biennial : A plant, which completes its life cycle in two seasons or two years.

Bolting : Flowering process in biennial plants, in which plant sends out a seed stalk and stops growing edible leaves.

Breeder's seed : Seed produced by a breeder under his own supervision.

Bud : A region of meristematic tissues with a potential to develop into leaves, shoots or flowers.

Bud break : Resumption of growth by the resting buds.

Bud scales : Leaves, which cover the bud.

Budding : A form of grafting in which a single vegetative bud of a desired variety is used for insertion onto the rootstock.

Budstick : A small shoot of current season's growth used for taking buds for budding.

Bulb : An underground vegetative structure, which consists of layers of fleshy scales overlapping each other.

Bulb scales : Leaf-like parts of a bulb, which surround the flower bud and are attached to the basal plate.

Bulblets : A small immature or miniature bulbs produced at the base of a mother bulb.

Callus : A mass of undifferentiated parenchymatous cells.

Cambium : A thin layer between the xylem and phloem tissues of dicot plants.

Cane : A mature shoot of a vine's previous season's growth.

Cell : A basic unit of life of all the living organisms.

Certified seed : A progeny either of foundation or breeder's seed, produced under the control of a State Seed Certifying Agency.

Clone : A population of asexually maintained individuals.

Cold frame : A propagating structure used for growing and protecting young plants in temperate climate.

Compatibility : Ability to unite with, grow and live together as the scion and rootstock in the grafting and budding processes.

Compatible : Different kinds of plants or varieties that would make a successful graft or bud union when intergrafted.

Complete flower : Flower with both male and female parts; sexually complete.

Compost : A mixture of organic waste and soil, which has been piled, moistened and allowed to decompose biologically. Sometimes, mineral fertilizers are also added to it.

Container grown : Growing plants to marketable size in a container.

Corm : A short, solid, vertical, enlarged underground stem, which contains undeveloped buds.

Cormels : Tiny corms produced around or near the base of the mother corm.

Cotyledons : The first leaves to appear on a plant; the seed leaves.

Cross-pollination : Transfer of pollen grains from stamen to the stigma of flowers borne on the same or another plant.

Cutting (noun) : A section of stem, root or other plant part used for propagation of plants.

Cytokinins : A group of plant growth hormones, which regulate nucleic acid and protein metabolism, cell division, organ initiation and delays senescence.

Damping off : A serious disease in the young nursery plants, causing them to die soon after seed germination.

Deciduous plant : A plant, which sheds its leaves during one or the other season.

Dicots : Plants having two cotyledons in their seeds.

Differentiation : Changes in structure and functions of cells and tissues during growth.

Dioecious : Plants species, which bear staminate (male) and female (pistilate) flowers on different plants.

Diploid : Individuals having double set of chromosomes.

Direct seeding : Planting seeds directly in a permanent place.

Disinfestation : Treating the seeds in chemicals or fungicides to avoid their infestation by pathogens.

Division : A method of propagation requiring the cutting and dividing of plants.

Dormancy : Lack of growth in any plant part due to one or the other reason.

Dormant period : A specific time period in which no growth occur in seed or plant.

Double fertilization : The process of sexual fertilization in which one nucleus from the male gametophyte fertilizes the egg nucleus to form the zygote and the second nucleus from the male gametophyte fertilizes two polar nuclei to form endosperm tissue.

Double working : Grafted twice, insertion of a mutually compatible interstock between the actual stock and the scion.

Embryo : A miniature plant within a seed, produced by the union of a male and female gamete.

Embryo culture : Growing seedlings from excised embryos on nutrient medium under controlled and aseptic conditions.

Endogenous : Produced within the tissue or internal use.

Endosperm : The food storage tissue in seeds of angiosperms, which provides nutrition to the developing embryo.

Entomophily : Pollination by insects.

Epicotyl : The part of the plant embryo axis situated above the cotyledons.

Epigeal germination : A type of seed germination in dicots in which the cotyledons rise above the soil surface.

Ethylene : A gaseous growth hormone, which regulates the fruit ripening process and abscission of different plant parts.

Etiolation : Exclusion of light.

Exogenous : Produced outside or external use.

Explant : A plant part to be used for various manipulations in tissue culture systems.

F_1 : First filial generation in a cross between two parents.

Fertilization : The union of an egg (female gamete) and a sperm (male gamete) to form the zygote.

Filament : The stalk-like part of a stamen.

Flower : A forerunner of a seed, the most attractive plant part.

Forcing : Any cultural practice to hasten bud break or flowering or growing plants outside their normal season.

Foundation seed : Seed produced from breeder's seed and handled carefully to maintain the genetic identity and purity of a cultivar.

Genetic purity : Treunesess of type, conforming to the characteristics of a variety as described by a breeder.

Germinate : To sprout or begin to grow.

Germination : A process in which seed embryo starts growing, which leads to the development of a seedling.

Gibbrellins : A group of natural growth hormones, responsible primarily for elongation of stems.

Girdling : A practice of removing a ring of bark around the periphery of a stem or a shoot.

Grafting : A method of inserting scion shoots on the stocks for raising new plants.

Grafting wax : A pliable, sticky, water-proof material made of bee's wax, resin and tallow, used for sealing the bud or graft union to avoid infection by pathogens and desiccation by wind and temperature.

Growth : An irreversible increase in cell size or cell number resulting in increase in the dry weight.

Gynoecium : The female reproductive part of a flower.

Hard seed : Seeds that remain hard and don't germinate because of hard seed coat.

Hardening-off : Forcing plants to adapt to outdoor conditions by withholding water, reducing the temperature or nutrient supply or combination of both.

Hardwood cutting : A cutting made from current season's growth, which is mature or hard.

Herbaceous : Plants that have soft, non-woody growth.

Heterosis : The increase in vigour, growth, size, yield or function of a hybrid progeny in comparison to either of the parents.

Heterozygous : Having different genes in the same cells or organisms.

Homozygous : Having similar genes in the same cell or organism.

Hormones : Growth regulating substances in plants.

Hotbed : A propagating structure used to germinate seeds, rooting of cuttings and growing other plants in temperate climate.

Hybrid : A progeny of two plants differing in one or more characters.

Hydrophily : Pollination by water.

Hypocotyl : Axis or stem of a seedling plant below the cotyledons.

Hypogeal germination : A type of seed germination in dicots wherein the cotyledons remain below the soil surface.

Imperfect flower : A flower lacking either androecium or gynoecium.

In vitro : In glass or in an artificial environment.

In vivo : In living; within the living organisms.

Incompatible : Different kinds or varieties that do not form successful graft or bud union.

Incomplete flower : Flower lacking either calyx, corrolla, or both.

Indexing : A technique used to test the vegetatively propagated plants for the presence or absence of viruses.

Inflorescence : A flowering system consisting of more than one flower.

Internode : The region of a stem between two successive nodes.

Isolation : To keep separate at a distance so as prevent the crossing among plant populations.

Isolation distance : A distance, which has to be maintained between the seed crop and the contaminant.

Lath-house : A propagation structure made up of wood, lath or plastic screen for protecting plants from excessive sunlight or frost.

Layering : A method of propagation of plants from parts that remain attached to the mother plant while rooting.

Marcottage : An alternative term used for air layering (*goottee*).

Mist propagation : A method of propagation in which water is continuously sprayed in the form of a mist to leafy cuttings in the rooting stage to reduce water loss thorough transpiration.

Monocots : Plants with one cotyledon in their seed.

Monoecious : A plant species bearing unisexual flowers, with flowers of both sexes on the same plant but at different places.

Node : An enlarged solid regions at stem, from where leaves and buds arise.

Nucellus : A tissue originally making up the major part of the young ovule, in which the embryo sac develops.

Nurse-graft : A rootstock that supports a scion until it roots itself.

Nursery : A place in an orchard or a locality kept for growing, handling and selling of the horticultural plants.

Nutrients : Plant food elements.

Offsets : Small new plants, developed from the parent plant, which can easily be separated and used as independent plants.

Off-type : Plant or seed, which is different from the true characteristics of a variety.

Organic : Dead and decaying plant parts such as green manure crops, peat moss, or animal manures, which improves the soil structure and water holding capacity of the soil.

Ornamental plants : The plants, which are grown for beautification and aesthetic purposes.

Ovary : The flask shaped part of pistil (gynoecium) of a flower, which contains ovules.

Ovule : The undeveloped seed.

Parthenocarpy : Development of seedless fruits without sexual fertilization.

Pedicel : The stem of the plant, which supports the flowers, the flower stalk.

Peduncle : A common leafless axis bearing several flowers.

Perfect flower : A flower, which contains both androecium and gynoecium.

pH level : The measure of a soil's acidity or alkalinity on a scale of 1-14.

Pinching : Removal of some plant parts by hand.

Pistillate : A flower having functional pistils but without or no functional stamens.

Plumule : The first shoot above the cotyledons of a newly germinated seedlings of a dicot plant.

Polarity : The tendency of a shoot to grow towards the top and roots to grow towards the bottom.

Pollen grains : The minute or microscopic yellow bodies that are produced by the anthers of a flower.

Pollination : The transfer of pollen grains from anthers to the stigma of the gynoecium.

Polyembryony : The presence of more than one embryo in a seed.

Primordia : Initials or beginning of structures like root primordia.

Propagate : To increase in number, to reproduce.

Radicle : The embryonic root of a germinating seed.

Relative humidity : The ratio of the weight of water vapour in a given quantity of air to the total weight of water vapour that quantity of air is capable of holding at a given temperature.

Reproduce : To increase in number.

Reproduction : The process of producing new individuals.

Rest period : Period in which viable seeds don't grow even in the presence of favourable environmental conditions.

Ringing : A method of removal of a narrow strip of bark around the periphery or the stem or a shoot.

Rooting : The development or growth of new roots.

Rooting hormone : A hormone, which helps the cuttings or layers to root faster.

Rooting media : A material in which the basal ends of cuttings are placed vertically for the development of roots.

Rootstock : A plant on which buds or scions are budded or grafted.

Scarification : The process of scratching or injuring the seed coat for facilitating seed germination.

Scion : A small shoot that is inserted by grafting or budding onto the rootstock.

Seed : The ripened ovule.

Seed coat : The outer covering of a seed.

Seedlings : Newly germinated young plants from seeds.

Seeds Act : Seeds act refers to an act, which regulates the quality of seeds offered for sale.

Self-pollination : Transfer of pollen grains from anthers of a flower to the stigma of the same flower.

Sexual propagation : Raising new plants from seeds, propagation by seeds.

Spadix : A racemose inflorescence in which the flowers are sessile and borne on a enlarged fleshy axis.

Sphagnum moss : A material formed from decayed bog plants; used as a growing media and for packing of plants.

Spike : A racemose inflorescence in which the flowers are sessile and borne on the same axis.

Staminate : A flower having functional stamens but without or non-functional pistil.

Stock-scion interaction : The effect of rootstock on a scion and *vice versa*.

Stolons : Runners by which plants reproduce vegetatively by sending out roots at each joint.

Stone fruit : A drupe or drupelet.

Stooling : A vegetative method of raising plants. It is also called as mound layering.

Strain : A variant in a clone.

Stratification : The practice of exposing imbibed seeds to 2 to 10°C for some period to break dormancy, which facilitates germination in them.

Syconus : A type of fruit in which achenes are borne inside the receptacle.

Symptoms : Some warning signals of plant growth initiated by problems related with nutrient deficiency or disease incidence.

Syncarpous : A flower having united carpels.

Testa : The outer, brown, comparatively thick seed coat.

Thalamus : The receprocle of a flower.

Transplanting : To shift plants from nursery to other location at permanent site.

Tuber : A modified stem, which is reproduced by growing roots from an eye, or a bud.

Tuberous root : A thick root containing large amounts of stored food, having roots located at one end and stem at the other.

Vegetative propagation : Raising new plants from vegetative (asexually) means.

Vermiculite : A light growing medium with neutral pH.

Weed : Any plant growing where it is not wanted.

Windbreak : Any natural or man-made screen to slow down the wind speed.

Zygote : A fertilized egg or cells arising from the fusion of male and female gametes.

• • •

Selected References

Adams, C.R., Brandford, K.M. and Early, M.P. (1993). Principles of horticulture. Butterworth Heinemann Ltd., Jordan Hills, Oxford.

Amen, R.D. (1968). A model of seed dormancy. *Botany Reviews,* **34** : 1-31.

Argles, G.K. (1969). Root formation by stem cuttings. Nurseryman and garden centre, Vol. 149, No. 3.

Bal, J.S. (1997). Fruit growing. Kalyani Publishers, Ludhiana, India.

Baudenditel, R.F. (1979). Horticulture : A basic awareness. Reston Publishing Company Inc., Reston, Virginia.

Beakbane, A.B. (1969). Relationships between structure and adventitious rooting. *Proc. Intl. Plant Prop. Soc.,* **19** :192-201.

Black, M. (1981). The role of endogenous hormones in germination and dormancy. *Israel J. Bot.,* **29** : 181-192.

Bose, T.K., Mitra, S.K. and Sanyal, D. (2001). Fruits : Tropical and subtropical. Naya Udyog Publishers, Kolkata (W.B.).

Bose, T.K., Mitra S.K., Sadhu, M.K. and Das, P. (2001). Propagation of tropical and subtropical Crops. Naya Prokash Publishers, Kolkata (W.B.).

Bose, T.K., Som, M.G. and Kabir, J. (1993). Vegetable Crops. Naya Prokash Publishers, Kolkata (W.B.).

Brison, F.R. and Adriance, G.W. (2000). Propagation of horticultural plants. Biotech Books, New Delhi.

Bunt, A.C. (1976). Modern potting composts : a manual on the preparation and use of growing media for pot plants. University Park, Pennsylvania State University, USA.

Carlson, R.F. and Mitchell, A.E. (1979). Budding and grafting fruit trees. Michigan Agri. Extension Bulletin, No. 508.

Cathey, H.M. and Campbell, L.E. (1980). Light and lighting systems for horticultural plants. *Hort. Reviews*, **2** : 491-537.

Chadha, K.L. (2001). Hand book of horticulture. DKMA, Indian Council of Agricultural Research, New Delhi.

Chattopadhyay, T.K. (2008). A text book on pomology. Vol. I : Fundamentals of fruit production. Kalyani Publishers, Ludhiana, India.

Chattopadhyay, T.K. (2008). A text book on pomology. Vol. II : Tropical fruits. Kalyani Publishers, Ludhiana, India.

Chattopadhyay, T.K. (2008). A text book on pomology. Vol. III : Sub-tropical fruits. Kalyani Publishers, Ludhiana, India.

Chattopadhyay, T.K. (2008). A text book on pomology. Vol. IV : Temperate Fruits. Kalyani Publishers, Ludhiana, India.

Davidson, H. and Mecklenburg, R. (1981). Nursery management : Administration and Culture. Englewood Cliffs, Prentice Hall, New York, USA.

Denisen, E.L. (1979). Principles of horticulture. 2nd edition, Macmillan Publishing Company, New York, USA.

Edmond, J.B. Senn, T.L., Andrews, F.S. and Halfacre, R.G. (1977). Fundamentals of horticulture. Tata Mac-Graw-Hill Publishing Company Ltd., New Delhi.

Frolich, E.F. (1961). Etiolation and rooting of cuttings. *American Nursery*, **73**(9) : 5-7.

Garner, R.J. (1979). The grafter's handbook. 4th ed., Oxford University Press, New York, USA.

Garner, V.R. (1986). Principles of horticultural production. Michigam State University Press, East Lansing, USA.

Gill, S.S., Bal, J.S. and Sandhu, A.S. (1985). Raising fruit nursery. Kalyani Publishers, Ludhiana, India.

Gray, D. (1981). Fluid drilling of vegetable seeds. *Hort. Reviews* **3** : 1-27.

Halfacre, R.G. and Barden, J.A. (1979). Horticulture. McGraw-Hill, New York, USA.

Hanan, J.J., Holley, W.D. and Goldsherry, K.L. (1978). Greenhouse management. Springer-Verlag, New York, USA.

Harrington, J.F. (1962). The effect of temperature on germination of several kinds of vegetable seeds. XVIth Intl. Hort. Cong., **2** : 435-441

Hartman, H.T., Kester, D.E. and Davies, F.T. (1993). Plant propagation. Prentice Hall of India Pvt. Limited, New Delhi.

Hartmann, H.T. and Beutel, A. (1979). Propagation of temerate zone fruit plants. Leaflet, California, Agri. Expt. Sta. California, USA.

Hartmann, H.T., Flocker, W.J. and Kofranek, A.M. (1981). Plant Science : Growth, development, and utilization of cultivated plants. Englewood Cliffs, N.J. : Prentice Hall, New York, USA.

Hayes, W.B. (1960). Fruit growing in India. Kitabistan, Allahabad.

Hecker, W.P. (1983). Phase change and intra-clonal variability. *Hort Science*, **18** : 22-24.

International Seed Testing Association (1978). International rules for seed testing. *Seed Sci. Technol.*, **4** : 114-177.

Janick, J. (1986). Horticultural Science, 4th Edition, U. H. Freeman and Company, San Francisco, USA.

Joiner, J.N. Poole R.T. and Conover, C.A. (1981). Propagation. *In :* Foliage plant production. Prentice Hall, Engle-wood Cliffs, N.Z.

Justice, O.L. (1972). Essentials of seed testing. *In* : Seed Biology. Vol. 3., T.T. Kizlowski, ed. Academic Press, New York, USA.

Khan, A.A. (1971). Cytokinins : Permissive role in seed germination. *Science*, **171** : 853-859.

Khan, A.A. (1981). Hormonal regulation of primary and secondary dormancy. *Israel J. Bot.*, **29** : 207-224.

Khan, A.A., Peck, N.H. and Samiry, C. (1981). Seed osmoconditioning : Physiological and biochemical changes. *Israel. J.Bot.*, **29** : 133-144.

Kumar, N., Khader, Md. Abdul JBM, Rangaswami, P. and Irulappan, I. (1993). Introduction to spices, plantation crops, medicinal and aromatic plants. Indian Council of Agricultural Research, New Delhi.

Laurie, A. and Ries, V.H. (1975). Floriculture : fundamentals and practices, 3rd edition, McGraw-Hill, New York, USA.

Leopard, A.C. and Kriedmann, P.E. (1975). Plant growth and development. 2nd edition, McGraw-Hill Company, New York, USA.

Malik, M.N. (2000). Horticulture. Biotech books, New Delhi.

Mayer, A.M. (1981). Control mechanism is seed germination, *Israel J. Bot.*, **29** : 1-4.

Mcmillan-Browse, P. (1979). Plant propagation. Simon and Schuster, New York, USA.

McMillan-Browse, P.D.A. (1978). Stratification – a detail of technique. *Proc. Intl. Plant Pro. Soc.*, **28** : 191-192.

Mengel, K. and Kirkby, E.A. (1987). Principles of plant nutrition. 4th edition. International Potash Institute, Worblaufen-Bern, Switzerland.

Miller, E.V. (1954). The natural origin of some popular varieties of fruits. *Economic Bot.*, **8** : 337-348.

Moore, R. and Walker, D.B. (1981). Graft compatibility-incompatibility in plants. *BioScience,* **31**(5) : 389-391.

Mukherjee, S.K. and Majumder, P.K. (1986). Propagation of tropical and Sub-tropical fruit crops. ICAR, New Delhi.

Nelson, P.V. (1985). Greenhouse operation and management, 3rd edition, Reston Publishing Company, New York, USA.

Nelson-Jones, W. (1969). Plant chimeras. Methuen and Company, London.

Nickell, L.G. (1982). Plant growth regulators: agricultural uses. Springer Verlag, New York, USA.

Patel, V. B., Singh, S.K., Dubey, A.K. (2008). A practical guide for plant propagation. Division of Fruits and Host Technology, IARI, New Delhi-12.

Pike, A.V. (1972). Propagation by roots. *Horticulture,* **50**(5) : 57-61.

Pruthi, J.S. (1996). Major spices and condiments : crop management and post harvest technology. Indian Council of Agricultural, Research, New Delhi.

Pruthi, J.S. (1998). Minor spices and condiments : crop management and post harvest technology. Indian Council of Agricultural Research, New Delhi.

Randhawa, G.S. and Mukhopadhyay, A. (1986). Floriculture in India. Allied Publishers Pvt. Limited, New Delhi.

Reddy, Y.T.N., Jankiram, T. and Reddy, D.S. (2001). Scientific nursery management. The House of Sapran (Media), Bangalore.

Rees, A.R. (1966). The physiology of ornamental bulbous plants. *Bot. Review*, **32** : 1-22.

Reiley, H.E. and Caroll, L.S. Jr. (1981). Introductory horticulture. 4th edition. Delmar Publishing Inc. Albany, New York, USA.

Reinert, J. and Bajaj, Y.P.S. (1997). Plant cell, tissue and organ culture, Springer-Verlag, Berlin, London.

Rice, R.W. and Rice, R.P. Jr. (1993). Practical horticulture. Englewood Cliffs, NJ, Prentice Hall, New Jersey, USA.

Sadhu, M.K. (1986). Plant propagation. New Age International Publishers, New Delhi.

Sharma, R.R. (2010). *Padav Pravardhan evam Podhshala Prabandhan* (Hindi). Commission for Scientific Technical Terms, Ministry of Human Resource Development, New Delhi.

Sharma, R.R. and Srivastav, M. (2004). Propagation and nursery management. Intl. Book Distributing Co., Lucknow, India .

Sharma, R.R. and Krishna, H. (2013). A textbook of plant propagation and nursery management. CBS Publishers, Delhi, India

Sharma, R.R. and Krishna, H. (2014). Fruit Production (Major Fruits). Daya Publishing House, Delhi, India.

Sharp, W.R., Larsen, P.O. Paddock, E.F. and Raghvan, V. (1979). Plant cell and tissue culture : Principles and applications. Ohio State University Press, Columbus, USA.

Singh, A. (1986). Fruit physiology and production. Kalyani Publishers, Ludhiana, India.

Street, H.E. (1977). Plant tissue and cell culture, 2nd ed. University of Calfornia Press, Berkeley, U.S.A.

Taylorson, R.B. and Hendricks, S.B. (1977). Dormancy in seeds. *Annual Review Plant Physiol.*, **28** : 231-254.

Thomson, J.R. (1979). An introduction to seed biology. John Wiley, New York, USA.

Thomson, J.R.(1979). An introduction to seed technology. John Wiley, New York, USA.

Thorpe, T.A. (1981). Plant tissue culture : Methods and applications in agriculture. Academic Press, New York, USA.

Tilney-Basset, R.A. (1963). The structure of periclinal chimeras. *Heredity*, **18** : 265-285.

Vasil, I. K. (1980). Perspectives in plant cell and tissue culture. *Intl. Rev. Cytol. supple.*, 11B, Academic Press, New York, USA.

Villiers, T.A. (1972) Seed dormancy. *In* : Seed biology, Vol. 2, (T.T. Kozlowski, Ed), Academic Press, New York, pp. 220-282.

Welch, H.J. (1970). Mist propagation and automatic watering, Faber and Faber, London.

• • •